Lowering Cholesterol in High-Risk Individuals and Populations

FUNDAMENTAL AND CLINICAL CARDIOLOGY

Editor-in-Chief

Samuel Z. Goldhaber, M.D.
Harvard Medical School
and Brigham and Women's Hospital
Boston, Massachusetts

Associate Editor, Europe

Henri Bounameaux, M.D.
University Hospital of Geneva
Geneva, Switzerland

1. *Drug Treatment of Hyperlipidemia*, edited by Basil M. Rifkind
2. *Cardiotonic Drugs: A Clinical Review, Second Edition, Revised and Expanded*, edited by Carl V. Leier
3. *Complications of Coronary Angioplasty*, edited by Alexander J. R. Black, H. Vernon Anderson, and Stephen G. Ellis
4. *Unstable Angina*, edited by John D. Rutherford
5. *Beta-Blockers and Cardiac Arrhythmias*, edited by Prakash C. Deedwania
6. *Exercise and the Heart in Health and Disease*, edited by Roy J. Shephard and Henry S. Miller, Jr.
7. *Cardiopulmonary Physiology in Critical Care*, edited by Steven M. Scharf
8. *Atherosclerotic Cardiovascular Disease, Hemostasis, and Endothelial Function*, edited by Robert Boyer Francis, Jr.
9. *Coronary Heart Disease Prevention*, edited by Frank G. Yanowitz
10. *Thrombolysis and Adjunctive Therapy for Acute Myocardial Infarction*, edited by Eric R. Bates
11. *Stunned Myocardium: Properties, Mechanisms, and Clinical Manifestations*, edited by Robert A. Kloner and Karin Przyklenk
12. *Prevention of Venous Thromboembolism*, edited by Samuel Z. Goldhaber
13. *Silent Myocardial Ischemia and Infarction: Third Edition*, Peter F. Cohn
14. *Congestive Cardiac Failure: Pathophysiology and Treatment*, edited by David B. Barnett, Hubert Pouleur, and Gary S. Francis
15. *Heart Failure: Basic Science and Clinical Aspects*, edited by Judith K. Gwathmey, G. Maurice Briggs, and Paul D. Allen

16. *Coronary Thrombolysis in Perspective: Principles Underlying Conjunctive and Adjunctive Therapy,* edited by Burton E. Sobel and Désiré Collen
17. *Cardiovascular Disease in the Elderly Patient,* edited by Donald D. Tresch and Wilbert S. Aronow
18. *Systemic Cardiac Embolism,* edited by Michael D. Ezekowitz
19. *Low-Molecular-Weight Heparins in Prophylaxis and Therapy of Thromboembolic Diseases,* edited by Henri Bounameaux
20. *Valvular Heart Disease,* edited by Muayed Al Zaibag and Carlos M. G. Duran
21. *Implantable Cardioverter-Defibrillators: A Comprehensive Textbook,* edited by N. A. Mark Estes III, Antonis S. Manolis, and Paul J. Wang
22. *Individualized Therapy of Hypertension,* edited by Norman M. Kaplan and C. Venkata S. Ram
23. *Atlas of Coronary Balloon Angioplasty,* Bernhard Meier and Vivek K. Mehan
24. *Lowering Cholesterol in High-Risk Individuals and Populations,* edited by Basil M. Rifkind
25. *Interventional Cardiology: New Techniques and Strategies for Diagnosis and Treatment,* edited by Christopher J. White and Stephen R. Ramee

ADDITIONAL VOLUMES IN PREPARATION

Lowering Cholesterol in High-Risk Individuals and Populations

edited by

Basil M. Rifkind

National Heart, Lung, and Blood Institute
National Institutes of Health
Bethesda, Maryland

Marcel Dekker, Inc.   New York • Basel • Hong Kong

Library of Congress Cataloging-in-Publication Data

Lowering cholesterol in high-risk individuals and populations / edited
by Basil M. Rifkind.
    p. cm. -- (Fundamental and clinical cardiology ; 24)
    Includes bibliographical references and index.
    ISBN 0-8247-9412-5 (hardcover : alk. paper)
    1. Hypercholesteremia--Treatment. 2. Hypercholesteremia-
-Prevention. I. Rifkind, Basil M. II. Series: Fundamental and
clinical cardiology ; v. 24.
    [DNLM: 1. Hypercholesterolemia--complications.
2. Hypercholesterolemia--therapy. 3. Coronary Disease--prevention
& control. 4. Risk Factors. W1 FU538TD v. 24 1995 / WD 200.5.H8
L917 1995]
RC632.H83L68   1995
616.1'36--dc20
DNLM/DLC   94-45081
For Library of Congress   CIP

This book was edited by Dr. Basil M. Rifkind in his private capacity. No official support or endorsement by the National Institutes of Health is intended and none should be inferred.

The publisher offers discounts on this book when ordered in bulk quantities. For more information, write to Special Sales/Professional Marketing at the address below.

This book is printed on acid-free paper.

Marcel Dekker, Inc.
270 Madison Avenue, New York, New York 10016

Current printing (last digit):
10 9 8 7 6 5 4 3 2 1

PRINTED IN THE UNITED STATES OF AMERICA

Series Introduction

Cholesterol reduction is a highly controversial, complex, and often confusing topic that centers itself squarely at the interface of primary and subspecialty care. Which patients should be treated? Which cholesterol-lowering agents should be prescribed? How long should patients be treated? (Is the prescription of pharmacotherapy, in most instances, the beginning of a lifelong commitment to drug-induced cholesterol lowering?) To what extent should we rely on dietary intervention?

These deceptively straightforward questions arise daily in my practice of general clinical cardiology. Fortunately, there has been an exponential increase in the available scientific basis for clinical decision making in this field. However, the vital information is scattered among a confusing array of widely varying subspecialty, general internal medicine, and government publications.

Dr. Basil M. Rifkind, a renowned authority in the field, has performed a public service by enlisting world leaders to analyze and synthesize the glut of information into an extremely well-organized and lucid book. The compilation ranges from summaries of advances in basic science to community approaches toward lowering cholesterol levels.

Lowering Cholesterol in High-Risk Individuals and Populations fulfills my goal as the Editor-in-Chief of the Fundamental and Clinical Cardiology series. The publisher and I strive to assemble the talents of individuals such as Dr. Rifkind to discuss virtually every area of cardiovascular medicine. The current book, number 24 in our series, could not possibly be more relevant or timely. I predict that Dr. Rifkind's book will be cited in articles and during clinical patient rounds well into the 21st century.

Samuel Z. Goldhaber

Preface

Since the Lipid Research Clinic's Coronary Primary Prevention Trial reported its findings in 1983, much effort has been devoted to the development of detailed guidelines for the management of high blood cholesterol, at the level of both the individual patient and the population as a whole. The various reports of the National Cholesterol Education Program have been especially noteworthy in providing guidance for the management of adults and children and for the U.S. population. National surveys have shown the guidelines to be widely known and to have been acted upon in many respects. The guidelines are ambitious in their scope, widespread in their potential impact, and costly in their implementation even if, in the long run, they result in substantial reductions in death and disability from coronary heart disease and in individual and public expenditures. The guidelines have many ramifications, given the number of people involved, the relevance of the guidelines to our life-style, and their application to healthy as well as to sick people. Public policy, the diet of Americans, and clinical practice have been and will continue to be substantially affected by them. The continued decline in the nation's serum cholesterol levels and in its total mortality rate may reflect the impact of these guidelines in conjunction with other forces.

As implementation proceeds, a variety of important issues have arisen, such as: Who should receive cholesterol-lowering drugs? What are the cost consequences of such programs? Is total mortality reduced by treatment? To what extent should other coronary risk factors be taken into account? In addition, scientific study continues to bring new and expanding insights into the pathogenesis, diagnosis, and treatment of high blood cholesterol and leads to the reshaping of guidelines. The present volume deals with many of these questions, issues, and advances. It is intended to be of interest to the several disciplines that relate to its various aspects, including primary care physicians, internal medicine practitioners, cardiologists, pediatricians, epidemiologists, public health physicians, clinical chemists, dietitians, and nurses involved in preventive care.

Basil M. Rifkind

Contents

Series Introduction   *iii*

Preface   *v*

Contributors   *ix*

1. **The High-Risk Strategy for Adults**   1
    Alan Chait

2. **Cholesterol Lowering and Total Mortality**   33
    David J. Gordon

3. **Secondary Prevention of Coronary Heart Disease**   49
    Jacques E. Rossouw

4. **Evolution of the Atherosclerotic Plaque**   69
    Antonio Fernández-Ortiz and Valentin Fuster

5. **Dyslipoproteinemia in Older People**   99
    Walter H. Ettinger, Jr., and William R. Hazzard

6. **Cholesterol Lowering in Women**   119
    John C. LaRosa

7. An Approach to Cholesterol Levels in Children and
    Adolescents   133
    Ronald M. Lauer

8. Population Strategy   149
    Thomas A. Pearson

9. The Community Approach   167
    Russell V. Luepker

10. The Cholesterol-Lowering Diet   183
    Margo A. Denke and Scott M. Grundy

11. Implementing Dietary Change   209
    Penny M. Kris-Etherton, Sharon L. Peterson,
    Madeleine Sigman-Grant, Lori Beth Dixon, Suzanne M. Jaax,
    and Lynne W. Scott

12. Drug Therapy   271
    Henry N. Ginsberg

13. Measuring Cholesterol and Other Lipids and Lipoproteins
    in Cardiovascular Risk Assessment   291
    G. Russell Warnick

14. The Cost-Effectiveness of Programs to Lower Serum
    Cholesterol   311
    David J. Cohen, Lee Goldman, and Milton C. Weinstein

15. Emerging Opportunities for Atherosclerosis Prevention:
    Beyond Cholesterol-Lowering Therapy   337
    Daniel Steinberg and Joseph L. Witztum

16. The Scandinavian Simvastatin Survival Study (4S): Editor's
    Summary   357
    Basil M. Rifkind

Index   *363*

Contributors

Allan Chait, M.D. Professor, Department of Medicine, University of Washington, Seattle, Washington

David J. Cohen, M.D. Instructor in Medicine, Harvard Medical School, and Cardiovascular Division, Beth Israel Hospital, Boston, Massachusetts

Margo A. Denke, M.D. Associate Professor, Department of Internal Medicine, and Center for Human Nutrition, University of Texas Southwestern Medical Center, Dallas, Texas

Lori Beth Dixon, Ph.D. Postdoctoral Fellow, Department of Nutrition, Pennsylvania State University, University Park, Pennsylvania

Walter H. Ettinger, Jr., M.D. Professor, Departments of Internal Medicine and Public Health Sciences, Bowman Gray School of Medicine, Wake Forest University, Winston-Salem, North Carolina

Antonio Fernández-Ortiz, M.D. Cardiopulmonary Department, Hospital Universitario San Carlos, Madrid, Spain

Valentin Fuster, M.D., Ph.D. Director, Cardiovascular Institute, and Vice Chairman, Department of Medicine, Department of Cardiology, Mount Sinai Medical Center, New York, New York

Henry N. Ginsberg, M.D. Tilden-Weger-Beiler Professor of Medicine, and Chief, Division of Preventive Medicine and Nutrition, Department of Medicine, Columbia University College of Physicians and Surgeons, New York, New York

Lee Goldman, M.D. Professor of Medicine, Harvard Medical School; Professor of Epidemiology, Harvard School of Public Health; and Chief Medical Officer, Brigham and Women's Hospital, Boston, Massachusetts

David J. Gordon, M.D., Ph.D. Senior Medical Officer, Division of Heart and Vascular Diseases, National Heart, Lung, and Blood Institute, National Institutes of Health, Bethesda, Maryland

Scott M. Grundy, M.D. Director, Center for Human Nutrition, and Professor of Internal Medicine and Biochemistry, University of Texas Southwestern Medical Center, Dallas, Texas

William R. Hazzard, M.D. Professor and Chairman, Department of Internal Medicine, Bowman Gray School of Medicine, Wake Forest University, Winston-Salem, North Carolina

Suzanne M. Jaax, M.S., R.D./L.D. Research Dietician, Department of Medicine, Baylor College of Medicine, and The Methodist Hospital, Houston, Texas

Penny M. Kris-Etherton, Ph.D., R.D. Professor, Department of Nutrition, Pennsylvania State University, University Park, Pennsylvania

John C. LaRosa, F.A.C.P. Chancellor and Professor of Medicine, Tulane University Medical Center, New Orleans, Louisiana

Ronald M. Lauer, M.D. Director, Division of Pediatric Cardiology, and Professor, Department of Pediatrics and Preventive Medicine, University of Iowa, Iowa City, Iowa

Russell V. Luepker, M.D., M.S. Professor and Head, Division of Epidemiology, School of Public Health, University of Minnesota, Minneapolis, Minnesota

Thomas A. Pearson, M.D., Ph.D. Director, Mary Imogene Bassett Research Insitute, Cooperstown, and Professor of Public Health, Columbia University, New York, New York

Sharon L. Peterson, R.D. Ph.D. Candidate, Graduate Program in Nutrition, Pennsylvania State University, University Park, Pennsylvania

Basil M. Rifkind, M.D. Senior Scientific Advisor, Vascular Research Program, Division of Heart and Vascular Diseases, National Heart, Lung, and Blood Institute, National Institutes of Health, Bethesda, Maryland

Jacques E. Rossouw, M.D. Office of Disease Prevention, National Institutes of Health, Bethesda, Maryland

Lynne W. Scott, M.A., R.D./L.D. Assistant Professor, Department of Medicine, Baylor College of Medicine, and The Methodist Hospital, Houston, Texas

Madeleine Sigman-Grant, Ph.D., R.D. Assistant Professor, Department of Food Science, Pennsylvania State University, University Park, Pennsylvania

Daniel Steinberg, M.D., Ph.D. Professor, Department of Medicine, University of California, San Diego, La Jolla, California

G. Russell Warnick, M.S., M.B.A. President, Pacific Biometrics, Inc., Pacific Biometrics Research Foundation, Seattle, Washington

Milton C. Weinstein, Ph.D. Henry J. Kaiser Professor of Health Policy and Management, Department of Health Policy and Management, Harvard School of Public Health, and Department of Medicine, Brigham and Women's Hospital, Boston, Massachusetts

Joseph L. Witztum, M.D. Professor, Department of Medicine, University of California, San Diego, La Jolla, California

1

The High-Risk Strategy for Adults

Alan Chait

University of Washington, Seattle, Washington

INTRODUCTION

Two complementary strategies have been developed to reduce coronary heart disease (CHD) risk by lowering blood cholesterol levels. The first, the population-based approach, aims to decrease the average cholesterol level of the population at large, thereby shifting the cholesterol distribution towards lower and more favorable levels. This public health approach relies on programs that educate the public about the relationship between blood cholesterol and CHD and the dietary changes that need to be undertaken to lower blood cholesterol levels. The success of the population-based approach is evident in the falling blood cholesterol levels during the past few decades (1) accompanied by a concomitant reduction in CHD mortality (2).

The second approach to CHD prevention is the high-risk approach. The strategy is to identify individuals at the highest risk of developing new or recurrent CHD by virtue of their lipid and lipoprotein levels, with or without other risk factors, and then target them for specific intervention. This approach screens potential candidates and then uses specific therapeutic diets with or without drugs for intervention.

To achieve these complementary strategies, the National Cholesterol Education Program (NCEP) of the National Heart Lung and Blood Institute of the National Institutes of Health was established in the 1980s. The NCEP was comprised of several panels, including a Population Panel to address the issue of the population-based approach, an Adult Treatment Panel (ATP) to provide guidelines for the management of high-risk adults, a Childhood Panel to provide guidelines for the identification and management of hyperlipidemia in children and adolescents, and a Laboratory Panel to improve and standardize the measurement of cholesterol and other lipids.

These panels' recommendations have increased cholesterol awareness among physicians and the public and have provided clinicians with useful guidelines for the evaluation and management of patients with hyperlipidemia. The first ATP recommendations (ATP I) were published in 1988 (3) and were widely praised for their simplicity and ease of application. However, with time it became apparent that the original guidelines needed to be updated. For example, therapy was often targeted to patients who were not at the highest risk (4), e.g., young females without other risk factors and lower-risk elderly subjects. Further, many dyslipidemic patients who were at very high risk were not identified by the original cutpoints (5). Finally, new research findings, especially regarding treatment of patients with established atherosclerotic disease, were taken into account in updating the guidelines for high-risk adults. In keeping with the more longstanding Joint National Committee for the Detection, Evaluation and Treatment of High Blood Pressure, which updates its recommendations at regular intervals, updated recommendations from the NCEP's Adult Treatment Panel, entitled ATP II, were published in June 1993 (6). The specifics of these updated guidelines and the reasons for these changes will be the focus of this chapter.

RECOMMENDATIONS OF ATP II

Who and How to Screen

As with the original guidelines, the goal of screening is to identify individuals at increased risk of cardiovascular disease by virtue of high LDL cholesterol levels. It is recommended that serum cholesterol be measured in all adults above the age of 20 and that this measurement be repeated at least every 5 years. In ATP II it is recommended that measurement of HDL cholesterol be included in the initial evaluation. Because neither serum total nor HDL cholesterol levels change to a great extent in the postprandial state, these initial measurements can be performed on blood samples obtained in the nonfasted state.

In the absence of clinical evidence of heart disease or other atherosclerotic disease, total cholesterol levels < 200 mg/dl are classified as "desirable," 200–239 mg/dl as "borderline high," and ≥ 240 mg/dl as "high" (Table 1). HDL cholesterol levels < 35 mg/dl are classified as "low." Low HDL is regarded as a cardiovascular risk factor that enters into follow-up and therapeutic decisions (see below).

Table 1 Stratification of Individuals by Cholesterol Levels

	Desirable	Borderline high	High
Serum cholesterol (mg/dl)	<200	200–239	≥240
LDL cholesterol (mg/dl)	<130	130–159	≥160

Follow-Up and Therapeutic Decisions

Follow-up and therapeutic decisions depend on whether there are other cardio-vascular risk factors and whether there is evidence of established atherosclerotic disease. Strategies to prevent the development and initial clinical manifestations of atherosclerotic disease are referred to as *primary prevention*. Treatment to prevent progression and further clinical manifestations in individuals with coronary artery or other atherosclerotic diseases is referred to as *secondary prevention*.

When desirable serum total and HDL cholesterol levels are present, no further short-term follow-up is indicated. It is recommended that such individuals be given advice and educational materials concerning the use of a prudent diet and other risk-reduction strategies for the population at large and have their lipid values repeated within 5 years. Full lipoprotein analysis is recommended (1) in individuals with desirable serum total cholesterol but low HDL cholesterol levels, (2) when total cholesterol levels are classified as high, and (3) in all individuals with established coronary artery disease (CAD) or other atherosclerotic diseases. Lipoproteins also should be evaluated (4) in individuals with borderline total cholesterol levels and either low HDL or two or more other risk factors (see below). Lipoprotein analysis must be performed in the fasted state since LDL cholesterol is estimated as LDL cholesterol − total cholesterol (measured) − HDL cholesterol (measured) − very low-density lipoprotein (VLDL) cholesterol (calculated as triglycerides/5) (7). Serum triglyceride values fluctuate markedly in the postprandial state leading to inaccurate VLDL and, therefore, LDL cholesterol values.

Therapeutic decisions are based on LDL cholesterol levels and the presence or absence of other risk factors and established atherosclerotic disease. LDL cholesterol values < 130 mg/dl are classified as "desirable," 130–159 mg/dl as "borderline high," and ≥ 160 mg/dl as "high" (Table 1). The primary-prevention approach to individuals with desirable LDL cholesterol levels should be the same as for individuals with desirable total cholesterol levels (see above). Individuals with borderline LDL levels (130–159 mg/dl) and fewer than two "net" risk factors (see below) should be provided with dietary information and reevaluated within 1 year. For individuals with borderline LDL cholesterol and two or more net risk factors, with high-risk LDL cholesterol levels (>160 mg/dl) in the absence of atherosclerotic disease, or with LDL cholesterol levels > 130 mg/dl in the presence of established atherosclerotic disease, diet therapy should be instituted. If after an adequate trial of diet therapy LDL cholesterol remains >190 mg/dl in the absence of risk factors, >160 mg/dl in their presence, or >130 mg/dl in subjects with established atherosclerotic disease, drug therapy is recommended.

Thus, there are three major decision steps for which cutpoints, based on the presence or absence of risk factors and established atherosclerotic disease, are useful (Table 2).

1. The suggested cutpoints for lipoprotein analysis are total cholesterol ≥ 240 mg/dl in the absence of risk factors and 200–239 mg/dl in their presence or HDL cholesterol < 35 mg/dl. Lipoprotein measurements are recommended in all individuals with established atherosclerotic disease.
2. Diet therapy is recommended for LDL cholesterol > 160 mg/dl in the absence of risk factors, > 130 mg/dl in their presence, and > 100 mg/dl in individuals with established disease.
3. Drug therapy is recommended for LDL cholesterol > 190 mg/dl in the absence of risk factors, > 160 mg/dl in their presence, and > 130 mg/dl in individuals with atherosclerotic disease.

Target LDL cholesterol levels during therapy are the same as those for institution of diet treatment.

Good clinical judgment is strongly recommended in the application of these guidelines. Some specific examples of when these guidelines should not be rigidly applied are provided in ATP II. For example, it is suggested that young individuals, i.e., men < 45 years of age and premenopausal women, who have no other risk factors should be considered for drug therapy some time in the future unless LDL cholesterol values exceed 220 mg/dl. Clearly, clinical judgment needs to be exercised in applying this recommendation. Another situation in which good clinical judgment is necessary is with elderly patients (see below).

Table 2 Action Points in ATP II (mg/dl)

Decision	No risk factors	Two or more risk factors	Established atherosclerotic disease
Lipoprotein analysis	TC > 240	TC > 200	All
Diet therapy	LDL-C > 160	> 130	>100
Drug therapy	LDL-C > 190	> 160	>130

TC = total serum cholesterol; LDL-C = low-density lipoprotein cholesterol.

Risk Factors

There are many cardiovascular risk factors. Only those for which the evidence is incontrovertible and that can easily and reliably be measured are included in ATP II. The positive risk factors that determine whether an individual patient will be stratified into the higher-risk group for therapeutic decisions include:

Age ≥ 45 years for males and age ≥ 55 years or premature menopause without the use of estrogen replacement therapy in females

A positive family history of premature CHD, i.e., the presence of a myocardial infarction or sudden death, coronary artery bypass surgery or angioplasty in first-degree male relatives < 55 years of age or in first-degree female relatives < 65 years of age

Current cigarette smoking

The presence of hypertension, defined as a blood pressure ≥ 140/90, or the use of antihypertensive medications for the treatment of hypertension

Diabetes mellitus

Low HDL, i.e., < 35 mg/dl in males or females

In the original guidelines, male gender was regarded as a positive risk factor. With the appreciation that cardiovascular disease is as common a cause of mortality in females, albeit it at a later age than in males, this risk factor has been replaced in ATP II with gender-specific age. Obesity, defined as >30% above ideal body weight, was designated a positive risk factor in the original guidelines for adults. With the increasing realization that the cardiovascular risk associated with obesity occurs predominantly when the adiposity is distributed centrally, especially viscerally (8,9), a different approach to the issue of obesity has been taken in ATP II. The central obesity/insulin resistance syndrome is associated with several cardiovascular risk factors that probably account for most, if not all, of the increased risk associated with obesity. These include dyslipidemia (high triglycerides, low HDL cholesterol levels, and the presence of small, dense LDL particles), hypertension, and glucose intolerance/non–insulin-dependent diabetes (see below). All improve when weight is lost and weight loss is maintained. Thus, low HDL cholesterol, hypertension, and diabetes remain risk factors in ATP II. However, rather than including obesity as a general risk factor, without regard to its type or consequences, weight control now becomes an important goal of therapy in ATP II. ATP II also contains a special section on the approach to the diagnosis and management of individuals with hypertriglyceridemia and diabetes, which often occur in subjects with central obesity.

Another new feature in ATP II is the notion of a "negative risk factor." The inverse relationship between HDL cholesterol levels and cardiovascular disease risk is well accepted (10,11). Individuals with high levels of HDL cholesterol are at lower risk than individuals with low HDL cholesterol levels, even in the presence of other risk factors. Thus, a high HDL cholesterol level will tend to offset the adverse effects of other risk factors such as high LDL cholesterol (12). Therefore, when assigning net risk to the individual patient, HDL cholesterol levels > 60 mg/dl have been designated a negative risk factor. Two or more *net* risk factors are required before assigning individuals to the "risk factor–positive" group. For example, a female over the age of 55 with hypertension whose HDL cholesterol is >60 mg/dl would not be classified as in the higher-risk category because only one net risk factor is present.

Diet Therapy

Diet therapy focuses on a nutritionally adequate diet that is low in fat (especially saturated fat) administered in a therapeutic setting. The goal is to permanently change eating habits by consumption of a diet rich in fruits, vegetables, and grains, which are low in fat. Weight control and physical activity also are stressed in ATP II. These measures not only lower blood cholesterol, but also are associated with modification of other risk factors such as hypertension, insulin resistance, hypertriglyceridemia, and low HDL. Physical activity has also been shown to be associated with increased cardiovascular fitness and reduced cardiovascular mortality beyond that which can be attributed to improvement in risk factors (13).

The two therapeutic diets, Step I and Step II, from the original guidelines are retained (Table 3). Fat should provide no more than 30% of total calories in both diets. Saturated fat intake should be limited to <10% of calories in the Step I diet

Table 3 Dietary Therapy for High Blood Cholesterol

Nutrient[a]	Recommended intake	
	Step I diet	Step II diet
Total fat	30% or less of total calories	
Saturated fatty acids	8–10% of total calories	<7% of total calories
Polyunsaturated fatty acids	≤10% of total calories	
Monounsaturated fatty acids	≤15% of total calories	
Carbohydrates	55% or more of total calories	
Protein	~15% of total calories	
Cholesterol	<300 mg/day	<200 mg/day
Total calories	To achieve and maintain desirable weight	

[a]Calories from alcohol not included.
Source: Ref. 6.

and to <7% of calories in the Step II diet. Dietary cholesterol intake should be <300 mg/day in the Step I diet and <200 mg/day in the Step II diet. A useful instrument is now included for rapid assessment of an individual's current eating habits and fat intake. In the event that the individual requiring diet therapy already is consuming a diet similar to the Step I diet, it is recommended that a Step II diet be initiated from the beginning. A Step II diet should also be prescribed if therapeutic goals are not achieved in spite of good adherence to the Step I diet.

It is recognized that the major cause of failure of dietary therapy is inadequate compliance. Therefore the use of a health care team that includes dietitians, nurses, physicians assistants, and office staff is strongly encouraged as a means of increasing dietary adherence.

Drug Therapy

Drug therapy is recommended when diet therapy fails to achieve the desired goal in very high-risk individuals. A trial of at least 6 months of adherence to the prescribed diet is recommended before proceeding to drug therapy. However, for individuals with familial forms of hyperlipidemia such as familial hyper-cholesterolemia, who almost invariably will require drug therapy, it is not necessary to wait 6 months before adding drugs to the therapeutic regime. The groups of lipid-lowering drugs currently available in the United States are shown in Table 4.

A new feature of ATP II is the consideration of HDL cholesterol levels in the choice of drug. When HDL is low, a drug that increases HDL, e.g., nicotinic acid, should be considered. The HMG-CoA reductase inhibitors were unavailable at the time of publication of the original guidelines. Their use is now put into perspective in ATP II. The use of combinations of drugs also is recommended in some circumstances, especially in secondary prevention when the lower therapeutic goals are not achieved by the use of a single drug. Low-dose combination therapy also is useful in primary prevention when therapeutic goals cannot be achieved by the starting dose of a single agent such as a reductase inhibitor or bile acid resin. Rather than increasing the dose of that agent, it often is more beneficial and cost-effective to add a low dose of the other agent to achieve adequate lowering of LDL cholesterol levels.

A new feature of ATP II is the recommendation to delay drug therapy in individuals who are not at excessively high risk by virtue of their LDL cholesterol levels. This applies particularly to the young, i.e., males < 45 or females either < 55 or premenopausal, who probably were overtreated in the past. It is now recommended that drug therapy be delayed for LDL cholesterol levels in the range of 190–220 mg/dl if no other risk factors are present. However, if other risk factors are present or if the individual has a familial form of hyperlipidemia such as familial hypercholesterolemia, drug therapy should not be delayed in the young,

Table 4 Lipid-Lowering Drugs Available in the United States

Class	Drug	Dose	Major effect	Side effects	Comments
Bile acid–binding resins	Cholestyramine	4–24 g/day	↓ LDL-C	GI distress	Difficult to maintain long-term
	Colestipol	5–30 g/day	↓ LDL-C	GI distress	Difficult to maintain long-term
Nicotinic acid	Various brands	1.5–6 g/day	↓ LDL-C ↓ TG ↑ HDL-C	GI distress Skin flushing Hepatotoxic Hyperglycemia	Need to start slowly Very useful for dyslipidemia
HMG-CoA reductase inhibitors	Lovastatin	10–80 mg/day	↓ LDL-C	Mildly hepatotoxic Myopathy	Few side effects
	Simvastatin	5–40 mg/day	↓ LDL-C	Mildly hepatotoxic Myopathy	Few side effects
	Pravastatin	10–40 mg/day	↓ LDL-C	Mildly hepatotoxic Myopathy	Few side effects
	Fluvastatin	20–80 mg/day	↓ LDL-C	Mildly hepatotoxic Myopathy	Few side effects
Fibrates	Gemfibrozil	600 mg bid	↓ TG ↑ HDL-C	Myopathy Gallstones	Very useful for hypertriglyceridemia
	Clofibrate	1 g bid	↓ TG ↑ HDL-C	Myopathy Gallstones	Very useful for hypertriglyceridemia
Estrogens	Multiple preparations		↓ LDL-C ↑ HDL-C	Endometrial CA	Useful for postmenopausal females
Probucol		500 mg bid	↓ LDL-C ↓ HDL-C	GI symptoms Prolonged QT interval	Antioxidant

since many very high-risk individuals develop clinical manifestations of atherosclerosis before these age cutpoints.

CHANGES FROM THE ORIGINAL GUIDELINES FOR ADULTS

The major changes between ATP I and ATP II occur largely in three areas. First, there is increased emphasis on HDL. Not only is it now recommended that HDL cholesterol levels be determined during the initial evaluation of patients (see above), but a high HDL cholesterol level is now regarded as a "negative risk factor." It also is recommended that the HDL cholesterol level be taken into account when making decisions about drug therapy. Second, with the increased realization that drug therapy should be reserved for the highest-risk individuals, there is an increased emphasis on diet, exercise, and other life-style measures in ATP II. Third, the most important new feature of ATP II is its increased emphasis on secondary prevention, i.e., treatment of the individual who already has evidence of established coronary artery or other atherosclerotic disease who is at very high risk of developing a further clinical event in the near to intermediate term. As a result of a considerable body of scientific evidence that has appeared since publication of the original guidelines, it is clear that this very high-risk group of individuals can benefit markedly from aggressive lipid-lowering therapy. As a result, new cutpoints and therapeutic goals have been established for subjects with established atherosclerotic disease. ATP II also deals with several lipid-related issues that were discussed in considerably less depth in the original report such as triglycerides, obesity, and diabetes, to name but a few of the areas of special interest to high-risk individuals. Finally, a cost-benefit analysis section and a discussion about the potentially harmful effects of very low blood cholesterol levels are included in ATP II.

RATIONALE FOR VARIOUS RECOMMENDATIONS AND AREAS OF CONTINUING CONTROVERSY

This section is an overview of the rationale for some of the new recommendations of ATP II and a discussion of some of the areas that still evoke controversy. The recommendations were a consensus of a large panel of individuals from many disciplines and backgrounds, and the guidelines were approved by a large number of organizations, societies, and groups. Therefore, not all issues were agreed upon uniformly, recommendations often represent compromise, and some areas remain to some extent controversial. Nonetheless, ATP II represents a simple and practical set of guidelines for the detection, evaluation, and management of individuals at increased risk of atherosclerotic complications by virtue of their lipids and lipoproteins. Clearly however, there is considerable room for good clinical judgment in the application of these guidelines.

LDL as the Target for Therapy

There is a good rationale to continue to use LDL cholesterol as the basis for therapeutic decisions. Total plasma cholesterol is a surrogate measure for LDL cholesterol. Recent analyses have provided additional support for the very strong relationship between serum cholesterol and CHD (reviewed in Ref. 14). Pooled data from 10 large cohort studies and three international comparisons confirmed the strength, consistency, and graded nature of the relationship between cholesterol levels and CHD mortality (16–18), making anything less than a causal relationship highly unlikely. Although cholesterol is but one of several cardiovascular risk factors, it is of interest that risk factors such as hypertension, cigarette smoking, and even diabetes have much less impact on CHD rates when serum cholesterol—and hence LDL cholesterol—levels are low (19).

In addition to these observational studies, analysis of 25 randomized trials of serum cholesterol reduction shows that the effect of cholesterol reduction in reducing CHD is even greater than previously believed, i.e., 10% reduction in cholesterol leads to 25% reduction in CHD (17). The effect of cholesterol-lowering therapy increases with time, with little benefit being observed in the first couple of years and benefit continuing to occur up till 5 years. Thus, previous meta-analyses that have not taken the duration of trials into account are likely to have underestimated the benefits of cholesterol lowering. The recent angiographic trials are particularly convincing (see below). Lesion progression is reduced, regression may occur, and clinical events are reduced when LDL cholesterol levels are reduced by aggressive diet (20) and drug therapy (21–24). LDL cholesterol can be reduced to target levels in most individuals with either diet or lipid-lowering drugs.

Evaluation of Adults Greater Than 20 Years of Age

It is recommended that all adults > 20 years of age be screened for the presence of high blood cholesterol levels. Screening should be performed in a case-finding mode, i.e., during visits to the physician for whatever reason. This also affords the opportunity to screen for risk factors other than abnormal lipids, emphasizing the multifactorial nature of atherosclerosis. This age cutpoint has been criticized because of the expense of screening such a large population. It has been suggested that screening might be delayed without much impact on development of heart disease, which seldom has its clinical onset before the age of 45. Nonetheless, the onset of complicated lesions occurs at an early age (25), and clinical disease does sometimes become manifest before the age of 45, especially in cases of familial forms of hyperlipidemia or when multiple risk factors are present. The hyperlipidemia associated with some genetic disorders, e.g., familial combined hyperlipidemia and remnant removal disease, usually does not become manifest before the third or fourth decade. Thus, absence of hyperlipidemia in the early twenties might lead to a false sense of security in some individuals who are at high risk by virtue of these genetic disorders. However, the guidelines do recommend repeating lipid measurements at various intervals depending on level and other risk factors, which should ultimately lead to the detection of these high-risk individuals. A less expensive approach might be to screen all adults > 20 who have other risk factors, particularly a positive family history of premature atherosclerotic disease, and to defer routine screening of other adults until they enter the medical system for some other ailment.

Cutpoints

The cutpoints chosen for use in the initial guidelines and expanded upon in ATP II are obviously somewhat arbitrary. The levels chosen are based on the epidemiological relationship between serum and LDL cholesterol and the development of CAD but also have been chosen to permit simple application nationwide. Thus, increments of 30 mg/dl separate the various categories. The new cutpoint for the use of drugs in low-risk young individuals also was intentionally set 30 mg/dl higher than in older individuals. While this approach is an intentional oversimplification, it has considerable merit in combining an approach based on sound epidemiological data while maintaining practicality for widespread application.

Age and Sex

Management of hyperlipidemia in females and in the elderly are areas still much debated. Much less is known about the relationship between hyperlipidemia and atherosclerosis in females than in males, because most studies performed during the past several decades have focused on middle-aged men at high risk. However, as more studies are being performed on females, it is apparent that heart disease is as important a cause of morbidity and mortality in women as it is in men (26–28), although the onset of the disease usually occurs about 10–15 years later in women than in men who are matched in all other respects. Less is known about the relationships between lipoproteins and cardiovascular disease in females than in males. Although there are not a lot of epidemiological data relating LDL levels to CAD risk in females (reviewed in Ref. 28), women with familial hypercholesterolemia have an increased risk of myocardial infarction, although the onset occurs later than in males (29,30). However, they appear to respond as well as men to aggressive lipid-lowering therapy (23). This implies that LDL is a modifiable risk factor in females and provides justification for the application of the ATP II guidelines to females, although more gender-related data clearly is required. Hypertriglyceridemia (28,31,32) and low HDL (28,31,33) may be more important risk factors in women than in men. It is not yet clear how this should affect strategies for prevention of atherosclerotic disease in females. However, measures that either decrease HDL levels or increase triglycerides should be undertaken with caution in females until further information is available.

The approach to diagnosis and management of hyperlipidemia in the elderly is even more controversial. The relationship between LDL cholesterol and CAD is less strong in the elderly than in young individuals (34), although the relationship is reasonably strong in some studies (35,36) and weak in others (37,38). Thus, the relative risk of heart disease is lower in the elderly than in the young. However, since many more elderly persons suffer complications of atherosclerotic disease than do young ones, the attributable risk is considerably higher in the elderly than in younger individuals (39). Limited information is available about the response to lipid-lowering therapy in the elderly. However, from the information that is available, there does not appear to be a different response between older and younger participants in lipid-lowering studies (40,41). Also, a dietary intervention study performed in elderly men (42) had results similar to those of dietary studies performed in younger individuals. Since age is an important independent cardio-vascular risk factor (36), the elderly are thus more likely to develop clinical manifestations of CAD in the short term than are the young. Therefore, cost-benefit analyses support intervention in this group of individuals (15), who might benefit from therapy after a short period of time.

However, if LDL is not a significant risk factor in the elderly, direct application of the ATP II guidelines to this large group of individuals may not be appropriate. Many elderly individuals have other diseases that may limit their lifespans. Therefore, these individuals should not undergo expensive lipid-lowering therapy. It also could be argued that use of the considerable resources required for a lipid-lowering program such as ATP II should be reserved primarily for those at risk of *premature* cardiovascular disease. There is probably no aspect of the ATP II guidelines that calls for good clinical judgment more than treatment of the elderly. A reasonable approach may be to use diet therapy for primary prevention of atherosclerosis in the elderly, while reserving drug treatment for (1) secondary prevention, (2) the rare individual with familial hypercholesterolemia who reaches old age, and (3) elderly subjects with marked elevations of LDL cholesterol levels not due to other disorders such as hypothyroidism.

Young individuals without other risk factors seldom develop premature vascular disease, which tends to occur mainly in individuals with multiple risk factors, especially in the face of a positive family history (43,44). Following publication of the original guidelines a disproportionate use of drug therapy occurred in young individuals without other risk factors (4). This group of individuals are likely to have a low cost-to-benefit ratio (15). To prevent inappropriate use of valuable resources, the updated guidelines have included higher cutpoints for the use of drug therapy (>220 mg/dl) in young men (>45 years) and females before the menopause. However, these higher cutpoints for the young should not preclude the use of drug therapy in high-risk young individuals with other risk factors, since serum cholesterol in the young predicts subsequent cardiovascular disease (45) and since individuals often develop clinical events below the age of 45 years (44).

Overemphasis on Lipids in CHD Prevention and Treatment of Atherosclerotic Disorders

The focus of the NCEP on lipids and lipoproteins has been felt by some to neglect the many other important cardiovascular risk factors that play a role in the genesis of this multifactorial disease. However, as mentioned earlier, atherosclerosis is not much of a problem in populations with low levels of LDL cholesterol, even when other risk factors are present (19). By categorizing individuals on the basis of other risk factors, including established atherosclerotic disease, ATP II now focuses on *overall cardiovascular risk status* as the major determinant of whether and how to treat. The algorithms developed emphasize the need to detect and treat risk factors other than lipids and lipoproteins. The increased focus in ATP II on life-style changes such as diet, exercise, and weight control also should benefit other risk factors such as central obesity, hypertension, non–insulin-dependent diabetes, and insulin resistance.

HDL

Inclusion of HDL in the Initial Evaluation

A consensus conference concerning the role of HDL and triglycerides as cardio-vascular risk factors was held at the National Institute of Health in 1992 (10). Consensus was reached that sufficient data was available concerning the predictive value of HDL as a cardiovascular risk factor from multiple studies throughout the world to justify its inclusion during the initial evaluation of lipids. Methods for measurement of HDL cholesterol had also become widely available and quality control had improved sufficiently to make the measurement of HDL cholesterol practicable.

High HDL as a Negative Risk Factor

Just as a low level of HDL cholesterol is predictive of the development of cardiovascular disease in the general population, so is a high HDL cholesterol level predictive of protection from myocardial infarction (46). Even in high-risk individuals, a relatively high level of HDL cholesterol is associated with less atherosclerotic disease (47). Therefore, it seems reasonable to designate high HDL cholesterol as a negative risk factor. However, the inverse relationship between HDL cholesterol and cardiovascular disease incidence is continuous and does not appear to have a threshold value. Thus, a value for HDL cholesterol that could be designated "high" had to be chosen analogous to the designation of values <35 mg/dl as "low." Choice of such a value therefore is somewhat arbitrary, but nonetheless has to be practical and stratify subjects into distinct categories with different risks. Despite the well-known gender differences in the distribution of HDL cholesterol levels (26), for reasons of simplicity and practicality, it was decided not to specify separate cutpoints for high HDL cholesterol levels for males and females, consistent with the choice of a single value designating low HDL cholesterol levels in both sexes.

HDL in Therapeutic Decisions

Therapeutic measures to increase HDL cholesterol levels still lag considerably behind those for lowering LDL cholesterol. Also, to date no intervention trials have specifically shown that increasing HDL will reduce cardiovascular disease. However, clinical trials (48,49) and angiographic studies (20–23) have strongly suggested that increases in HDL levels have a beneficial effect on outcomes. Thus, it was felt that to be important to identify individuals with low HDL cholesterol levels who might respond to lifestyle changes such as weight reduction and increased physical activity. Further, when selecting drugs for treatment of high LDL cholesterol levels, nicotinic acid can be a good drug of choice if HDL cholesterol levels are low concurrently. Finally, since it is easier to reduce LDL than to increase HDL levels, it seems reasonable to be more aggressive in treating raised LDL cholesterol in individuals who also have low HDL cholesterol values.

Emphasis on Family History

A genetic disorder of lipoprotein metabolism is present in more than half the cases of premature CAD (50). A positive family history of premature atherosclerotic disease is a very important cardiovascular risk factor, independent of measurement of cholesterol levels (51). Individuals with familial forms of hyperlipidemia, such as familial hypercholesterolemia and familial combined hyperlipidemia, are unlikely to achieve and maintain target LDL cholesterol levels with diet therapy alone and usually require the use of lipid-lowering drugs. Therefore ATP II stresses the importance of family history during evaluation of risk factors. Since the parents of the individual being evaluated might not yet have clinical manifestations of atherosclerotic disease or might have had other diseases that led to death prior to the onset of symptoms of CAD, the family history should be more extensive than mere evaluation of first-degree relatives. Inquiry about grandparents and uncles and aunts, particularly on the mother's side of the family, can provide considerable information that points to the presence of a familial form of hyperlipidemia. This information is useful in applying clinical judgment in difficult cases or where the decision to treat or use drugs is not clear-cut. Because of the markedly increased risk associated with these familial disorders, it is reasonable to use a lower threshold for intervention than when the family history clearly is negative.

Small, Dense LDL

As alluded to earlier, the presence of small, dense LDL is associated with increased cardiovascular risk. LDL particles tend to be distributed in two major subpopulations, which are in part genetically determined (52). About 20–25% of the population have small, dense LDL as their major LDL subpopulations—so-called pattern B. When the number of these small, dense particles is increased in plasma for whatever reason, cardiovascular risk is increased (53). Because these small, dense particles have a low ratio of cholesterol to protein, plasma LDL cholesterol levels sometimes are within the normal range despite an increase in the number of these particles and might not be elevated by NCEP criteria. Cardio-vascular risk may nonetheless be increased (53).

Small, dense LDL occur as part of a pattern of distribution of lipoprotein particles that has been termed "dyslipidemia." VLDL also tend to small and dense, and there is some accumulation of remnants of the triglyceride-rich lipo-proteins (54). HDL levels tend to be low, although often not <35 mg/dl (54). This dyslipidemic pattern is characteristic of familial combined hyperlipidemia (55) and non–insulin-dependent diabetes (54). It also is seen in the insulin resistance syndrome (56), which is characterized by central obesity, dyslipidemia, hyperten-sion, and glucose intolerance, a constellation that has been termed the "deadly quartet" (57).

Measurement of LDL particle size and density is a research procedure that is unlikely to become widely available for clinical use in the foreseeable future. Thus, evaluation of LDL particle size and density was not included in ATP II. However, many of the features associated with the presence of small, dense LDL particles will result in patients with small, dense LDL being stratified into high-risk categories, particularly since these features, e.g., hypertension, diabetes, and low HDL, tend to coexist in the same individual, who thus is likely to end up with two or more net risk factors.

Obesity and the Insulin Resistance Syndrome

There are two major forms of obesity: gynoid obesity, in which the excess adipose tissue is distributed primarily over the lower body (buttocks and thighs), and android obesity, in which the excess adiposity has a more central distribution (9,58). The central deposition of adipose tissue is associated with the features of the insulin resistance syndrome alluded to earlier, which often cluster in the same individual (54). These metabolic derangements do not occur in gynoid obesity, which may explain why cardiovascular risk is increased in central but not gynoid obesity. The features of the insulin resistance syndrome and increased cardio-vascular risk appear to be even more pronounced when the central distribution of adiposity is intra-abdominal (visceral) rather than subcutaneous.

The central distribution of adiposity usually is easy to detect by simple clinical evaluation. Measurement of waist-to-hip ratios provides a number that has been useful in research studies (59), but that does not have much utility in clinical practice. More precise evaluation of adipose tissue distribution can be made by computerized tomography, but the very high cost of this procedure makes its clinical use quite impractical.

Although obesity per se is no longer included as a risk factor in ATP II, individuals with the dangerous central form of obesity nonetheless are likely to be detected by the ATP II algorithms by virtue of the various risk factors that occur in association with central obesity. Further, individuals with the more benign form of gynoid obesity, in whom these various metabolic abnormalities do not exist (58), will no longer be assigned to a high-risk category as might have occurred with the earlier guidelines. The increased emphasis on weight control and physical activity in ATP II also is likely to have a beneficial impact, since all of the associated metabolic abnormalities and risk factors, including hypertriglyceridemia (see below), improve with weight loss in central obesity (60). Unfortunately, some studies have suggested that weight loss does not necessarily improve cardiovascular risk because of the weight cycling that frequently accompanies it (61,62).

Hypertriglyceridemia

The role of hypertriglyceridemia as a cardiovascular risk factor is becoming increasingly clear. A 1992 consensus conference on triglycerides and HDL (10) provided some of the background for the recommendations regarding triglycerides included in ATP II. Epidemiological studies have been relatively consistent in showing a positive relationship between triglycerides and CAD using univariate analysis (63). However, when using multivariate analyses that include HDL, this independent relationship often is lost because of the interdependence of triglycerides and HDL (63). The relationship between triglycerides and CAD is stronger in women (28,32,63). Not all forms of hypertriglyceridemia impart equal cardiovascular risk. The hypertriglyceridemia associated with familial combined hyperlipidemia (64,65) and some forms of familial hypertriglyceridemia (65) appear to increase cardiovascular disease risk, whereas most families with familial hypertriglyceridemia are not at increased risk (64). Hypertriglyceridemia associated with diabetes mellitus (66), chronic renal failure (67,68), and the insulin resistance syndrome (54,56) also appear to increase risk, whereas most other secondary forms of hypertriglyceridemia, e.g., those due to ethanol, estrogens, and high-carbohydrate diets, do not (69). The hypertriglyceridemic states that are associated with increased risk tend to be characterized by the dyslipidemic pattern referred to earlier, with small, dense VLDL, remnant lipoproteins, small, dense LDL, and low HDL levels. Thus, hypertriglyceridemia may simply be a marker that heralds the presence of other subtle abnormalities of lipoprotein composition and concentration that may be atherogenic.

In ATP II, triglyceride values of 200–400 mg/dl are classified as borderline high, 400–1000 mg/dl as high, and >1000 mg/dl as very high. Since triglycerides may be a marker of some underlying disorder that imparts increased cardiovascular risk, the clinical utility of stratification of triglyceride levels into groups is not obvious. Rather, the approach to the management of the individual with hypertriglyceridemia requires a careful and detailed family history and an attempt to ascertain the cause of the hypertriglyceridemia. Individuals in whom hypertriglyceridemia is associated with a positive family history of premature atherosclerotic disease, diabetes, or insulin resistance syndrome are probably best treated more aggressively than when none of these situations is present. Therapy should include treatment of all the associated risk factors.

Triglyceride values > 1000 mg/dl, however, present a special situation and usually signify the coexistence of genetic and secondary forms of hypertriglyceridemia (70). With triglyceride values > 2000 mg/dl there is an increased risk of pancreatitis and other features of the chylomicronemia syndrome (71). Marked hypertriglyceridemia is a special situation that requires vigorous attempts at triglyceride reduction by recognition and treatment of underlying secondary forms of hypertriglyceridemia, elimination of drugs that increase triglyceride levels, and often specific drugs, such as fibrates or nicotinic acid, to reduce plasma triglyceride levels.

Diabetes

Management of the patient with diabetes requires special consideration in view of the marked increase in cardiovascular risk seen in both insulin-dependent and non–insulin-dependent diabetes (reviewed in Ref. 72). Further, hyperlipidemia or dyslipidemia occurs frequently in individuals with diabetes (reviewed in Ref. 73). Although diabetes is considered a single risk factor for the purpose of calculation of net risk factors, some authorities consider diabetes to be the equivalent of established atherosclerotic disease with respect to cardiovascular risk. Hence, lower thresholds for therapy, analogous to those used for the management of established atherosclerotic disease, have been suggested by some to be appropriate for use in patients with diabetes.

Hypertriglyceridemia is the most common plasma lipid abnormality seen in diabetes (73). The risk of coronary heart disease is increased in diabetic subjects with high triglyceride levels (66). HDL levels are normal or even increased in insulin-dependent diabetes (73) and are low in non–insulin-dependent diabetes as part of the dyslipidemic pattern that characterizes this disorder (73,74). LDL levels, on which therapeutic decisions are made in ATP II, usually are within the normal range, although often towards the high end of normal (73). Hypertension also occurs frequently in patients with diabetes, so they may be targeted for intervention by virtue of borderline high LDL levels and multiple risk factors.

However, many diabetic patients will not fulfill the criteria for therapy, especially drug therapy, because LDL levels are not particularly elevated in diabetes, unless complications such as the nephrotic syndrome are present or another familial or secondary form of hypercholesterolemia coexists. ATP II does not offer specific guidelines for the management of hypertriglyceridemia in diabetes. Information also is lacking as to whether therapy specifically directed towards lowering triglyceride levels will reduce the alarming incidence of cardiovascular disease in this disorder. Despite this lack of information, many authorities are using medications such as reductase inhibitors (75) or fibrates (76) in diabetic subjects with hyperlipidemia or dyslipidemia. Unfortunately, LDL cholesterol levels often increase with the use of fibrates (76), and reductase inhibitors often fail to normalize triglyceride levels (75), thus raising the question of whether a combination of a fibrate and a reductase inhibitor should be used despite the increased incidence of myopathy that occurs with this combination (77). Clearly, additional information concerning atherosclerosis prevention in diabetes is needed, including the role of tight glycemic control, which has been suggested to be important by the Diabetes Control and Complications Trial (78).

Apo B and Lp(a) as Designated Risk Factors in ATP II

Increased plasma levels of apolipoprotein B (apo B) are observed when LDL levels are elevated (79) and could therefore potentially be used as a surrogate marker for high LDL. A high plasma level of apo B also is a cardiovascular risk factor even when total plasma and LDL cholesterol levels are not clearly elevated (80). This situation is due to the presence of an increased number of small, dense LDL particles, which have a low ratio of cholesterol to protein (79). Thus, "hyperapo B" can occur in states characterized by the presence of small, dense LDL and be a marker of atherogenic disorders such as familial combined hyperlipidemia (81).

To date however, measurement of apo B has not been shown to have better predictive power than total plasma or LDL cholesterol (82), although recent data suggests that both apo B and Lp(a) may be of predictive value as assays improve (83). Further, assays for the measurement of apo B have not been adequately standardized, and their availability is not yet sufficiently widespread for its incorporation into the ATP guidelines. As more data regarding apo B become available and assays improve, utilization of apo B measures might be incorporated into future update of NCEP guidelines.

Lp(a), an LDL particle to which apo(a) is linked by a disulfide bridge, is another cardiovascular risk factor being intensively studied at present. Because of sequence homology with plasminogen, it competes with this protein, thereby favoring both thrombogenesis and atherogenesis (84). Although some recent studies have failed to show it to be predictive of CAD (85,86), subject selection (86) or problems with storage of samples (85–87) might explain the discrepancies with other cross-sectional (83,88) and prospective studies (83,89) that suggest that Lp(a) is an important risk factor. Levels of Lp(a) isoform distribution are different between whites and blacks and may signify less cardiovascular risk in blacks than in whites (90) for reasons that are not yet understood. Lp(a) levels also are resistant to many diet and drug therapies currently in use (91). Lp(a) measurement is difficult, in part because many of the antibodies that have been used in the past cross-react with plasminogen. Current assays also require further standardization due to the multiple isoforms of apo(a) reacting differently with many of the antibodies used (87). Therefore, measurement of Lp(a) levels was not included in ATP II, although it too may find a place in future revisions of these guidelines.

Established Atherosclerotic Disease

One of the major changes in ATP II is the inclusion of subjects with established atherosclerotic disease as a separate category, with lower target levels and cut-points for intervention with both diet and drugs. There are two major reasons for this change. First, it is apparent that the strongest predictor of a future myocardial infarct is a previous infarction or the presence of other forms of atherosclerotic disease (92,93). Second, several angiographic studies published during the past few years have shown conclusively that aggressive lipid-lowering therapy can delay the progression of CAD and even cause regression (21–23). While most of these studies have relied on drug therapy, similar results recently have been obtained using a cholesterol-lowering diet (20). Even though the magnitude of the change in luminal diameter with treatment often was not very impressive, a disproportionate and rapid decrease in clinical events was noted in several studies (22,94,95).

Recently it has become apparent that the atherosclerotic lesions that become occluded and result in myocardial infarctions are not the largest, most occlusive lesions. From angiograms obtained after myocardial infarction, it appears that thromboses that result in heart attacks are most likely to occur in lesions that occlude the lumen by only 40–60%. These lesions are believed to be unstable in that they undergo necrosis, fissuring, hemorrhage, and thrombosis, all of which can lead to acute occlusion and myocardial infarction (96). It is hypothesized that the reduction in clinical events associated with lipid-lowering therapy is due to stabilization of these potentially dangerous unstable atherosclerotic lesions (97,98). Slowing of atherosclerosis progression and induction of progression is related to the extent of lipid lowering, however achieved (97). It also appears to be maximal when total plasma cholesterol levels < 200 mg/dl or LDL cholesterol levels < 100 mg/dl are achieved (99). Because of the very high likelihood of the recurrence of clinical events in individuals with established disease, and because of the well-documented protection that can be afforded these individuals by aggressive lipid-lowering therapy, this secondary prevention approach is relatively cost-effective (15). Its inclusion in the revised guidelines is an important advance in targeting therapy to those at the highest risk.

Diet

Diet remains a mainstay of therapy in ATP II, since it is inexpensive, safe, and can be effective if properly applied. Yet diet therapy has frequently been criticized as being ineffective in achieving meaningful reductions in serum cholesterol levels. Indeed, the magnitude of cholesterol lowering in a recent well-controlled study was disappointingly low at 5% (100).

There are several reasons why individuals fail to respond adequately to diet therapy. The first and most common is poor adherence to the prescribed diet. Second, the prescribed diet may not differ sufficiently from that habitually consumed by the individual, particularly since health-conscious individuals in the United States tend to limit their dietary intake of saturated fat and cholesterol. Indeed, the nationwide consumption of total fat, saturated fat, and cholesterol continues to decline according to the most recent data from NHANES III (1). This may in part explain why a Step II diet was not very effective in the study alluded to earlier (100) and why a similar diet markedly reduced LDL cholesterol levels and atherosclerosis progression in Great Britain (20), where the population is less cholesterol-conscious and average consumption of fat is higher than in the United States. Third, diet therapy alone is unlikely to achieve marked cholesterol reduction in individuals with severe genetic forms of hyperlipidemia, although its effect is additive to and independent of drug therapy (100). Finally, some individuals may have true resistance to diet therapy on a genetic basis, analogous with that observed in inbred strains of nonhuman primates (101). However, genetic markers are not yet available to determine the existence or frequency of diet responsivity or resistance.

With these considerations in mind, the diet section of ATP II has been extensively revised, although it still retains the Step I and II diets, with their graded reductions in the consumption of calories from fat (especially saturated fat) and cholesterol. The emphasis is on consumption of a nutritionally adequate diet rich in fruit, vegetables, grains, cereals, and legumes. It recommends consumption of fish, poultry, low-fat meats, and low- or nonfat dairy products rather than foods that are rich in saturated fats and cholesterol. Saturated fatty acids continue to be identified as the nutrient most likely to raise blood cholesterol and are specifically targeted for reduction.

ATP II now strongly recommends the concerted and continued efforts of physicians, dietitians, nurses, physician-assistants, and other health care professionals to achieve successful and long-term dietary intervention. Frequent follow-up, reinforcement, encouragement, and education of the patient and his or her family are perceived as being a vital component of successful diet therapy. A useful instrument for rapid dietary assessment is included that should help to ascertain the individual's baseline diet. If an individual already is consuming a diet that approached Step I, it would be logical to move right on to a Step II diet. Sample menus are included for a number of different ethnic groups. Not only can these be useful for members of those ethnic groups, they can also be applied more generally to increase variety and palatability of a low-fat diet for the population at large. Exercise and weight control now is included under diet therapy and will be dealt with separately below.

This more practical approach to diet therapy should allow many subjects to achieve target LDL cholesterol levels with diet alone, thereby obviating the need for drug therapy. Further, long-term adherence to diet therapy can lower the dose of drug required (100), thereby reducing the cost and potential toxicity associated with lipid-lowering drugs.

Exercise and Weight Control

Maintenance of weight loss has proved difficult in practice (102). Successful maintenance of weight loss is aided by participation in an exercise program (103). The magnitude of reduction of LDL cholesterol levels can be greatly increased by concurrent weight loss (103,104). Further, plasma triglyceride levels fall and HDL cholesterol levels rise with sustained weight loss (103,105). HDL cholesterol levels also are increased by exercise independent of their effect on body weight (106). Therefore, exercise and body weight control have beneficial effects on several aspects of plasma lipids and lipoproteins.

Exercise also has beneficial effects on cardiovascular risk over and above its effect on serum lipid and lipoprotein levels. There appears to be a beneficial effect of exercise on cardiovascular disease risk that cannot be attributed to its effect on other risk factors (13) but that may relate to changes in the thrombolytic system (107) or be related to general cardiovascular fitness. Exercise also has a beneficial effect on blood pressure and blood glucose levels by reducing insulin resistance (108). Therefore, the use of exercise as a therapeutic modality has been included as an integral part of ATP II.

Drug Therapy

There are several differences in the approach to drug therapy between the original guidelines and ATP II. First and foremost, the guidelines have been revised so as to try and reserve the use of drugs for those individuals who clearly are at the highest risk by virtue of multiple risk factors, a strongly positive family history, or established atherosclerotic disease. Since drug therapy is relatively expensive, limiting its use to the highest-risk individuals should improve the cost-benefit ratio associated with lipid-lowering therapy. Second, experience with the HMG-CoA reductase inhibitors has become much greater since the original guidelines were published. Recommendations as to their use, including in combination therapy, are now included. The use of combinations of low doses of reductase inhibitors and bile acid resins or of reductase inhibitors and nicotinic acid appears to be more cost-effective than the use of high doses of reductase inhibitors, particularly since all four reductase inhibitors on the market have a curvilinear dose-response curve, with quantitatively greater responses at lower doses of the drug. Third, HDL cholesterol levels are taken into consideration in the choice of drug used. For example, nicotinic acid is the most effective drug currently available for raising HDL cholesterol. It is therefore a good choice when HDL cholesterol levels are low concurrently with elevated levels of LDL cholesterol. It also reduced plasma triglyceride levels and therefore is an excellent choice for the individual with dyslipidemia, although it frequently raises plasma glucose levels and therefore is difficult to use in patients with diabetes (109). Strategies to reduce the nuisance side effects associated with the use of nicotinic acid are included in ATP II, including the need to slowly titrate up to the maintenance dose, to take the medication with food, and to avoid its consumption concurrent with hot drinks or alcohol. The fibrates can also raise HDL cholesterol levels, although they usually are effective only if triglyceride levels are elevated concurrently (110). Since LDL cholesterol levels often increase with the use of fibrates, they need to be monitored carefully. The fibrates remain the drug of choice for marked hypertriglyceridemia (71). Finally, estrogen replacement therapy is now considered as a primary option in treating the hypercholesterolemic postmenopausal female, since this therapy can reduce LDL cholesterol and raise HDL cholesterol levels (111) as well as having beneficial effects on cardiovascular disease morbidity and mortality (112,113) and osteoporosis prevention (114).

Dangers of Low Blood Cholesterol

The potential danger of having too low a blood cholesterol level has been commented on. A U-shaped relationship exists between blood cholesterol level and mortality, i.e., mortality increases at the lowest levels of blood cholesterol (115). The reasons for this increase at the lowest cholesterol levels are not clear but are due in part to the fact that diseases associated with increased mortality, such as cancer and liver disease, cause low cholesterol levels (115). However, some studies have suggested that this U-shaped relationship persists even when individuals with these disorders have been eliminated from consideration (115). Some studies have suggested that this relationship is strongest among people of poor socioeconomic status due to old age or declining health (116,117). Even if this relationship is true, the increased mortality is only seen at total serum cholesterol levels below about 140 mg/dl, a level that occurs in about 2% of the population and is seldom achieved during lipid-lowering therapy for hypercholesterolemia. There is no data to suggest that lowering cholesterol levels into this range has the same effect. However, several of the intervention studies have shown a small increase, often not statistically significant, in noncardiac mortality in subjects receiving intervention for high cholesterol levels (reviewed in Ref. 115). No plausible mechanism for these deaths, which appear to be from a variety of causes, has been suggested. A NIH-sponsored consensus conference a few years ago concluded that there was little evidence to support an adverse effect of cholesterol lowering but that further data was required (115). A recent analysis of the largest cohort studies and randomized trials concluded that there is no evidence that low or reduced serum cholesterol levels increase mortality from any cause other than hemorrhagic stroke (18), which clearly does appear to be associated with very low cholesterol levels (117). Thus, there appears to be little cause for concern about potential adverse effects of applying the ATP II guidelines for cholesterol lowering.

THE FUTURE

As further knowledge regarding the relationship of lipids and lipoproteins to cardiovascular disease is obtained and as advances in the management of hyperlipidemia become available, periodic updates of these guidelines are likely, as has occurred with the guidelines for the management of hypertension. As methods become available for rapid and accurate detection of genetic forms of hyperlipidemia, these will no doubt be used to screen individuals from high-risk families. With the increasing appreciation of the importance of the dyslipidemic pattern of lipoprotein distribution, additional approaches to its diagnosis and management are likely to appear. Since central obesity appears to be a common underlying feature of many of the cardiovascular risk factors, it is likely that there will be intensive research efforts to understand the genetic and metabolic basis of central obesity and the mechanisms whereby it leads to accelerated atherosclerosis. Advances in the understanding of appetite regulation and body weight control are also likely to occur in the next few years. Such advances will no doubt continue to include an exercise component. When strategies to effectively control body weight and insulin resistance become available, many of the risk factors, including hyperlipidemia/dyslipidemia, might be manageable by a single approach. Availability of effective weight control measures also is likely to bring the management of hyperlipidemia, hypertension, and diabetes more closely together, since these disorders frequently coexist in the same subject.

REFERENCES

1. Sempos CT, Cleeman JI, Carroll MD, Johnson CL, Bachorik PS, Gordon DJ, Burt VL, Briefel RR, Brown CD, Lippel K, et al. Prevalence of high blood cholesterol among US adults. An update based on guidelines from the second report of the

National Cholesterol Education Program Adult Treatment Panel. JAMA 1993; 269: 3009–3014.

2. Sytkowski PA, Kannel WB, D'Agostino RB. Changes in risk factors and the decline in mortality from cardiovascular disease. The Framingham Heart Study. N Engl J Med 1990; 322:1635–1641.

3. Members of the National Cholesterol Education Program Expert Panel. Report of the National Cholesterol Program Expert Panel on detection, evaluation, and treatment of high blood cholesterol in adults. Arch Intern Med 1988; 148:36–69.

4. McIsaac WJ, Naylor CD, Basinski A. Mismatch of coronary risk and treatment intensity under the National Cholesterol Education Program Guidelines. J Gen Intern Med 1991; 6:518–523.

5. Miller M, Seidler A, Kwiterovich PO, Pearson TA. Long-term predictors of subsequent cardiovascular events with coronary artery disease and "desirable" levels of plasma total cholesterol. Circulation 1992; 86:1165–1170.

6. Members of the National Cholesterol Education Program Expert Panel. Summary of the second report of the National Cholesterol Education Program (NCEP) Expert Panel on detection, evaluation, and treatment of high blood cholesterol in adults (Adult Treatment Panel II). JAMA 1993; 269:3015–3023.

7. Friedewald WT, Levy RI, Fredrickson DS. Estimation of the concentration of low-density lipoprotein cholesterol in plasma, without use of the preparative ultracentrifuge. Clin Chem 1972; 18:499–502.

8. Despres JP, Moorjani S, Lupien PJ, Tremblay A, Nadeau A, Bouchard C. Regional distribution of body fat, plasma lipoproteins, and cardiovascular disease. Arteriosclerosis 1990; 10:497–511.

9. Bjorntorp P. Abdominal fat distribution and disease: an overview of epidemiological data. Ann Med 1992; 24:15–18.

10. NIH Consensus Development Panel on Triglyceride, High-Density Lipoprotein, and Coronary Heart Disease. Triglyceride, high-density lipoprotein, and coronary heart disease. JAMA 1993; 269:505–510.

11. Wilson PWF, Abbott RD, Castelli WP. High density lipoprotein cholesterol and mortality: the Framingham Heart Study. Arteriosclerosis 1988; 8:737–741.

12. Goldbourt U, Yaari S. Cholesterol and coronary heart disease mortality: a 23-year follow-up study of 9902 men in Israel. Arteriosclerosis 1990; 10:512–519.

13. Sandvik L, Erikssen J, Thaulow E, Erikssen G, Mundal R, Rodahl K. Physical fitness as a predictor of mortality among healthy, middle-aged Norwegian men. N Engl J Med 1993; 328:533–537.

14. Joint Statement by the American Heart Association and the National Heart, Lung, and Blood Institute. The cholesterol facts: a summary of the evidence relating dietary fats, serum cholesterol, and coronary heart disease. Circulation 1990; 81: 1721–1733.

15. National Cholesterol Education Program. Second report of the expert panel on detection, evaluation, and treatment of high blood cholesterol in adults (Adult Treatment Panel II). NIH Publication No. 93-3095, 1993.

16. Law MR, Wald NJ, Wu T, Hackshaw A, Bailey A. Systematic underestimation of

association between serum cholesterol concentration and ischaemic heart disease in observational studies: data from the BUPA study. Br Med J 1994; 308:363–366.

17. Law MR, Wald NJ, Thompson SG. By how much and how quickly does reduction in serum cholesterol concentration lower risk of ischaemic heart disease? Br Med J 1994; 308:367–373.

18. Law MR, Thompson SG, Wald NJ. Assessing possible hazards of reducing serum cholesterol. Br Med J 1994; 308:373–379.

19. People's Republic of China-United States Cardiovascular and Cardiopulmonary Epidemiology Research Group. An epidemiological study of cardiovascular and cardiopulmonary disease risk factors in four populations in the People's Republic of China: baseline report from the PRC-USA. Circulation 1992; 85:1083–1096.

20. Watts GF, Lewis B, Brunt JNH, et al. Effects on coronary artery disease of lipid-lowering diet, or diet plus cholestyramine, in the St. Thomas' Atherosclerosis Regression Study (STARS). Lancet 1992; 339:563–569.

21. Blankenhorn DH, Nessim SA, Johnson RL, Sanmarco ME, Azen SP, Cashin-Hemphill L. Beneficial effects of combined colestipol-niacin therapy on coronary atherosclerosis and coronary venous bypass grafts. JAMA 1987; 257:3233–3240.

22. Brown G, Albers JJ, Fisher LD, et al. Regression of coronary artery disease as a result of intensive lipid-lowering therapy in men with high levels of apolipoprotein B. N Engl J Med 1990; 323:1289–1298.

23. Kane JP, Malloy MJ, Ports TA, Phillips NR, Diehl JC, Havel RJ. Regression of coronary atherosclerosis during treatment of familial hypercholesterolemia with combined drug regimens. JAMA 1990; 264:3007–3012.

24. Rossouw JE, Lewis B, Rifkind BM. The value of lowering cholesterol after myocardial infarction. N Engl J Med 1990; 323:1112–1119.

25. Pathobiological Determinants of Atherosclerosis in Youth (PDAY) Research Group. Natural history of aortic and coronary atherosclerotic lesions in youth. Findings from the PDAY Study. Arterioscler Thromb 1993; 13:1291–1298.

26. Bush TL, Fried LP, Barrett-Connor E. Cholesterol, lipoproteins, and coronary heart disease in women. Clin Chem 1988; 34:B60–B70.

27. Kannel WB. Metabolic risk factors for coronary heart disease in women: Perspective from the Framingham Study. Am Heart J 1987; 114:413–419.

28. Moreno GT, Manson JE. Cholesterol and coronary heart disease in women: an overview of primary and secondary prevention. Coronary Artery Dis 1993; 4:580–587.

29. Rosenberg L, Kaufman DW, Helmrich SP, Miller DR, Stolley PD, Shapiro S. Myocardial infarction and cigarette smoking in women younger than 50 years of age. JAMA 1985; 253:2965–2969.

30. Slack J. Risks of ischaemic heart disease in familial hypercholesterolemic states. Lancet 1969; 2:1380–1382.

31. Gordon WP, Hjortland MC, Kannel WB, Sawber TR. High density lipoprotein as a protective factor against coronary heart disease. Am J Med 1977; 62:707–713.

32. Heyden S, Heiss G, Hames CG, Bartel AG. Fasting triglycerides as predictors of CHD mortality in Evans County, Georgia. J Chron Dis 1980; 33:275–282.

33. Jacobs DR, Mebane IL, Bangdiwala SI, Criqui MH, Tyroler HA. High density lipoprotein cholesterol as a predictor of cardiovascular disease mortality in men and women: the follow-up study of the Lipid Research Clinics Prevalence Study. Am J Epidemiol 1990; 131:32–34.

34. Manolio TA, Pearson TA, Wenger NK, Barrett-Connor E, Payne GH, Harlan WR. Cholesterol and heart disease in older persons and women. Review of an NHLBI Workshop. Ann Epidemiol 1992; 2:161–176.

35. Benfante R, Reed D. Is elevated serum cholesterol level a risk factor for coronary heart disease in the elderly? JAMA 1990; 263:393–396.

36. Rubin SM, Sidney S, Black DM, Browner WS, Hulley SB, Cummings SR. High blood cholesterol in elderly men and the excess risk for coronary heart disease. Ann Intern Med 1990; 113:916–920.

37. Castelli WP, Wilson PWF, Levy D, Anderson K. Cardiovascular risk factors in the elderly. Am J Cardiol 1989; 63:12H–19H.

38. Welborn TA, Wearne K. Coronary heart disease incidence and cardiovascular mortality in Busselton with reference to glucose and insulin concentrations. Diabetes Care 1979; 2:154–160.

39. Malenka DJ, Baron JA. Cholesterol and coronary heart disease. The importance of patient-specific attributable risk. Arch Intern Med 1988; 148:2247–2252.

40. Canner PL, Berge KG, Wenger NK, Stamler J, Friedman L, Prineas RJ, et al. Fifteen year mortality in coronary drug project patients: long-term benefit with niacin. J Am Coll Cardiol 1986; 8:1245–1255.

41. Lipid Research Clinics Program. The Lipid Research Clinics Coronary Primary Prevention Trial results: II. The relationship of reduction in incidence of coronary heart disease to cholesterol lowering. JAMA 1984; 251:365–374.

42. Dayton S, Pearce ML, Goldman H, Harnish A, Plotkin D, Shickman M, Winfield M, Zager A, Dixon W. Controlled trial of a diet high in unsaturated fat for prevention of atherosclerotic complications. Lancet 1968; 2:1060–1062.

43. Burke GL, Savage PJ, Sprafka JM, Selby JV, Jacobs Jr DR, Perkins LL, Roseman JM, Hughes GH, Fabsitz RR. Relation of risk factor levels in young adulthood to parental history of disease. The CARDIA Study. Circulation 1991; 84:1176–1187.

44. Tornvall P, Bavenholm P, Landou C, de Faire U, Hamsten A. Relation of plasma levels and composition of apolipoprotein B-containing lipoproteins to angiographically defined coronary artery disease in young patients with myocardial infarction. Circulation 1993; 88:2180–2189.

45. Klag MJ, Ford DE, Mead LA, He J, Whelton PK, Liang K-Y, Levine DM. Serum cholesterol in young men and subsequent cardiovascular disease. N Engl J Med 1993; 328:313–318.

46. Wilson PWF, Abbott RD, Castelli WP. High density lipoprotein cholesterol and mortality: the Framingham Heart Study. Arteriosclerosis 1988; 8:737–741.

47. Sharp SD, Williams RR, Hunt SC, Schumacher MC. Coronary risk factors and the severity of angiographic coronary artery disease in members of high-risk pedigrees. Am Heart J 1992; 123:279–285.

48. Lipid Research Clinics Program. The Lipid Research Clinics Coronary Primary

Prevention Trial results. I. Reduction in incidence of coronary heart disease. JAMA 1984; 251:351–364.

49. Manninen V, Tenkanen L, Koskinen P, et al. Joint effects of serum triglyceride and LDL cholesterol and HDL cholesterol concentrations on coronary heart disease risk in the Helsinki Heart Study: implications for treatment. Circulation 1992; 85:37–45.

50. Genest Jr JP, Martin-Munley SS, McNamara JR, Ordovas JM, Jenner J, Myers RH, Silberman SR, Wilson PWF, Salem DN, Schaefer EJ. Familial lipoprotein disorders in patients with premature coronary artery disease. Circulation 1992; 85:2025–2033.

51. Shea S, Ottman R, Gabrieli C, Stein Z, Nichols A. Family history as an independent risk factor for coronary artery disease. J Am Coll Cardiol 1984; 4:793–801.

52. Austin MA, King MC, Vranizan KM, Newman B, Krauss RM. Inheritance of low-density lipoprotein subclass patterns: results of complex segregation analysis. Am J Human Genet 1988; 43:838–846.

53. Austin MA, Breslow JL, Hennekens CH, Buring JE, Willett WC, Krauss RM. Low-density lipoprotein subclass patterns of risk of myocardial infarction. JAMA 1988; 260:1917–1921.

54. Krauss RM. The tangled web of coronary risk factors. Am J Med 1991; 90: 36S–41S.

55. Austin MA, Brunzell JD, Fitch WL, Krauss RM. Inheritance of low density lipoprotein subclass patterns in familial combined hyperlipidemia. Arteriosclerosis 1990; 10:520–530.

56. Selby JV, Austin MA, Newman B, Zhang D, et al. LDL subclass phenotypes and the insulin resistance syndrome in women. Circulation 1993; 88:381–387.

57. Kaplan NM. The deadly quartet: upper-body obesity, glucose intolerance, hyper-triglyceridemia, and hypertension. Arch Intern Med 1989; 149:1514–1520.

58. Kissebah AH. Insulin resistance in visceral obesity. Int J Obesity 1991; 15: 109–115.

59. Freedman DS, Jacobsen SJ, Barboriak JJ, Sobocinski KA, Anderson AJ, Kissebah AH, Sasse EA, Gruchow HW. Body fat distribution and male/female differences in lipids and lipoproteins. Circulation 1990; 81:1498–1506.

60. Fagerberg B, Berglund A, Andersson OK, Berglund G. Weight reduction versus antihypertensive drug therapy in obese men with high blood pressure: effects upon plasma insulin levels and association with changes in blood pressure and serum lipids. J Hypertension 1992; 10:1053–1061.

61. Blair SN, Shaten J, Brownell K, Collins G, Lissner L. Body weight change, all-cause mortality, and cause-specific mortality in the Multiple Risk Factor Intervention Trial. Ann Intern Med 1993; 119:749–757.

62. Jeffery RW, Wing RR, French SA. Weight cycling and cardiovascular risk factors in obese men and women. Am J Clin Nutr 1992; 55:641–644.

63. Austin MA. Plasma triglyceride and coronary heart disease. Arterioscler Thromb 1991; 11:2–14.

64. Brunzell JD, Schrott HG, Motulsky AG, Bierman EL. Myocardial infarction in the familial forms of hypertriglyceridemia. Metabolism 1976; 25:313–320.

65. Goldstein JL, Schrott HG, Hazzard WR, Bierman EL, Motulsky AG. Hyper-lipidemia in coronary heart disease. II. Genetic analysis of lipid levels in 176 families and delineation of a new inherited disorder, combined hyperlipidemia. J Clin Invest 1973; 52:1544–1568.

66. West KM, Ahuja MMS, Bennett PH, Czyzky A, Mateo de Acosta O, Fuller JH, Grab B, Grabauskas V, Jarrett RJ, Kosaka K, Keen H, Krolewski AB, Miki E, Schliak V, Teuscher A, Watkins PJ, Stober JA. The role of circulating glucose and triglyceride concentrations and their interactions with other "risk factors" as determinants of arterial disease in nine diabetic population samples from the WHO multinational study. Diabetes Care 1983; 6:361–369.

67. Ma KW, Greene EL, Raij L. Cardiovascular risk factors in chronic renal failure and hemodialysis populations. Am J Kidney Dis 1992; 19:505–513.

68. Tschope W, Koch M, Thomas B, Ritz E. Serum lipids predict cardiac death in diabetic patients on maintenance hemodialysis. Results of a prospective study. The German Study Group Diabetes and Uremia. Nephron 1993; 64:354–358.

69. Chait A, Brunzell JD. Acquired hyperlipidemia (secondary dyslipoproteinemias). Endocrinol Metab Clin North Am 1990; 19:259–278.

70. Chait A, Brunzell JD. Severe hypertriglyceridemia: role of familial and acquired disorders. Metabolism 1983; 32:209–214.

71. Chait A, Brunzell JD. Chylomicronemia syndrome. Adv Intern Med 1992; 37: 249–273.

72. Chait A, Bierman EL. Pathogenesis of macrovascular disease in diabetes. In: Kahn CR, Weir GC, eds. Joslin's Diabetes Mellitus. 13th ed. Philadelphia: Lea & Febiger, 1994:648–664.

73. Brunzell JD, Chait A, Bierman EL. Plasma lipoproteins in human diabetes mellitus. In: Alberti KG, Krall LP, eds. The Diabetes Annual I. Amsterdam: Elsevier Science Publishers, 1985:463–479.

74. Barakat HA, McLendon VD, Marks R, Pories W, Heath J, Carpenter JW. Influence of morbid obesity and non-insulin-dependent diabetes mellitus on high-density lipoprotein composition and subpopulation distribution. Metabolism 1992; 41: 37–41.

75. Garg A, Grundy SM. Lovastatin for lowering cholesterol levels in non-insulin-dependent diabetes mellitus. N Engl J Med 1988; 318:81–86.

76. Garg A, Grundy SM. Gemfibrozil alone and in combination with lovastatin for treatment of hypertriglyceridemia in NIDDM. Diabetes 1989; 38:364–372.

77. Glueck CJ, Oakes N, Speirs J, Tracy T, Lang J. Gemfibrozil- lovastatin therapy for primary hyperlipoproteinemias. Am J Cardiol 1992; 70:1–9.

78. The Diabetes Control and Complications Trial Research Group. The effect of intensive treatment of diabetes on the development and progression of long-term complications in insulin-dependent diabetes mellitus. N Engl J Med 1993; 329: 977–986.

79. Sniderman A, Shapiro S, Marpole D, Skinner B, Teng B, Kwiterovich Jr PO. Association of coronary atherosclerosis with hyperapobetalipoproteinemia [increased protein but normal cholesterol levels in human plasma low density (beta) lipoproteins]. Proc Natl Acad Sci USA 1980; 77:604–608.

80. Sniderman AD, Wolfson C, Teng B, Franklin FA, Bachorik PS, Kwiterovich Jr PO. Association of hyperapobetalipoproteinemia with endogenous hypertriglyceridemia and atherosclerosis. Ann Intern Med 1982; 97:833–839.

81. Austin MA, Horowitz H, Wijsman E, Krauss RM, Brunzell J. Biomodality of plasma apolipoprotein B levels in familial combined hyperlipidemia. Atherosclerosis 1992; 92:67–77.

82. Stampfer MJ, Sacks FM, Salvini S, Willett WC, Hennekens CH. A prospective study of cholesterol, apolipoproteins, and the risk of myocardial infarction. N Engl J Med 1991; 325:373–381.

83. Rader DJ, Hoeg JM, Brewer Jr HB. Quantitation of plasma apolipoproteins in the primary and secondary prevention of coronary artery disease. Ann Intern Med 1994; 120:1012–1025.

84. Scanu AM. Lipoprotein(a): a potential bridge between the fields of atherosclerosis and thrombosis. Arch Pathol Lab Med 1988; 112:1045–1047.

85. Jauhiainen M, Koskinen P, Ehnholm C, et al. Lipoprotein (a) and coronary heart disease risk. A nested case-control study of the Helsinki Heart Study participants. Atherosclerosis 1991; 89:59–67.

86. Ridker PM, Hennekens CH, Stampfer MJ. A prospective study of lipoprotein (a) and the risk of myocardial infarction. JAMA 1993; 270:2195–2199.

87. Albers JJ, Marcovina SM, Lodge MS. The unique lipoprotein(a): properties and immunochemical measurement. Clin Chem 1990; 36:2019–2026.

88. Dahlen GH, Guyton JR, Altar M, Farmer JA, Kautz JA, Gotto AM. Association of levels of lipoprotein(a), plasma lipids, and other lipoproteins with coronary artery disease documented by angiography. Circulation 1986; 74:758–765.

89. Schaefer EJ, Lamon-Fava S, Jenner JL, McNamara JR, Ordovas JM, David E, Abolafia JM, Lippel K, Levy RI. Lipoprotein(a) levels and risk of coronary heart disease in men: The Lipid Research Clinics Coronary Primary Prevention Trial. JAMA 1994; 271:999–1003.

90. Marcovina SM, Zhang ZH, Gaur VP, Albers JJ. Identification of 34 apolipoprotein(a) isoforms: differential expression of apolipoprotein(a) alleles between American blacks and whites. Biochem Biophys Res Commun 1993; 191:1192–1196.

91. Scanu AM, Lawn RM, Berg K. Lipoprotein(a) and atherosclerosis. Ann Intern Med 1991; 115:209–218.

92. Pekkanen J, Linn S, Heiss G. et al. Ten-year mortality from cardiovascular disease in relation to cholesterol level among men with and without preexisting cardiovascular disease. N Engl J Med 1990; 322:1700–1707.

93. Criqui MH, Langer RD, Fronek A, et al. Mortality over a period of 10 years in patients with peripheral arterial disease. N Engl J Med 1992; 326:381–386.

94. Pearson TA, Marx HJ. The rapid reduction in cardiac events with lipid-lowering therapy: mechanisms and implications. Am J Cardiol 1993; 72:1072–1073.

95. The Pravastatin Multinational Study Group for Cardiac Risk Patients. Effects of pravastatin in patients with serum total cholesterol levels from 5.2 to 7.8 mmol/liter (200 to 300 mg/dl) plus two additional atherosclerotic risk factors. Am J Cardiol 1993; 72:1031–1037.

96. Davies MJ, Richardson PD, Woolf N, Katz DR, Mann J. Risk of thrombosis in

human atherosclerotic plaques: role of extracellular lipid, macrophage, and smooth muscle cell content. Br Heart J 1993; 69:377–381.

97. Brown BG, Zhao XQ, Sacco DE, Albers JJ. Atherosclerosis regression, plaque disruption, and cardiovascular events: a rationale for lipid lowering in coronary artery disease. Annu Rev Med 1993; 44:365–376.

98. Brown BG, Zhao XQ, Sacco DE, Albers JJ. Lipid lowering and plaque regression. New insights into prevention of plaque disruption and clinical events in coronary disease. Circulation 1993; 87:1781–1791.

99. Stewart BF, Brown BG, Zhao XQ, Hillger LA, Sniderman AD, Dowdy A, Fisher LD, Albers JJ. Benefits of lipid-lowering therapy in men with elevated apolipoprotein B are not confined to those with very high low density lipoprotein cholesterol. J Am Coll Cardiol 1994; 23:899–906.

100. Hunninghake DB, Stein EA, Dujovne CA, Harris WS, Feldman EB, Miller VT, Tobert JA, Laskarzewski PM, Quiter E, Held J, Taylor AM, Hopper S, Leonard SB, Brewer BK. The efficacy of intensive dietary therapy alone or combined with lovastatin in outpatients with hypercholesterolemia. N Engl J Med 1993; 328:1213–1219.

101. McGill Jr HC, McMahan CA, Mott GE, Marinez YN, Kuehl TJ. Effects of selective breeding on the cholesterolemic responses to dietary saturated fat and cholesterol in baboons. Arteriosclerosis 1988; 8:33–39.

102. NIH Technology Assessment Conference Panel. Methods for voluntary weight loss and control. Consensus Development Conference March 30–April 1, 1992. Ann Intern Med 1993; 119:764–770.

103. Wood PD, Stefanick ML, Williams PT, Haskell WL. The effects of plasma lipoproteins of a prudent weight-reducing diet, with or without exercise, in overweight men and women. N Engl J Med 1991; 325:461–466.

104. Caggiula AW, Christakis G, Farrand M, et al. The multiple risk factor intervention trial (MRFIT). IV. Intervention on blood lipids. Prev Med 1981; 10:443–475.

105. Wood PD, Stefanick ML, Dreon DM, et al. Changes in plasma lipids and lipoproteins in overweight men during weight loss through dieting as compared with exercise. N Engl J Med 1988; 319:1173–1179.

106. Schieken RM. Effect of exercise on lipids. Ann NY Acad Sci 1991; 623:269–274.

107. Chandler WL, Veith RC, Fellingham GW, Levy WC, Schwartz RS, Cerqueira MD, Kahn SE, Larson VG, Cain KC, Beard JC, et al. Fibrinolytic response during exercise and epinephrine infusion in the same subjects. J Am Coll Cardiol 1992; 19:1412–1420.

108. Kahn SE, Larson VG, Beard JC, Cain KC, Fellingham G, Schwartz RS, Veith RC, Stratton JR, Cerqueira MD, Abrass IB. Effect of exercise on insulin action, glucose tolerance, and insulin secretion in aging. Am J Physiol 1990; 258:E937–E943.

109. Stern MP, Haffner SM. Dyslipidemia in type II diabetes. Implications for therapeutic intervention. Diabetes Care 1991; 14:1144–1159.

110. Zimetbaum P, Frishman WH, Kahn S. Effects of gemfibrozil and other fibric acid derivatives on blood lipids and lipoproteins. J Clin Pharmacol 1991; 31:25–37.

111. Lobo RA. Effects of hormonal replacement on lipid and lipoproteins in postmenopausal women. J Clin Endocrinol Metab 1991; 73:925–930.

112. Barrett-Connor E, Bush TL. Estrogen and coronary heart disease in women. JAMA 1991; 265:1861–1867.
113. Grady D, Rubin SM, Petitti DB, et al. Hormone therapy to prevent disease and prolong life in postmenopausal women. Ann Intern Med 1992; 117:1016–1037.
114. Odell WD, Heath 3rd H. Osteoporosis: pathophysiology, prevention, diagnosis and treatment. Dis Mon 1993; 39:789–867.
115. Jacobs D, Blackburn H, Higgins M, Reed D, Iso H, McMillan G, Neaton J, Nelson J, Potter J, Rifkind B, et al. Report of the Conference on Low Blood Cholesterol: mortality associations. Circulation 1992; 86:1046–1060.
116. Harris T, Feldman JJ, Kleinman JC, et al. The low cholesterol-mortality association in a national cohort. J Clin Epidemiol 1992; 45:595–601.
117. Yano K, Reed DM, Maclean CJ. Serum cholesterol and hemorrhagic stroke in the Honolulu Heart Program. Stroke 1992; 20:1460–1465.

2

Cholesterol Lowering and Total Mortality

David J. Gordon

National Heart, Lung, and Blood Institute, National Institutes of Health, Bethesda, Maryland

INTRODUCTION

The etiological role of cholesterol in coronary atherogenesis is well established (1). However, while the effectiveness of cholesterol-lowering therapy in reducing the incidence of myocardial infarction (MI) and other clinical sequelae of atherosclerotic coronary heart disease (CHD) has been demonstrated repeatedly in randomized clinical trials using a wide variety of cholesterol-lowering treatments, the beneficial effect of these treatments on CHD mortality in these trials has been offset by increased mortality from causes other than CHD, and overall mortality has not been reduced. Doubts about whether cholesterol lowering improves mortality have been reinforced by observational epidemiological studies, which have raised similar questions about excess non-CHD mortality in persons at the lowest end of the cholesterol distribution (2). Although this effect was observed at cholesterol levels far below those attained in cholesterol-lowering trials and may have been explained at least in part by the association of some preexisting diseases (cancer, liver diseases, etc.) with low cholesterol levels, the possibility that lowering cholesterol may, at least under some circumstances, do no more than trade one cause of death for another remains a concern.

Even comprehensive quantitative meta-analyses of randomized clinical trials of cholesterol lowering have failed to resolve the controversy surrounding this issue (3–10). For example, Holme (3) reported a strong correlation across trials between reductions in cholesterol level and mortality and emphasized that mortality tended to be reduced in trials in which cholesterol was lowered by more than 9%. At the other end of the spectrum, Ravnskov (6), looking at essentially the same trials, rejected altogether the notion that cholesterol lowering reduces even CHD mortality, while ascribing favorable trends in CHD death and nonfatal MI to reporting bias. This chapter will present an updated quantitative review of published (as of October 1993) cholesterol-lowering trials, which is intended primarily to highlight the substantial areas of *agreement* in previous meta-analyses of these trials and thereby to dispel some of the existing confusion of conflicting interpretations and opinions.

WHICH TRIALS ARE RELEVANT?

Nearly 50 controlled clinical trials in which each of nearly 85,000 participants received one of nearly 60 active cholesterol-lowering treatment regimens have been published since the 1961 (3–32). The objectives and experimental designs of these trials vary widely. Most were randomized, but some used matched controls. Most followed the intent-to-treat principle, but some excluded participants after randomization. Most trials focused exclusively on cholesterol lowering, but others incorporated cholesterol lowering into a multiple-risk factor intervention program. Some trials were conducted in patients without known CHD (primary prevention), others in patients with known CHD (secondary prevention), and still others in patients in which another disease (diabetic retinopathy, breast cancer) was the main target of intervention. The cholesterol-lowering interventions (alone and in combination) included various diets, partial ileal bypass surgery, resins, niacin, fibrates, hormones (estrogen, dextrothyroxine), and, most recently, tamoxifen and HMG CoA reductase inhibitors. The number of participants (treated plus control) in any one of these trials varied from 44 to nearly 60,000, their durations varied from less than a year to nearly 10 years, and their degrees of cholesterol lowering varied from less than 1% to more than 20%. While all of these trials may tell us something about cholesterol lowering, their great diversity presents a formidable obstacle to gleaning a coherent message.

Although reasonable people might differ as to which trials are germane to the analysis of cholesterol lowering and mortality, certain criteria have been adopted for the purpose of this review as represented in the following sections.

Trials Must Be Randomized and Must Report Their Results by "Intent-to-Treat"

Virtually all trialists agree that other means of obtaining a control group (self-selection, matching, etc.) are subject to serious biases that may invalidate the results of any trial that uses them. Since the exclusion of randomized participants from analysis because of noncompliance or intercurrent illnesses during the trial is tantamount to the absence of randomization, a full count of *all* events that occurred during the trial in each randomized group is imperative. This count must include even participants who discontinued their assigned treatment regimen during the trial for whatever reason (i.e., analysis by intent-to-treat). While all published meta-analyses have at least implicitly used randomization as an inclusion criterion, a few nonrandomized trials and non–intent-to-treat results from randomized studies have slipped through. For example, in one prominent trial included in all meta-analyses, intent-to-treat results were not published until 14 years after the initial publication of trial results (16,17).

Each Trial Must Provide an Unconfounded Test of the Question of Interest, i.e., Whether Cholesterol Lowering Is Beneficial in the Primary and Secondary Prevention of CHD

Although the application of this principle to decide what can reasonably be called a trial of cholesterol lowering is the source of most differences among meta-analyses with regard to trials included, this issue has not been dealt with explicitly and comprehensively. The operational definition proposed here is that cholesterol lowering must be the dominant known mechanism by which the particular intervention is thought to affect the endpoints of interest in the study population. This definition implies both that the cholesterol-lowering effect of the intervention must be nontrivial and that the potential non–cholesterol-mediated effects of the treatment regimen on the endpoint of interest must be unimportant by comparison. To protect against the biases inherent in discarding trials based on post hoc data, the application of this criterion should be based only on data available to the original planners of the trial, not on the hindsight of the meta-analyst.

The following groups of trials, included in some other meta-analyses, have been excluded by this criterion from the present one:

1. Multifactor trials, in which co-interventions on nonlipid risk factors were administered along with cholesterol reduction to the active-treatment group but not to the control group. In such trials, it is impossible to distinguish the effects of cholesterol lowering [less than 2% in the three largest multifactor intervention trials (8)] from the concurrent effects of antihypertensive drugs, smoking cessation, etc. Note that factorial designs, in which a similar fraction of each cholesterol-lowering treatment arm is randomized to receive one or more nonlipid interventions, would not be excluded by this criterion.
2. Studies in which cholesterol was lowered by less than 4%. While the cholesterol response to treatment is technically a postrandomization result, the danger of bias from using this criterion is minimal, since the cholesterol-lowering potency of the planned intervention is generally known *before* a trial is undertaken and in any case is established within the first month of a trial, well before the vast majority of clinical endpoints are recorded.

3. Some meta-analysts have excluded trials that used hormone-based interventions as potentially confounded by the nonlipid effects of the intervention (3,5,8–10). While this may be true, the same could be said of other systemic cholesterol-lowering drugs like niacin and the fibrates. While only one estrogen trial met criterion 2 for cholesterol lowering, the two trials using dextrothyroxine achieved cholesterol reductions comparable to those achieved by most other interventions used. While the use of these regimens for cholesterol lowering has now been discredited (at least in men), there is no compelling a priori reason to single out these trials as confounded.

4. Some single-factor trials using cholesterol-lowering interventions may be inappropriate to the study of the effect of cholesterol lowering on mortality because of the population studied. For example, a recent chemotherapy trial of tamoxifen (which significantly lowers cholesterol) in breast cancer patients could not provide an unconfounded test of the beneficial effect of cholesterol lowering on mortality, since the dominant impact of the intervention on mortality was determined by its effect on breast cancer recurrence, which accounted for 15 times as many deaths as did CHD in this study (32).

Duration of the Trial Must Be Sufficient to Allow the Impact of Cholesterol Lowering to Become Clinically Manifest

Since cholesterol lowering theoretically works by slowing or slowly reversing a chronic pathological process—atherosclerosis—most well-designed clinical trials have allowed for prolonged follow-up on treatment. It has been pointed out that the duration of even the longest clinical trials ought not be expected to be adequate to determine the excess risk associated with hypercholesterolemia, especially since many of the clinical events that determine a trial's outcome occur early in the trial and the mean duration of treatment preceding a trial's clinical events is generally only about half the total duration of the trial (33). Some meta-analysts have nevertheless included trials designed to assess the acute impact of treatment on intermediate endpoints (cholesterol levels, side effects, coronary angiography, etc.) in which the mean duration of treatment in patients with clinical events was far too short (a year or less) to be expected to have a significant impact on risk of major morbidity and mortality. Since the large number of short trials that have been conducted for various purposes contribute little to our understanding of the long-term effect of cholesterol lowering on morality, trials with a planned duration of less than 3 years have been excluded.

META-ANALYSIS OF 22 CHOLESTEROL-LOWERING TRIALS

The 22 treatment arms from 18 trials published before October 1993 that meet the above-mentioned criteria are shown in Table 1. These trials include 15,847 participants and 86,660 person-years of follow-up on treatment. Their mean cholesterol lowering was 10%, and their mean duration of follow-up on treatment was 5.5 years. Only four qualifying trials addressed primary prevention; the remaining 18 trials addressed secondary prevention. Seven of the trials, representing 58% of the person-years of treatment, used a fibrate (clofibrate or gemfibrozil). Three other trials, representing 17% of the person-years, used resins. Although six trials used diet alone, these were mostly smaller studies (averaging 200 on active treatment) and represented only 7% of the follow-up. The remaining 18% of the person-years of follow-up derived from six trials using niacin, ileal bypass surgery, or hormones (dextrothyroxine or estrogen). No published trials using HMG CoA reductase inhibitors were of sufficient duration to qualify for this analysis.

Meta-analyses of mortality due to CHD, non-CHD causes, and all causes combined and of the combined incidence of CHD death and nonfatal MI (CHD incidence) were performed for the trials listed in Table 1, using the Mantel-Haenszel procedure as modified by Yusuf et al. (34). This procedure, which can accommodate even small trials with few events, entails a comparison of observed versus "expected" (by chance) numbers of deaths in each trial. The difference between observed and expected events (O − E) was divided by its estimated variance (V) to obtain an estimate of the log odds ratio for each trial. The values of O-E and V for each trial were summed over all trials to estimate the log of the overall odds ratio ($\Sigma(O - E)/\Sigma V$) and its variance (ΣV). For each trial that compared more than one qualifying active treatment arm with a common control group, the active treatment arms were combined before this summation to avoid double-counting their control groups. However, the active treatment arms of multiarmed trials were considered separately when subgroups defined by intervention, cholesterol lowering, etc. were examined. The nominal two-tailed p-values reported here, which are based on these Mantel-Haenszel odds ratios, do not take into account when the events in each trial occurred; thus, significance levels for individual trials may not agree precisely with those that appeared in the original trial reports.

Despite a 17% reduction in CHD incidence ($p < 0.001$), total mortality was essentially unchanged by cholesterol lowering (Table 1). The latter finding reflected the offsetting effects of a 9% decrease in CHD mortality ($p = 0.03$) and a 24% increase in non-CHD mortality ($p = 0.001$). This result is basically similar to results reported in other meta-analyses examining differing groups of trials.

Even after culling out all the confounded and short-term trials, the trials selected for this meta-analysis still represent considerable heterogeneity in their target populations and interventions. It is therefore instructive to compare the results in various subgroups of trials to gain insight into why the significant reduction in CHD incidence associated with cholesterol-lowering treatment has not translated into an overall reduction in mortality.

The first subgroup analysis compares the results in primary versus secondary prevention trials (Table 2). Although 18 of the 22 trials were in the latter category,

Table 1 Randomized Trials of Cholesterol Lowering and CHD Prevention[a]

Trial	# Persons Trt.	# Persons Ctrl.	FU (yr)	Percent cholesterol reduction	Mortality Total Trt.	Total Ctrl.	Total OR	CHD Trt.	CHD Ctrl.	CHD OR	Non-CHD Trt.	Non-CHD Ctrl.	Non-CHD OR	CHD incidence Trt.	CHD incidence Ctrl.	CHD incidence OR
Fibrates (50,333 person-yr)																
Clofibrate																
CDP (1975)	1,103	2,789	6.2	6.4	288	723	1.01	221	583	0.95	67	140	1.23	309	839	0.91
Newcastle (1971)	244	253	3.6	13.0	31	51	**0.58**	25	44	**0.55**	6	7	0.89	52	81	**0.58**
Scottish physicians (1971)	350	367	3.4	14.0	42	48	0.91	34	35	1.02	8	13	0.64	61	72	0.87
Stockholm (1988)[b]	279	276	5.0	13.0	**61**	**82**	**0.66**	**47**	**73**	**0.57**	14	9	1.56	**72**	**100**	**0.61**
WHO (1978)	5,331	5,296	5.3	9.0	**236**	**181**	**1.31**	91	77	1.18	**145**	**104**	**1.39**	**167**	**208**	**0.79**
Gemfibrozil																
Helsinki (1987)	2,051	2,030	5.0	9.7	45	42	1.06	14	19	0.73	31	23	1.34	**56**	**84**	**0.65**
Helsinki ancillary (1993)	311	317	5.0	8.5	19	12	1.64	17	8	2.16	2	4	0.52	35	24	1.54
Resins (14,491 person-yr)																
LRC-CPPT (1984)	1,907	1,899	7.4	9.0	68	71	0.95	32	44	0.72	36	27	1.33	155	187	0.81
NHLBI Type II (1984)	59	57	5.0	18.7	5	7	0.67									
STARS (1992)	26	28	3.3	19.8	0	3	0.13	0	3	0.13	0	0		1	5	0.25
Niacin (7365 person-yr)																
CDP (1975)	1,119	2,789	6.2	9.9	277	723	0.94	215	583	0.90	62	140	1.11	**287**	**839**	**0.81**

USVA Drug-Lipid (1968)	145	284	3.2	14.0	31	54	1.16	28	48	1.18	3	6	0.98	49	93	1.05
Diet (6356 person-yr)																
Los Angeles Veterans Domiciliary (1969)	424	422	7.0	12.7	174	177	0.96	41	50	0.80	133	127	1.06	54	71	0.72
MRC low fat (1965)	123	129	3.0	9.3	20	24	0.85	17	20	0.87	3	4	0.78	27	27	1.06
MRC Soya (1968)	199	194	4.0	14.2	28	31	0.86	25	25	0.97	3	6	0.49	45	51	0.82
Oslo (1966)	206	206	5.0	14.4	41	55	0.68	37	50	0.69	4	5	0.80	57	74	0.68
STARS (1992)	27	28	3.3	11.0	1	3	0.36	1	3	0.36	0	0		2	5	0.40
Sydney (1978)	221	237	5.0	4.6	39	28	1.59									
Surgery (4084 person-yr)																
POSCH (1990)	421	417	9.7	22.3	49	62	0.76	32	44	0.70	**41**	**65**	**1.67**	**82**	**125**	**0.57**
Hormones (4031 person-yr)																
CDP (D-thyroxine, 1972)	1,110	2,789	3.0	11.4	160	339	1.22	119	274	1.10	17	18	0.93	197	449	1.13
Oliver, Boyd (estrogen, 1961)	50	50	5.0	16.1	17	12	1.62	13	10	1.40	4	2	2.02	18	18	1.00
USVA drug-lipid (D-thyroxine, 1968)	141	284	3.2	7.0	23	54	0.83	18	48	0.73	5	6	1.76	41	93	0.84
All trials (86,660 person-yr)	15,847	18,324	5.5	10.0	1655	**2056**	1.01	**1027**	**1455**	**0.91**	**576**	**566**	**1.24**	**1767**	**2601**	**0.83**

[a]Meeting the following criteria: (1) analyzed by intent to treat, (2) nonconfounded, (3) at least 3 years duration, (4) at least 4% cholesterol reduction. Primary prevention trials are indicated in italic type.

[b]Some participants also received niacin.

Trt. = Active treatment group; Ctrl. = Control group; OR = odds ratio (Trt. vs. Ctrl.).

Statistically significant results shown in boldface type.

Table 2 Cholesterol-Lowering Trials: Primary Versus Secondary Prevention

Population	No. trials	No. treated	Person-years	Mean cholesterol reduction (%)	Percent change in risk[a]			
					Mortality			CHD incidence[b]
					Total	CHD[b]	Non-CHD[b]	
Primary prevention	4	9,713	55,589	9	12.6	−7.1	**26.0**	**−23.2**
Secondary prevention	18	6,134	31,071	11	−3.4	**−9.5**	**21.3**	**−14.2**
Total	22	15,847	86,660	10	1.0	**−9.1**	**23.8**	**−16.8**

[a]Statistically significant results are indicated in boldface type.
[b]Two secondary prevention trials, NHLBI Type II and Sydney, did not report cause–specific mortality or nonfatal MI for each treatment group.

many of them were quite small, and the primary prevention trials actually contributed 64% of the total person-years on treatment. The relative risk for all-cause mortality, though not statistically significant in either group of trials, was distinctly more favorable in the secondary prevention (3% risk reduction) than in the primary prevention (13% risk increase) setting. This difference was due entirely to the fact that 80% of deaths in the secondary prevention trials versus only 37% in the primary prevention trials were attributed to CHD. The cause-specific relative risks for CHD (7–10% decrease) and non-CHD mortality (21–26% increase) considered separately were very similar in the primary and secondary prevention setting (Table 2). Mortality from nonmedical causes (accidents, suicide, homicide) was increased by 47% ($p = 0.08$) among the primary prevention trials, in general agreement with the report of Muldoon et al. (4), largely due to the influence of the CPPT (18) and Helsinki Heart Study (16). However, this trend fell short of statistical significance and was not reproduced in the four secondary prevention trials for which separate tallies of nonmedical and other non-CHD deaths were published (7% increase, $p = 0.8$) (11,12,19,29).

A second subgroup analysis, in which the 22 trials were grouped according to whether cholesterol reductions of greater or less than the median (12%) were achieved, indicates that the degree of cholesterol lowering achieved was an important determinant of risk reduction (Table 3). Among the 11 trials that exceeded this criterion, highly significant reductions were observed in CHD incidence (31%; $p < 0.001$), CHD mortality (27%; $p < 0.001$), and all-causes mortality (20%; $p = 0.002$). Non-CHD mortality was increased by 11% ($p = 0.4$). In the remaining 11 trials, CHD incidence was reduced by only 11% ($p = 0.003$), which though statistically significant did not translate into reductions in CHD or all-causes mortality. In this group of trials, there was a 30% increase in non-CHD mortality ($p < 0.001$) and a 10% increase in all-causes mortality ($p = 0.03$).

The finding of a "dose-response" relationship between reductions in CHD morbidity and mortality among participants in trials of cholesterol lowering and the cholesterol-lowering efficacy of the treatment they received, which has been reported by several other meta-analyses (3,8–10), lends credence to the attribution of reductions in CHD risk to cholesterol lowering per se. By the same token, the absence of a similar "dose-response" relationship between non-CHD mortality and the cholesterol-lowering efficacy of the treatment regimen, which appear to be related *inversely*, suggests that increases in non-CHD mortality are likely to be intervention-specific and not intrinsic to their cholesterol-lowering activity.

The final subgroup analysis compares the results achieved by different classes of cholesterol-lowering interventions (Table 4). Except for the three trials that used hormones (estrogen or dextrothyroxine) to lower cholesterol, significant reductions in CHD incidence were observed for all intervention classes, ranging from 18% for the seven fibrate trials to 43% for the single partial ileal bypass surgery trial. Nonsignificant reductions in CHD mortality were also observed for all but the hormone trials. As might be expected from Table 3, the greatest reduction in

Table 3 Cholesterol-Lowering Trials By Success in Lowering Cholesterol

Cholesterol reduction	No. trials	No. treated	Person-years	Mean cholesterol reduction (%)	Percent change in risk[a]			
						Mortality		CHD incidence[b]
					Total	CHD[b]	Non-CHD[b]	
At least 12%	11	2,399	12,974	15	**-19.7**	**-27.1**	10.8	**-31.0**
Less than 12%	11	13,448	72,770	9	**9.6**	-1.5	**29.7**	**-11.3**
Total	22	15,847	86,660	10	1.0	**-9.1**	**23.8**	**-16.8**

[a]Statistically significant results are indicated in boldface type.
[b]The NHLBI Type II (19% cholesterol reduction) and Sydney (5% cholesterol reduction) trials did not report cause-specific mortality or nonfatal MI for each treatment group.

Table 4 Cholesterol-Lowering Trials By Intervention Class

Intervention	No. trials	No. treated	Person-years	Mean cholesterol reduction (%)	Percent change in risk[a]			
					Total	Mortality CHD[b]	Non-CHD[b]	CHD incidence[b]
Surgery	1	421	4,084	22	−24.5	−30.0	−6.7	**−43.1**
Resins	3	1,992	14,491	9	−10.7	−32.2	33.1	**−20.6**
Diet	6	1,200	6,356	11	−6.0	−20.9	0.4	**−23.8**
Niacin	2	1,264	7,365	10	−4.1	−7.5	8.1	**−17.0**
Fibrates[c]	7	9,669	50,333	9	3.3	−8.0	**32.4**	**−18.4**
Hormones	3	1,301	4,031	11	18.0	5.5	77.0	7.3
Total	22	15,847	86,660	10	1.0	**−9.1**	**23.8**	**−16.8**
(without fibrates or hormones)	12	4,877	32,295	12	−6.9	**−14.7**	6.4	**−22.1**

[a]Statistically significant results are indicated in boldface type.
[b]The NHLBI Type II (resin) and Sydney (diet) trials did not report cause–specific mortality or nonfatal MI for each treatment group.
[c]The Stockholm trial, which used both clofibrate and niacin, is counted as a fibrate trial.

CHD was attained by the intervention (surgery) that reduced cholesterol the most. The resin, diet, niacin, and fibrate trials all reported similar and far smaller reductions in cholesterol levels (9–11%) and CHD rates (18–24%).

The significant overall increase in non-CHD mortality was largely a product of the trials using fibrates and hormones. The WHO and CDP clofibrate trials were responsible for most of the excess non-CHD mortality in the fibrate trial category (Table 1). This excess was attributed mainly to cancer and diseases of the hepatobiliary system in the WHO trial (16,17) and to non-CHD cardiovascular and unknown causes in the CDP (11,12). The fibrates have long been known to alter bile composition such that the risk of gallstones is increased. In the dextrothyroxine arm of the CDP, which accounted for nearly all of the non-CHD deaths in the hormone trial category, most of the excess deaths were ascribed to cardiovascular diseases other than CHD (30). This phenomenon was taken as evidence of a direct toxic effect of the drug on the myocardium of these patients, all of whom had suffered a prior infarction.

The trial subgroups using surgery, resins, niacin, and diet all showed nonsignificant trends toward reduced all-causes mortality (Table 4). As for CHD incidence and mortality, the trend was most favorable for the surgery trial, in which a 22% cholesterol reduction was achieved and in which the reduction in CHD mortality was not offset by an adverse trend in non-CHD mortality. In the fibrate trials, an 8% reduction in CHD mortality ($p = 0.2$) was more than offset by a 32% increase in non-CHD mortality ($p = 0.002$), with a net 3% increase in all-causes mortality ($p = 0.5$). In the three hormone trials, where increases in both CHD and non-CHD mortality were associated with treatment, there was a net 18% increase in all-causes mortality ($p = 0.09$).

The extent to which the observed neutralization of the beneficial effects of cholesterol-lowering treatment on CHD by apparent adverse effects on non-CHD mortality is attributable to specific drugs rather than generic to cholesterol lowering is best appreciated by looking at the trials that used neither a fibrate nor a hormone (Table 4). In these 12 trials, the significant 15% decrease in CHD mortality ($p = 0.02$) more than offset the nonsignificant 6% increase in non-CHD mortality ($p = 0.5$) associated with treatment, and all-causes mortality was reduced by 7% ($p = 0.19$). This finding is fundamentally in agreement with a meta-analysis by Gould et al. (10), which excluded hormone trials at the outset and simultaneously analyzed the effects of intervention and cholesterol reduction on mortality.

INTERPRETATION OF META-ANALYSIS

Meta-analysis, particularly as applied to clinical trials of cholesterol lowering, has been a focus of controversy. While its quantitative methodology provides a veneer of objectivity, it is the meta-analyst who decides the criteria for the selection of relevant trials and the weight given to each result. These subjective decisions, generally made in full knowledge of the findings of each individual trial, offer ample leeway for advocates of a particular viewpoint to use these analyses to make the best possible case for their own interpretation and to criticize those of their opponents. However, the following two points should not be lost in the sound and fury of this debate.

1. No published comprehensive meta-analysis of cholesterol-lowering trials has failed to show a significant overall reduction of CHD incidence.
2. No published comprehensive meta-analysis of cholesterol-lowering trials has succeeded in showing a significant overall reduction in all-causes mortality.

While some meta-analysts have emphasized one of these findings and/or minimized the other, their objective results, like those presented here, are in fundamental agreement in these two regards, despite substantive differences in trial selection and methodology.

This meta-analysis, like others, finds evidence that non-CHD mortality is increased among trial participants who received cholesterol-lowering treatment. However, this increase, which shows no dose-response relation to cholesterol lowering, is largely attributable to trials that used hormone- or fibrate-based regimens, which comprise 63% of all the person-years of follow-up among the trials analyzed. The trials that used clofibrate, particularly the WHO trial, are primarily responsible for the adverse trend in non-CHD mortality associated with fibrate use, although the two gemfibrozil trials report similar trends (Tables 1 and 4). The observation that trials using other regimens, some of which produced far greater reductions in cholesterol levels than did the hormones and fibrates, exhibited only a small, nonsignificant increase in non-CHD mortality argues against the hypothesis that an increase in non-CHD mortality is a necessary consequence of cholesterol lowering per se.

Nevertheless, the question of whether the reduction in CHD associated with cholesterol lowering is ultimately of benefit to the patient or merely shifts mortality from one cause to another cannot be definitively resolved based on trials that have been completed to date. While post hoc subgroup analyses of these trials, like those in Tables 2–4, may provide important insights, they will always be subject to the criticism that the groupings, however reasonable they appear, are influenced by the analyst's prior knowledge of the results. However, these analyses are most useful for generating hypotheses that may be tested in meta-analyses of trials whose results are not yet known.

The results of meta-analyses of completed trials, particularly the powerful dose-response relationship between cholesterol reduction and reductions not only in CHD incidence but in all-causes mortality, suggest that any real hope of resolving the unanswered questions about the impact of cholesterol lowering on mortality should be invested in ongoing trials using potent cholesterol-lowering interventions without known major adverse non-CHD effects. To this end, a meta-analysis of ongoing and planned trials using selection criteria similar to those described above but with a far more stringent requirement for cholesterol reduction [e.g., at least a 20–25% reduction in low-density lipoprotein (LDL) cholesterol] is proposed. The HMG CoA reductase inhibitors, which are now being used in several large, long-term trials scheduled to be completed between 1994 and 2001, are likely to comprise the bulk of such an analysis, although new trials using other similarly potent interventions (LDL apheresis, partial ileal bypass, etc.) might also qualify. Collaborative meta-analyses that take full advantage of the statistical power of these planned and ongoing cholesterol-lowering trials are currently in an advanced stage of planning. At this stage, this prospective approach would appear to have far more to offer than continuing to look back at trials that did not lower cholesterol very much and/or used drugs that may have had offsetting toxic effects, especially in view of the impossibility of freeing retrospective analyses of known trials from our biases.

In conclusion, cholesterol-lowering trials completed to date have clearly confirmed that lowering cholesterol levels reduces CHD risk but have failed to establish a broader benefit in terms of mortality. Further analysis of these results suggests that the failure by past trials in their totality to demonstrate a net reduction in all-causes mortality may have been due to a combination of insufficient efficacy in lowering cholesterol levels and possible adverse effects of some drug classes most commonly used in these trials. If this surmise is correct, the new generation of cholesterol-lowering trials, using far more potent cholesterol-lowering agents with (as yet) no known major adverse effects on non-CHD mortality, may provide a more pure and powerful test of the effect of cholesterol lowering on mortality. By specifying hypothesis and selection criteria *before the results of these new trials are known*, one can hope for an unbiased resolution that is free of the contentiousness that currently prevails.

REFERENCES

1. Consensus conference statement on lowering blood cholesterol to prevent heart disease. JAMA 1985; 253:2080–2086.
2. Jacobs D, Blackburn H, Higgins M, et al. Report of the conference on low blood cholesterol mortality associations. Circulation 1992; 86:1046–1060.
3. Holme I. An analysis of randomized trials evaluating the effect of cholesterol reduction on total mortality and coronary heart disease incidence. Circulation 1990; 82:1916–1924.
4. Muldoon MF, Manuck SB, Mathews KA. Lowering cholesterol concentrations and mortality: a quantitative review of primary prevention trials. Br Med J 1990; 301: 309–314.
5. Rossouw JE, Lewis B, Rifkind BM. The value of lowering cholesterol after myocardial infarction. N Engl J Med 1990; 323:1112–1119.

6. Ravnskov U. Cholesterol lowering trials in coronary heart disease: frequency of citation and outcome. BMJ 1992; 305:15–19.
7. Davey Smith G, Song F, Sheldon TA. Cholesterol lowering and mortality and mortality: the importance of considering initial level of risk. BMJ 1993; 306:1367–1373.
8. Holme I. Relation of coronary heart disease incidence and total mortality to plasma cholesterol reduction in randomised trials: use of meta-analysis. Br Heart J 1993; 69(suppl):S42–S47.
9. Thompson SG. Controversies in meta-analysis: the case of the trials of serum cholesterol reduction. Stat Meth Med Res 1993; 2:173–192.
10. Gould AL, Rossouw J, Santanello NC, Heyse JF, Furberg C. Cholesterol lowering confers clinical benefit: a new look at old data. Circulation 1995 (in press).
11. Coronary Drug Project Research Group. The Coronary Drug Project: clofibrate and niacin in coronary heart disease. JAMA 1975; 231:360–381.
12. Canner PL, Berge KG, Wenger NK, Stamler J, Friedman L, Prineas RJ, Friedewald W. Fifteen year mortality in Coronary Drug Project patients: long-term benefit with niacin. J Am Coll Cardiol 1986; 18:1245–1255.
13. Group of Physicians of the Newcastle Upon Tyne Region. Trial of clofibrate in the treatment of ischaemic heart disease. Five year study. BMJ 1971; 4:767–775.
14. Research Committee of the Scottish Society of Physicians. Ischaemic heart disease: a secondary prevention trial using clofibrate. BMJ 1971; 4:775–784.
15. Carlson LA, Rosenhamer G. Reduction of mortality in the Stockholm ischaemic heart disease secondary prevention study by combined treatment with clofibrate and nicotinic acid. Acta Med Scand 1988; 223:405–418.
16. Report from the Committee of Principal Investigators. A cooperative trial in the primary prevention of ischaemic heart disease using clofibrate. Br Heart J 1978; 40:1069–1103.
17. Heady JA, Morris JN, Oliver MF. WHO clofibrate/cholesterol trial: clarifications. Lancet 1992; 340:1405–1406.
18. Frick MH, Elo O, Heinonen O, Heinsalmi P, Helo P, Huttunen JK, Kaitaniemi P, Koskinen P, Manninen V, Mäenpää H, Mälkönen M, Mänttäri M, Norola S, Pasternack A, Pikkarainen J, Romo M, Sjöblom T, Nikkilä E. Helsinki Heart Study: primary-prevention trial with gemfibrozil in middle-aged men with dyslipidemia. Safety of treatment, changes in risk factors, and incidence of coronary heart disease. N Engl J Med 1987; 317:1237–1245.
19. Frick MH, Heinonen OP, Huttunen JK, Koskinen P, Manttari M, Manninen V. Efficacy of gemfibrozil in dyslipidemic subjects with suspected heart disease. An ancillary study in the Helsinki Heart Study frame population. Ann Med 1993; 25:41–45.
20. Lipid Research Clinics Program. The Lipid Research Clinics Coronary Primary Prevention Trial Results I. Reduction in incidence of coronary heart disease. JAMA 1984; 251:351–364.
21. Levy RI, Brensike JF, Epstein SE, Kelsey SF, Passamani ER, Richardson JM, Loh IK, Stone NJ, Aldrich RF, Battaglini JW, Moriarty DJ, Fisher ML, Friedman L, Friedewald W, Detre KM. The influences of changes in lipid values induced by cholestyramine and diet on progression of coronary artery disease: Results of the NHLBI Type II Coronary Intervention Study. Circulation 1984; 69:325–337.

22. Watts GF, Lewis B, Brunt JNH, Lewis ES, Coltart DJ, Smith LDR, Mann JI, Swan AV. Effects on coronary artery disease of lipid-lowering diet, or diet plus cholestyramine, in the St. Thomas' Atherosclerosis Regression Study (STARS). Lancet 1992; 339:563–569.

23. Schock HK. The U.S. Veterans Administration cardiology drug-lipid study: an interim report. Adv Exp Med Biol 1968; 4:405–420.

24. Dayton S, Pearce, ML, Hashimoto S, Dixon WJ, Tomiyasu U. A controlled clinical trial of a diet high in unsaturated fat in preventing complications of atherosclerosis. Circulation 1969; 39, 40 (suppl 2).

25. Research Committee. Low-fat diet in myocardial infarction: a controlled trial. Lancet 1965; 2:501–504.

26. Research Committee to the Medical Research Council. Controlled trial of soybean oil in myocardial infarction. Lancet 1968; 2:693–700.

27. Leren P. The effect of plasma cholesterol lowering diet in male survivors of myocardial infarction. A controlled clinical trial. Acta Med Scand 1966: 466(suppl):1–92.

28. Woodhill JM, Palmer AJ, Leelerthaepin B, McGilchrist C, Blacket RB. Low fat, low cholesterol diet in secondary prevention of coronary heart disease. Adv Exp Med Biol 1978; 109:317–330.

29. Buchwald H, Varco RL, Matts JP, Long JM, Fitch LL, Campbell GS, Pearce MB, Yellin AE, Edmiston A, Smink RD, Sawin HS, Campos CT, Hansen BJ, Tuna N, Karnegis J, Sanmarco ME, Amplatz K, Castaneda-Zunida WR, Hunter DW, Bissett JK, Weber FJ, Stevenson JW, Leon AS, Chalmers TC, and the POSCH Group. Effect of partial ileal bypass surgery on morbidity and mortality from coronary heart disease in patients with hypercholesterolemia. Report of the Program on the Surgical Control of Hypercholesterolemia (POSCH). N Engl J Med 1990; 323:946–955.

30. Coronary Drug Project. Findings leading to further modifications of its protocol with respect to dextrothyroxine. JAMA 1972; 220:996–1008.

31. Oliver MF, Boyd GS. Influence of reduction of serum lipids on prognosis of coronary heart disease. A five-year study using oestrogen. Lancet 1961; 2:499–505.

32. Rutqvist LE, Mattsson A for the Stockholm Breast Cancer Study Group. Cardiac and thromboembolic morbidity among post-menopausal women with early stage breast cancer in a randomized trial of adjuvant tamoxifen. J Natl Cancer Inst 1993; 85:1398–1406.

33. Collins R. Cholesterol lowering trials in their epidemiologic context. Postgrad Med J 1993; 69(suppl 1):S3–S7.

34. Yusuf S, Peto R, Collins R, Sleight P. Beta blockade during and after myocardial infarction: an overview of randomized trials. Prog Cardiovasc Dis 1985; 27:335–371.

3

Secondary Prevention of Coronary Heart Disease

Jacques E. Rossouw

Office of Disease Prevention, National Institutes of Health, Bethesda, Maryland

INTRODUCTION

The most direct and convincing evidence that lowering cholesterol reduces coronary heart disease (CHD) risk comes from clinical trials. The clinical trials of cholesterol lowering are conventionally categorized as primary prevention if performed in ostensibly healthy individuals, secondary prevention if performed in individuals with preexisting clinical CHD, and trials in which the main outcome is the angiographic appearance of the coronary vessels. These distinctions are somewhat artificial in that many of the "healthy" middle-aged and older healthy primary trial participants (particularly those who developed clinical symptoms during the trial) almost surely had underlying coronary atherosclerosis, and conceptually the inferences drawn from the primary and secondary trials are rather similar. The angiographic trials clearly address the same patient population as the secondary prevention trials, but typically the interventions have been more aggressive and of a shorter duration.

The various types of trial evidence can best be regarded as complementary. Nevertheless, there is some merit to examining them separately, in that the numbers of outcomes are much higher in secondary prevention trials (due to higher event rates) than in primary prevention trials, and in the angiographic trials every participant provides angiographic outcome information. The secondary prevention and angiographic trials are therefore potentially more efficient at testing the cholesterol hypothesis, since adequate statistical power will be achieved with much smaller numbers of participants, given a similar magnitude of treatment effect. The higher absolute risk of CHD in patients with existing coronary disease leads naturally to a more intensive therapeutic approach to elevated cholesterol in such patients, so there is some clinical relevance in evaluating secondary prevention trials separately (1).

NARRATIVE AND META-ANALYTIC OVERVIEWS

This chapter reviews cholesterol lowering for secondary prevention of CHD and for angiographic change. Two techniques are useful in such reviews: the traditional narrative review and the more recent meta-analytic review. A narrative review looks at each trial separately, describes the patient population, the treatments, and the outcomes. An attempt is then made to draw inferences from the group of trials. The narrative review has certain advantages in that it is easily understood and can describe clinical differences in patient populations and differences in trial design that may affect the outcomes of interest in each particular trial. However, this kind of review is subject to potential bias on the part of the reviewer, since inclusion and exclusion of particular trials, and the interpretation of their results, may influence the conclusions of the reviewer. The results are often ambiguous because individual trials may have limited statistical power, particularly for outcomes other than the primary outcome for which the trial was designed.

Meta-analysis offers a statistically valid way of pooling the results of a group of trials by summing the treatment effects and variances of each trial in order to obtain an estimated average treatment effect. A carefully performed meta-analysis should provide a more objective evaluation of the effects observed in a given set of trials, however, selection of trials to be included may again bias the results. To avoid this bias, meta-analyses should be as inclusive as possible, though not at the expense of introducing a different bias by inclusion of trials that have serious design flaws (e.g., nonrandomized design, incomplete or unblinded ascertainment of outcomes) or are confounded (e.g., multifactorial interventions). The main advantage of meta-analysis is the increase in statistical power achieved by the combined analysis over the individual trials. This allows for more stable estimates of treatment effect in regard to the primary outcome (CHD) and also allows exploration of secondary outcomes and of subgroup hypotheses. The main disadvantage of meta-analyses are, first, that they do not account sufficiently for clinical and statistical heterogeneity of the individual trials and, second, that large trials have an unduly large influence on the summary estimate of treatment effect. In spite of these potential problems, meta-analysis is a very efficient means of summarizing a large body of data.

INCLUSION CRITERIA

For inclusion in the present review, clinical outcome trials had to be randomized (in order to reduce the possibility of selection bias in the treatment and control groups), CHD had to be a prespecified outcome of interest, the ascertainment of CHD outcome data had to be as complete (by intent-to-treat analysis) and objective (blinded) as possible, and cholesterol lowering had to be the single intervention. In addition, trials that were likely to be noncontributory by reason of lack of cholesterol lowering or very short duration (less than 2 years) were excluded, as were trials using drugs that are no longer in clinical use (thyroxin, estrogen in males). Twelve secondary prevention trials (with a total of 14 active treatment arms) met the criteria (Table 1) (2–14).

The angiographic trials had to meet similar criteria (randomized, complete and blinded outcome ascertainment, unconfounded by co-interventions), but the degree of cholesterol lowering obtained in these trials is two- to threefold greater than in clinical outcome trials, and the primary outcome is angiographic change, which can be expected to antedate change in clinical outcomes, so that a shorter duration of treatment (more than 1 year) was regarded as acceptable. Seven angiographic trials (with nine active treatment arms) met the inclusion criteria (Table 2) (2,15–20). Though the angiographic trials were not primarily designed to measure treatment effect on clinical outcomes such as myocardial infarction, these were recorded and are susceptible to meta-analysis.

SECONDARY PREVENTION TRIALS

The Veterans Administration Cardiology Drug-Lipid Study

This was one of the first generation of cholesterol-lowering interventions studies. A number of drugs (estrogen, thyroxin, nicotinic acid) thought to be potentially protective against CHD were tested in a double-blind randomized partial factorial design (14). Follow-up was almost complete because of the setting of the study in the VA system. Only the placebo and aluminum nicotinate groups will be considered here. The patient population was comprised of men who had had a transmural myocardial infarction in the preceding 12 months. Seventy percent had their infarction less than 3 months prior to enrollment, therefore this was a group at very high risk for reinfarction. Mean cholesterol levels at baseline were 241 mg/dl.

The men on 4 g of aluminum nicotinate daily reduced their cholesterol levels by 14% at 12 months, while no change was observed in the placebo group. Over the 3 years in the trial, 13 out of 77 (16.9%) men on aluminum nicotinate developed a fatal CHD event (i.e., myocardial infarction) versus 23 out of 143 (16.1%) on placebo; nonfatal CHD events occurred in 12 (15.6%) and 19 (13.3%) men, respectively. Though the numbers are clearly too small, no beneficial trend was

Table 1 Summary Narrative Review of Secondary Prevention Trials

Trial	Intervention	Number of participants (treated/control)	All CHD events (treated/control)	Significance of treatment effect All CHD	Nonfatal CHD	Fatal CHD	Ref.
VA Drug-Lipid	Niacin	77/143	25/42	NC	NC	NC	14
Newcastle	Clofibrate	244/253	52/81	→	NC	→	8
Edinburgh	Clofibrate	350/367	54/72	NC	NC	NC	12
CDP[a]	Clofibrate	1103/2789	384/1018	NC	→	NC	4,5
	Niacin	1119/2789	352/1018	→	NC	→	
Stockholm	Clofibrate + niacin	279/276	72/100	→	NC	NC	6
Helsinki Ancillary	Gemfibrozil	311/317	35/24	NC	NC	NC	7
Rose	Corn oil	28/26	12/6	NC	NC	NC	13
	Olive oil	26/26	9/6	NC	NC	NC	
MRC	Low-fat diet	199/194	45/51	NC	NC	NC	10
MRC	Soybean oil	123/129	30/31	NC	NC	NC	11
Oslo	Diet	206/206	61/81	→	NC	NC	9
DART	Diet	1018/1015	132/144	NC	→	NC	3
POSCH	Partial ilial bypass	421/417	82/125	→	NC	NC	2

[a]CDP nonfatal CHD (4), fatal CHD (5), all CHD sum of nonfatal and fatal CHD.
NC = No change.
→ = Significantly reduced, $p < 0.05$.

discernible. Side effects were frequent with aluminum nicotinate, and one third discontinued the drug while almost half of the remaining patients reduced their dosage.

The Newcastle and Edinburgh Trials of Clofibrate in the Treatment of Ischaemic Heart Disease

Both of these secondary prevention trials were unusual in that they included patients with angina as well as those with previous myocardial infarctions, and they also included a small proportion of women participants (8,12). Participants were unselected in regard to lipid levels, except that in Newcastle those with serum cholesterol levels exceeding 400 mg/dl or with xanthomata were excluded. Participants were less than 65 years of age in Newcastle and 49–69 years old in Edinburgh. Both trials were double-blind placebo-controlled studies of rather small size: there were 244 participants on clofibrate and 253 on placebo in the Newcastle trial and 350 on clofibrate and 367 on placebo in the Edinburgh trial. The dose of clofibrate was 1.5–2.0 g per day, and duration was 5 years in Newcastle and 6 years in Edinburgh.

The total cholesterol levels in Newcastle fell by 10% compared to controls, and in Edinburgh by 15%. Significantly fewer (-33%) CHD events occurred in the clofibrate compared to the placebo group in Newcastle (52 vs. 81, $p < 0.01$). The reduction in fatal CHD events (but not in nonfatal CHD) was significant. In Edinburgh the 21% reduction in CHD was not significant (54 vs. 72). In both studies the subgroups who had angina at entry had significantly fewer sudden deaths on treatment than the controls, and there were more such patients in Newcastle than in Edinburgh, possibly accounting for the more favorable results in regard to all myocardial events in the Edinburgh trial.

The reports of these early studies created some interest at the time of their publication because both concluded that the beneficial effects in regard to CHD were relatively independent of the initial level of cholesterol and of the response to treatment. If true, this would suggest a nonlipid mode of action of clofibrate. However, these studies were far too small to allow for meaningful conclusions to be drawn from such subgroup analyses, so that they could not adequately test this hypothesis. In the much larger World Health Organization (WHO) clofibrate primary prevention trial, benefit was more marked in those with highest levels of serum cholesterol at baseline and in those who had the greatest fall on treatment (21).

The Coronary Drug Project

In the Coronary Drug Project (CDP), men aged 30–64 with at least one documented previous myocardial infarction were recruited in 53 centers (4,5). They were unselected in regard to serum cholesterol level. Originally they were randomized into five treatment groups and a placebo group; however, the two estrogen groups and the dextrothyroxine group were discontinued early because of an unacceptably high incidence of toxic effects. The clofibrate- and niacin-active treatment groups completed the planned 6 years of the double-blind, placebo-controlled trial.

Clofibrate was given at a total daily dose of 1.8 g to 1103 men. Compared to the 2789 men in the placebo group, clofibrate reduced total cholesterol by 6% and triglycerides by 22%. CHD events decreased by 5% (384 in clofibrate group vs. 1018 in placebo group, not significant). Overall, this large and well-conducted study failed to provide evidence in favor of using clofibrate in unselected patients with CHD, perhaps because of the modest decrease in serum cholesterol obtained.

The 1119 patients who received as total of 3 g of niacin per day experienced net decreases (compared to the 2789 men on placebo) of 10% in total cholesterol and 26% in triglycerides. At the end of 6 years CHD events had decreased by 14% (352 on niacin vs. 1018 on placebo, $p < 0.05$). The decrease in nonfatal CHD was significant, whereas that for fatal CHD was not. Other cardiovascular endpoints also tended to decrease, significantly so for stroke and new angina, except that there was a significant excess of cardiac arrhythmias.

The Stockholm Ischemic Heart Disease Secondary Prevention Study

This trial was a small, randomized comparison of a combination of clofibrate (2 g/day) plus niacin (3 g/day) with no treatment in consecutive survivors of myocardial infarction less than 70 years of age (6). There were 279 participants in the active treatment group and 279 in the control group. One fifth were women, but these were not analyzed separately. The treatment period was 5 years. Compliance with clofibrate was very good; that with niacin was poor.

The net decrease in total cholesterol was 13% in the actively treated group, while serum triglycerides declined by 19%. CHD events decreased by 28% (72 in the treated group, 100 in the control group, $p < 0.05$). The reduction for fatal CHD was significant, that for nonfatal CHD was not. Unique among the trials, in-trial all-cause mortality was also significantly lower in the treated group (61 vs. 82, $p < 0.05$). Subgroup analyses seemed to indicate that individuals who had elevated triglycerides and those who showed the greatest decline in triglycerides had the greatest reductions in CHD. No such relationships were noted for total cholesterol. The findings might suggest that the role of triglycerides in the etiology of CHD needs to be reassessed. While provocative, the subgroup analyses are not sufficient to make the case due to the small numbers in each of the subgroups.

Helsinki Heart Study Secondary Prevention Ancillary Study

The screening for the Helsinki Heart Study primary prevention trial cohort identified 683 middle-aged men who met the lipid criteria (non-HDL cholesterol in excess of 200 mg/dl) but were ineligible by reason of preexisting CHD (7). These men were offered the opportunity to participate in a randomized secondary prevention trial of gemfibrozil 600 mg twice daily or placebo, and 628 men accepted. The 311 men in the active-treatment group and 317 in the placebo group were followed for 5 years. During the trial total cholesterol was about 10% lower, LDL cholesterol was 7% lower, HDL cholesterol was 9% higher, and triglycerides were 38% lower in the active-treatment group compared to the control group.

About one third of participants discontinued medications in each of the groups, and the clinical outcome data were reported by both the on-treatment and intent-to-treat approaches. None of the clinical outcomes were significantly different, however, by the intent-to-treat analysis there were numerically more total CHD events (35 vs. 24, a 49% increase) in the active-treatment compared to the placebo group, more nonfatal CHD events (21 vs. 17), more fatal CHD events (14 vs. 7), more cardiac deaths including fatal CHD (17 vs. 8), and more total deaths (19 vs. 12). Vascular surgery was less common in the active-treatment group (5 vs. 15), but abdominal surgery was more common (16 vs. 1). Analyses using the on-treatment approach found no indications of differences between the groups, so that most of the events accounting for the apparent differences occurred in participants who were not on study medication.

The study was an afterthought of the investigators and did not have the design power to obtain significant results (the power for CHD was retrospectively calculated to be 24%). The results therefore may be due to chance, as is the case in many of the smaller studies reviewed here. The complete absence of a favorable trend in any clinical outcome category is nonetheless worrisome, and the data are certainly suitable for inclusion in a meta-analysis with other similar studies.

Corn Oil in the Treatment of Ischemic Heart Disease

This early dietary secondary prevention trial randomized patients with myocardial infarction to one of three groups: control ($n = 26$), olive oil supplement ($n = 26$), and corn oil supplement ($n = 28$) (13). Patients in each of the actively treated groups received a daily supplement of 80 g of oil, and they were advised to avoid fried and fatty foods. Control patients received no dietary advice. Over the next 2 years the total cholesterol levels in the control group did not change, that of the olive oil group dropped by 2%, and that of the corn oil group dropped by 10%. Reinfarctions occurred in 6 (1 fatal) of the control patients, 9 (3 fatal) on olive oil, and 12 (5 fatal) on corn oil. None of these results was significant. There were no deaths from noncardiac causes.

The Medical Research Council Low-Fat Diet Trial

Myocardial infarction survivors were randomized to low-fat diet (40 g of fat daily, unrestricted as to type of fat or cholesterol) or usual diet (10). It was a small study

(123 in treatment group, 129 in control group) that lasted for 3 years. The reduction in total cholesterol was modest (9%), and there was no change in the rate of CHD (30 in treatment group, 31 in control group).

The Medical Research Council Trial of Soybean Oil

This trial was slightly larger (though still grossly underpowered), with 199 patients in the treatment group and 194 in the control group (11). The treatment diet was supplemented with soybean oil and was otherwise unrestricted. The total cholesterol fell by 13% over the 4 years of the study and the rate of CHD events by 14% (45 in treatment group, 51 in control group, not significant).

The Oslo Diet-Heart Study

This study randomized male myocardial infarction survivors to a diet substituting polyunsaturated fats (including substantial amounts of fish) for saturated fats and low in cholesterol or to a control (usual) diet (9). There were 206 participants in each group, and the trial lasted 5 years. The mean reduction in total cholesterol was 16%, and CHD events were reduced by 25% (61 in diet group vs. 81 in control group, $p < 0.05$).

Diet and Reinfarction Trial

The Diet and Reinfarction Trial (DART) is the largest published study of secondary prevention by diet ($n = 2033$) (3). The study was designed to have the power to detect a 30% reduction in CHD events. Men aged less than 70 who had recently been hospitalized for acute myocardial infarction were randomized to one group ($n = 1018$) that received fat advice (designed to lower total fat to 30% or less of energy intake and increase the P/S ratio to 1.0) and one group that received no fat advice ($n = 1015$). They were also randomized in a factorial design to fatty fish advice versus no fish advice, and to fiber advice versus no fiber advice; however, only the fat advice comparison will be discussed here, as it was the only intervention that was designed to lower cholesterol.

Adherence to the intervention diet fell short of the target. At 2 years, fat as a percentage of energy was 32% in the fat-advice group versus 35% in the no-fat-advice group, and the P/S ratio was 0.78 versus 0.44. The changes in serum cholesterol were correspondingly small: a 4% reduction in the fat-advice group compared to the no-fat-advice group. HDL cholesterol did not change in either group.

The numbers of CHD events were similar in the two groups (132 fatal plus nonfatal in the fat-advice group, 144 in the no-fat-advice group, a nonsignificant reduction of 8%), and the numbers of CHD deaths were identical in both groups (97). Total mortality was also very similar (111 in the fat-advice group versus 113 in the no-fat-advice group). Because of poor adherence to the diet and the small changes in blood cholesterol, the study did not have the power to test the efficacy of dietary cholesterol-lowering therapy. However, it does point to the relative ineffectiveness of standard dietary intervention in bringing about changes in behavior.

The Program on the Surgical Control of the Hyperlipidemias

This clinical outcome trial (the primary endpoint was total mortality) was unique in many respects (2). The intervention was partial ileal bypass, which is the surgical equivalent of administering a bile acid–binding resin with near-complete compliance. The duration of intervention was long—almost 10 years. This study also recorded the angiographic appearance of the coronary vessels at intervals and therefore provided the opportunity of directly comparing angiographic and clinical responses to treatments. The angiographic outcomes are described with the other angiographic studies.

There were 421 patients (90% men) 30–64 years of age in the surgery group and 417 in the control group. The mean reduction in total cholesterol was 23%, in LDL cholesterol 38%, and the HDL cholesterol was increased by 4% in the surgery compared to the control group. Though the 28% reduction in fatal CHD was not significant, the 35% reduction in fatal plus nonfatal CHD was highly significant (82 in surgery group, 125 in control, $p < 0.001$). The incidences of coronary revascularization, unstable angina, and peripheral vascular disease were also significantly reduced.

NARRATIVE AND META-ANALYTIC OVERVIEWS OF THE SECONDARY PREVENTION TRIALS

In overview of the secondary prevention trials, some interesting differences compared to primary prevention trials emerge. In addition to higher event rates, two thirds of the CHD events were fatal, compared to about one third in five primary prevention trials with the same inclusion criteria (21–26). In spite of the high event rates, less than half of the secondary prevention trials produced significant results for combined fatal and nonfatal CHD (6 out of 13 active treatments), and only 2 active treatments had significant reductions in either fatal or nonfatal CHD separately. Thus, on narrative review the case for efficacy of cholesterol lowering to reduce CHD is not very convincing.

Meta-analysis of these same secondary prevention trials, on the other hand, provided unequivocally favorable results (Fig. 1): the odds ratios for all CHD events was 0.83 (95% CI 0.76–0.90, $p < 0.001$), for nonfatal CHD 0.77 (0.68–0.87, $p < 0.001$), and for fatal CHD 0.89 (0.81–0.99, $p < 0.05$). For the primary prevention trials the corresponding odds ratios were 0.80 (0.71–0.89, $p < 0.001$),

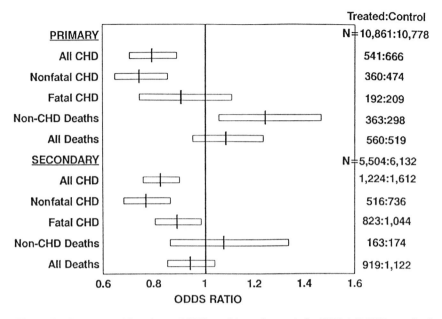

Figure 1 Summary odds ratios and 95% confidence intervals for CHD (all CHD, nonfatal CHD, and fatal CHD), non-CHD mortality, and all-cause mortality in 5 primary and 12 secondary prevention trials of cholesterol lowering. Details of trials are given in text. Meta-analysis was performed by the method of Mantel and Haenzel (31), as modified by Yusuf et al. (32). Numbers (N) of participants in treated and control groups in the pooled data are indicated to the right of the diagram, as are numbers of events.

0.75 (0.65–0.86, $p < 0.001$), and 0.91 (0.75–1.11, NS). These odds ratios were of the same magnitude and direction as those for secondary prevention trials, even though fatal CHD events were not significantly reduced in primary prevention trials. These results are a demonstration of the ability of meta-analysis to pool a number of studies in order to provide a clearer indication of the overall trend than can be obtained from narrative review. They suggest similar benefit in regard to coronary events for secondary and primary prevention trials.

The odds ratios for noncardiac deaths and total mortality were 1.07 (0.87–1.34, NS) and 0.94 (0.85–1.04, NS), respectively, in the secondary prevention trials, and 1.25 (1.06–1.47, $p < 0.05$) and 1.08 (0.95–1.24, NS), respectively, in the primary prevention trials. The overall results do not suggest noncardiac adverse effects attributable to cholesterol lowering in secondary prevention trials, whereas there is some suggestion of an increase in noncardiac mortality in primary prevention trials. However, these summary results need to be treated with caution, since the aggregation of data may obscure treatment-specific adverse trends. Also, the analysis does not take into account the degree of cholesterol lowering obtained in particular trials and hence may underestimate reduction of cardiac events (and thus total mortality) that could be obtained by more effective cholesterol-lowering therapies. An analysis that simultaneously accounts for both choice of treatment and degree of cholesterol lowering obtained has indicated that cholesterol lowering confers benefit for cardiac and total mortality, while noncardiac mortality is unrelated to degree of cholesterol lowering itself but is related to choice of treatment (27). Fibrates and hormones (thyroxin, estrogens) appear to have inherent toxicities, absent in other interventions, which lead to increases in noncardiac and total mortality.

DESCRIPTION OF ANGIOGRAPHIC STUDIES

NHLBI Type II Coronary Intervention Study

This trial selected patients 21–55 years of age (81% males, 19% females) with elevated LDL cholesterol levels (i.e., Type II hyperlipoproteinemia) and angiographic evidence of coronary artery disease (17). The mean level of total cholesterol of the patients was very high at 323 mg/dl. During the trial period of 5 years, total cholesterol levels were 11% lower, LDL cholesterol 19% lower, and HDL cholesterol 5% higher on cholestyramine than on placebo. Though lesions progressed in both groups, a nonsignificant trend towards less progression on angiography was seen in the active-treatment group. Progression was seen in 32% of the cholestyramine group and 49% of the placebo group, and regression was seen in 7% of both groups. Eight of the 71 patients randomized to 24 g of cholestyramine daily developed myocardial infarction, compared with 12 out of 72 on placebo (NS).

Cholesterol-Lowering Atherosclerosis Study

This study enrolled nonsmoking men aged 40–59 who had previously undergone coronary artery bypass grafting and who had demonstrated good compliance and good response in serum cholesterol to the study medications (15). Participants were thus a more favorable group in which to intervene than those in the NHLBI study. Unlike the NHLBI patients, the CLAS patients were unselected in regard to baseline lipids, and they had lower mean serum cholesterol levels (246 mg/dl in drug group, 243 mg/dl in placebo group) and a wide range of levels (185–350 mg/dl). Patients were randomly allocated to groups (94 in each group) receiving either colestipol 30 g/day plus niacin 3–12 g/day or placebos. All patients were placed on a cholesterol-lowering diet.

During the 2 years of the study, the mean total cholesterol in the drug group was 22% lower, LDL cholesterol 39% lower, and HDL cholesterol 37% higher than in the placebo group. On average, lesions progressed in both groups, however, global

Table 2 Summary Narrative Review of Angiographic Regression Studies

Study	Intervention	Number of participants (treated/control)	Significance of treatment effect Progression	Regression
NHLBI Type II	Cholestyramine	59/57	NC	NC
CLAS	Colestipol + niacin	80/82	↓	↑
POSCH	Partial ilial bypass	333/301	↓	↑
FATS	Colestipol + lovastatin	38/46	↓	↑
FATS	Colestipol + niacin	36/46	NC	↑
SCOR	Cholestyramine + niacin + lovastatin	40/42	NC	↑
STARS	Diet	26/24	↓	↑
STARS	Diet + Cholestyramine	24/24	↓	↑
MARS	Lovastatin	123/124	NC	↑

NC = No change.
↓ = Significantly reduced, $p < 0.05$.
↑ = Significantly increased, $p < 0.05$.

change score by visual panel was significantly more favorable (less progression) in the active-treatment group. Progression was seen in 33% of the active-treatment group versus 55% in the placebo group, and regression was seen in 19% and 7%, respectively. Myocardial infarctions occurred in 1 out of 94 active-treatment patients and 5 out of 94 placebo-treatment patients (NS).

Familial Atherosclerosis Treatment Study

To be eligible for this trial, men less than 62 years of age had to have a family history of premature coronary disease, elevated apolipoprotein B levels, and angiographic coronary artery disease, but not have undergone coronary artery bypass grafting (18). Two types of combination therapy were assessed: colestipol (30 g/day) plus niacin (4 g/day) ($n = 48$) and colestipol plus lovastatin (40 mg/day) ($n = 46$). The control group ($n = 52$) received placebo (and in some cases, colestipol). All patients were placed on a cholesterol-lowering diet. The trial duration was 2.5 years.

In-trial total cholesterol levels decreased by 17%, LDL cholesterol levels by 20%, and HDL cholesterol levels increased by 37% in the colestipol plus niacin group compared to controls, and in the colestipol plus lovastatin group the changes in total, LDL, HDL cholesterol levels were -28%, -34%, and 2%, respectively.

This was the first unifactorial angiographic trial to show significant regression within the active-treatment groups. On average, lesions in the control group progressed, and the differences between groups were highly significant. Progression occurred in 25% of patients on colestipol plus niacin, 21% on colestipol plus lovastatin, and in 46% on placebo. The corresponding rates of regression were 39% and 32% in the two active-treatment groups and 11% in the placebo group. There were no clinical CHD events in the colestipol-plus-niacin group, 2 in the colestipol-plus-lovastatin group, and none in the placebo group.

Specialized Center of Research Study

To qualify for this study, patients had to have heterozygous familial hypercholesterolemia and quantifiable lesions on coronary angiography but did not need to have symptomatic coronary disease (19). The study was also unusual in that 55 of the 97 patients were women. The treated group was initially prescribed up to 30 g of colestipol and up to 7.5 g of niacin daily, and when lovastatin became available, this was added to the regimen. Not all patients took all three drugs, but most were on two- or three-drug combinations during the course of the trial. The control patients were treated with diet and were offered low-dose colestipol (15 g/day); about half elected to take it.

Compared to the control group, total and LDL cholesterol declined by 22% and 26%, respectively, and HDL cholesterol increased by 16%. The study found significant regression within the active-treatment group and significant differences between active and control groups. When analyzed separately, women fared better than men. Progression was seen in 20% of the active-treatment group versus 41% of the controls, and regression was seen in 33% and 13%, respectively. None of the 48 actively treated patients, and 1 of the 49 control patients, developed clinical CHD events.

The St. Thomas Atherosclerosis Regression Study

This study of British men assessed the effect of dietary reduction of plasma cholesterol on coronary atherosclerosis and also the effect of the addition of cholestyramine (16 g/day) to diet (20). There were 30 patients on diet only, 30 on diet plus cholestyramine, and 30 on usual care. In order to be eligible, men had to have plasma cholesterol levels between 232 and 387 mg/dl and coronary atherosclerosis demonstrated by quantitative coronary angiography. Mean trial duration was 3 years and 3 months.

Total cholesterol levels in the diet an diet-plus-cholestyramine groups decreased by 11% and 20%, respectively, compared to the usual-care group. LDL cholesterol levels declined by 12% and 28%, and HDL cholesterol levels also declined by 6% and 2%. Within the active-treatment groups, angiographic regression predominated (significant for diet-plus-cholestyramine group), and differences between active and control groups were significant. Progression was seen in 15% of the diet group, 12% of the diet-plus-cholestyramine group, and 46% of controls. Regression was seen in 38% and 33% of the two active-treatment groups compared to 4% of the control group. CHD events were significantly ($p < 0.05$) less frequent in the combined active-treatment groups (3 out of 60) than in the control group (5 out of 30).

Program on the Surgical Control of the Hyperlipidemias

This study has been described above. Although primarily designed as a clinical outcome study, the investigators did measure the angiographic appearance at baseline, 3, 5, 7, and 10 years (2). Overall, lesions tended to progress in both the surgery and control groups, but progression was significantly less at 3 years and at every subsequent evaluation in the surgery compared to the control group. At 5 years 38% of the surgery group versus 65% of the control group had progressed, and 13% versus 5% had regressed. The differences between groups did not widen further after 5 years. Interestingly, angiographic progression in the first 3 years was strongly related to subsequent CHD events in individuals, lending strong support to the idea that angiographic trials are suitable surrogates for clinical outcome trials (28).

Monitored Atherosclerosis Regression Study

This is the first published angiographic trial of drug monotherapy (lovastatin 40 mg/day) for lowering cholesterol (16). Participants had to have angiographically defined disease but were not selected by lipid level. All but 9% of the 270 patients were male, and ages ranged from 35 to 67 years. All patients received dietary advice; 134 were randomized to lovastatin and 136 to placebo. The trial was designed to be evaluated at 2 years, with a 2-year extension conditional on the initial results. The primary outcome was average per-patient percent diameter stenosis as measured by quantitative coronary angiography (QCA). The trial was stopped at 2 years by an external review board primarily because of significantly favorable results as measured by the secondary outcome measure of global change score by visual panel ($p < 0.002$) and favorable results by QCA in the subgroup with severe lesions (>50% stenosis, $p < 0.01$).

Lovastatin treatment reduced total cholesterol by 32%, LDL cholesterol by 43%, and triglycerides by 25%, and it increased HDL cholesterol by 4%. For the overall active-treatment group, change in percent stenosis by QCA showed average progression, and although progression was less in the lovastatin group than in the placebo group, the difference was not significant. Progression was seen in 24% of the lovastatin group and 33% of controls (NS), and regression was seen in 19% of the lovastatin group and 10% of controls ($p = 0.04$). As measured by global change score, the lovastatin group did significantly better than the control group ($p < 0.002$). Five lovastatin patients and 7 placebo patients had coronary events (NS).

NARRATIVE AND META-ANALYTIC OVERVIEWS OF ANGIOGRAPHIC STUDIES

Narrative review indicates that there were significantly fewer patients with overall progression in five out of the nine comparisons of actively treated groups with control groups. On average, lesions progressed in the control groups and very few patients in these groups showed regression. In contrast, regression occurred on average in five out of nine actively treated groups, and compared to the control groups significantly more patients showed regression (eight out of nine comparisons). Thus, it is apparent that progression can be slowed or inhibited and actual regression of lesions can be obtained by cholesterol-lowering therapy.

These conclusions are confirmed by the meta-analysis (Fig. 2), which showed

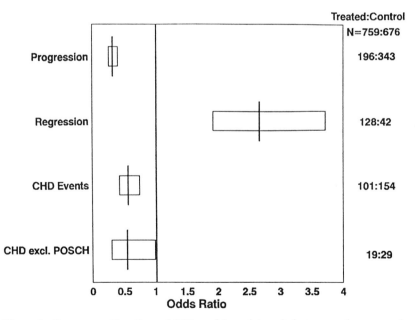

Figure 2   Summary odds ratios and 95% confidence intervals for progression, regression, and all CHD events in angiographic regression studies of cholesterol lowering. Details of trials are given in text. Meta-analysis was performed by the method of Mantel and Haenzel (31), as modified by Yusuf et al. (32). Numbers (N) of participants in treated and control groups in the pooled data are indicated to the right of the diagram, as are numbers of outcomes.

that the odds of progression was reduced by 68% in active treatment compared to control groups (odds ratio 0.32, 95% CI 0.26–0.40, $p < 0.0001$). On the other hand, regression was 267% more likely to occur (odds ratio 2.67, 95% CI 1.92–3.71, $p < 0.0001$).

The meta-analysis comparing the incidence of CHD in the angiographic studies is revealing, in spite of the small numbers of clinical events in the individual angiographic trials. Excluding the Program on Surgical Control of Hyper-lipidemias (POSCH) data, in the 8 active treatment arms, 19 out of 502 (3.8%) of patients developed clinical CHD events compared with 29 out of 432 (6.7%) in the 6 control groups. The meta-analysis indicates that a risk reduction of 45% can be expected from intensive short-term lipid-lowering therapy (odds ratio 0.55, 95% CI 0.30–0.99, $p < 0.05$). Confidence in the estimate is enhanced by adding the 208 MI events from the POSCH study: the odds ratio is unchanged at 0.56, but the confidence interval is narrower (95% CI 0.43–0.75, $p < 0.001$).

These estimates indicate that a substantial improvement in the prognosis of patients with existing coronary disease is achievable by lipid-lowering treatment. Untreated, such patients have a very high rate of cardiovascular events, perhaps seven times higher than those who do not have existing disease. The angiographic changes appear to be relevant to clinical outcomes, as shown by the concomitant reduction of angiographic severity and clinical events in the aggregated data and by the results within the POSCH study where angiographic change in the first 3 years predicted clinical events in the subsequent years (28). At first blush the risk reductions in the angiographic trials appear to be unexpectedly large. However, the magnitude of risk reduction for myocardial infarction is quite consistent with the large reductions in cholesterol obtained by more efficacious therapy. Benefit within a relatively short period of 2–3 years may relate to the ability of cholesterol reduction to stabilize smaller lipid-rich plaques, which are particularly prone to fissuring, thrombosis, and the sudden occlusion of vessels that leads to myocardial infarction (29,30). On the available evidence, angiographic studies appear to be suitable surrogates for clinical endpoint trials. At a minimum they complement the evidence of benefit that has accrued from the clinical outcome trials.

CONCLUSIONS AND IMPLICATIONS

The clinical trials and angiographic studies of cholesterol lowering have been of decisive importance in persuading scientific and public opinion that elevated serum cholesterol is a causal element in the chain of events leading to CHD and that treatment by diet and drugs is effective in lowering the risk of CHD. The appropriateness of these opinions is well illustrated by the analyses of the combined trials, which show that the clinical event rate can be lowered by about 20% if cholesterol levels are lowered by 10%. The reduced risk for CHD applies to both primary and secondary prevention. Furthermore, the angiographic studies have now demonstrated that more vigorous lipid-lowering therapy leads to improvements in the angiographic appearance of coronary vessels, which are accompanied by large reductions in CHD risk, and that these reductions in risk can be obtained within a relatively short space of time.

Diet and a variety of drugs appear to modify the risk of CHD. The results of studies using combinations of drugs, for example, bile acid–binding resins with either niacin or HMG-CoA reductase inhibitors, are particularly impressive. The primary purpose of treatment remains the reduction of total and LDL cholesterol; however, the possibility of an additional benefit from improving other aspects of the lipid profile (such as raising HDL cholesterol levels) at the same time should not be ignored. In many instances combinations of drugs will be needed to achieve optimal lowering of serum cholesterol or to treat all elements of the lipid disorder.

Though the treatment of high-risk but apparently healthy individuals should not be neglected, it would be particularly appropriate to institute intensive diet and combination drug therapy in patients with existing CHD, in view of their high risk of (often fatal) reinfarction if left untreated. The secondary prevention trials provide evidence that clinical events can be reduced in such patients.

In secondary prevention there is no indication of an increase in noncardiac mortality, and because of the preponderance of cardiac deaths the trend for total mortality is favorable though not quite significant. The benefit is underestimated in these analyses because the degree of cholesterol lowering obtained has not been taken into account. The angiographic studies strongly suggest that large reductions in cholesterol to much lower levels (in-treatment LDL cholesterol levels below 100 mg/dl were frequently observed) than those achieved in the secondary prevention trials will markedly reduce the rate of coronary events in patients with existing disease.

REFERENCES

1. Rossouw JE, Lewis B, Rifind BM. The value of lowering cholesterol after myocardial infarction. N Engl J Med 1990; 323:1112.
2. Buchwald H, Varco RL, Matts JF, et al. Effect of partial ileal bypass surgery on mortality and morbidity from coronary artery disease in patients with hypercholesterolemia. N Engl J Med 1990; 323:946.
3. Burr ML, Gilbert JF, Holliday RM, Sweetnam PM, Elwood PC, Deadman NM. Effects of changes in fat, fish, and fibre intakes on death and myocardial reinfarction: diet and reinfarction trial (DART). Lancet 1989; ii:757–761.
4. Canner PL, Berge KH, Wenger NK, et al. Fifteen year mortality in Coronary Drug Project patients: long-term benefit with niacin. J Am Coll Cardiol 1986; 8:1245.
5. Coronary Drug Project Research Group. Clofibrate and niacin in coronary heart disease. JAMA 1975; 231:360–380.
6. Carlson LA, Rosenhammer G. Reduction in mortality in the Stockholm Ischaemic

Heart Disease Secondary Prevention Study by combined treatment with clofibrate and nicotinic acid. Acta Med Scand 1988; 223:405.

7. Frick MH, Heinonen OP, Huttunen JK, Koskinen P, Manttari M, Manninen V. Efficacy of gemfibrozil in dyslipaemic subjects with suspected heart disease. An ancillary study in the Helsini heart study frame population. Ann Med 1993; 25: 41–45.

8. Group of Physicians of the Newcastle upon Tyne Region. Trial of clofibrate in the treatment of ischaemic heart disease. Br Med J 1971; 4:767.

9. Leren P. The Oslo Diet-Heart Study. Eleven-year report. Circulation 1970; 42:935.

10. Research Committee to the Medical Research Council. Low-fat diet in myocardial infarction—a controlled trial. Lancet 1965; 2:501.

11. Research Committee to the Medical Research Council. Controlled trial of soya-bean oil in myocardial infarction. Lancet 1968; 2:693.

12. Research Committee of the Scottish Society of physicians. Ischaemic Heart Disease: a secondary prevention trial using clofibrate. Br Med J 1971; 4:775.

13. Rose GA, Thomson WB, Williams RT. Corn oil in treatment of ischaemic heart disease. BMJ 1965; i:1531–1533.

14. Schoch HK. The US Veterans Administration cardiology drug-lipid study: and interm report. Adv Exp Med Biol 1968; 4:405–420.

15. Blankenhorn DH, Nessim SA, Johnson RL, Sanmarco ME, Azen SP, Cashin-Hemphill L. Beneficial effects of combined colestipol-niacin therapy on coronary atherosclerosis and coronary venous bypass grafts. JAMA 1987; 257:3233.

16. Blankenhorn DH, Azen SP, Kramsch DM, et al. Coronary angiographic changes with lovastatin therapy. The Monitored Atherosclerosis Regression Study (MARS). Ann Intern Med 1993; 119:969–976.

17. Brensike JF, Levy RI, Kelsey SF, et al. Effects of therapy with cholestyramine on progression of coronary arteriosclerosis: results of the NHLBI Type II Coronary Intervention Study. Circulation 1984; 69:313.

18. Brown BG, Albers JJ, Fisher LD. Regression of coronary artery disease as a result of intensive lipid-lowering therapy in men with high levels of apolipoprotein B. N Engl J Med 1990; 323:1289.

19. Kane JP, Malloy MJ, Ports TA, Phillips, NR, Diehl JC, Havel RJ. Regression of coronary atherosclerosis during treatment of familial hypercholesterolemia with combined drug regimens. JAMA 1990; 264:3007.

20. Watts GF, Lewis B, Brunt JNH, et al. Effects of coronary artery disease of lipid-lowering diet, or diet plus cholestyramine, in the St Thomas' Atherosclerosis Regression Study (STARS). Lancet 1992; 339:563.

21. Committee of Principal Investigators. A cooperative trial in the primary prevention of ischaemic heart disease using clofibrate. Br Heart J 1978; 40:1069.

22. Heady JA, Morris JN, Oliver MF. WHO clofibrate/cholesterol trial: clarifications. Lancet 1993; 340:1405.

23. Dayton S, Pearce ML, Hashimoto S, Dixon WJ, Tomiyasu U. A controlled clinical trial of a diet high in unsaturated fat in preventing complications of atherosclerosis. Circulation 1969; 40(suppl 2):1.

24. Dorr AE, Gunderson K, Schneider JC, Spencer TW, Martin WB. Colestipol hydro-

chloride in hypercholesterolemic patients: effect on serum cholesterol and mortality. J Chron Dis 1978; 31:5–14.

25. Frick MH, Elo O, Haapa K, et al. Helsinki Heart Study: primary-prevention trial with gemfibrozil in middle-aged men with dyslipidemia. Safety of treatment, changes in risk factors, and incidence of coronary heart disease. N Engl J Med 1987; 317:1237.

26. Lipid Research Clinics Program. The Lipid Research Clinics Coronary Primary Prevention Trial results. I. Reduction in incidence of coronary heart disease. JAMA 1984; 251:351.

27. Gould AL, Rossouw JR, Furberg CD, Santanello NC, Heyse JF. Cholesterol reduction yields clinical benefit: a new look at old data. Circulation. Submitted for publication.

28. Buchwald H, Matts JP, Fitch LL, et al., for the Program on the Surgical Control of the Hyperlipidemias (POSCH). Changes in sequential coronary arteriograms and subsequent coronary events. JAMA 1992; 268:1429–1433.

29. Fuster V, Badimon L, Badimon J, Chesebro J. The pathogenesis of coronary artery disease and the acute coronary syndromes I. N Engl J Med 1992; 326:242–250.

30. Fuster V, Badimon L, Badimon J, Chesebro J. The pathogenesis of coronary artery disease and the acute coronary syndromes II. N Engl J Med 1992; 326:310–318.

31. Mantel N, Haenszel W. Statistical aspects of the analysis of data from retrospective studies of disease. J Natl Cancer Inst 1959; 22:719.

32. Yusuf S, Peto R, Collins R, Sleight P. Beta blockade during and after myocardial infarction: an overview of randomized trials. Progr Cardiovasc Dis 1985; 27:355.

4

Evolution of the Atherosclerotic Plaque

Antonio Fernández-Ortiz

Hospital Universitario San Carlos, Madrid, Spain

Valentin Fuster

Mount Sinai Medical Center, New York, New York

INTRODUCTION

Atherosclerosis is a disease by no means unique to twentieth-century humans. Advanced calcified lesions were found in ancient Egyptian mummies. However, progress in our comprehension of the initiation and evolution of atherosclerotic plaques has been slow and recent. While most of the morphological facts about the process of atherosclerosis are now known, it is unclear how these facts fit into an overall framework that fully explains the pathogenesis and progression of the disease.

Atherosclerosis is primarily a focal intimal disease of arteries ranging from the size of the aorta down to the size of tertiary branches of coronary arteries (1 mm diameter). By definition, atherosclerosis is characterized by atherosis (soft "gruel") and sclerosis (hard, collagenous). A few patients have only hard or only soft plaques, but most patients have plaques containing variable amounts of both hard and soft components (1–3). The connection, if any, between the hard collagenous and the soft atheromatous components is uncertain, and most attempts to reconstruct a dynamic sequence in the evolution of atherosclerotic plaques based on the morphology and composition of advanced lesions have been disappointing. Most observers believe that plaques soften with time: the older the individual (4–8) and the larger the plaque (5,6,9–15), the more gruel. Others, however, suggest that softening occurs early during the evolution of fibrous plaques (16,17) and that hard lesions could represent the late, sclerotic phase of the disease (18).

Recent advances in cellular and molecular biology have opened a window of opportunity to better understand factors influencing atherosclerotic plaque initiation, progression, or even regression. The purpose of this chapter is to give a connecting overview of morphological features and biological mechanisms involved in atherosclerosis.

INITIATION AND LOCALIZATION OF ATHEROSCLEROSIS

Identification of the earliest events in lesion development is important if we are to understand the critical pathogenic mechanisms that become obscure as the lesions progress. During the nineteenth century, two major hypotheses were used to explain the pathogenesis of atherosclerosis: the "incrustation" hypothesis and the "lipid" hypothesis. The incrustation theory, proposed by Rokitansky in 1852 (19), suggested that intimal thickening resulted from fibrin deposition with subsequent organization by fibroblast and secondary lipid accumulation. The lipid theory, proposed by Virchow in 1856 (20), suggested that blood lipids transudate into the arterial wall, interact with proteoglycans, and promote intimal proliferation.

Vascular Response-to-Injury Hypothesis

The prevalent view of atherogenesis proposed by Ross integrates both theories into a more complex "response-to-injury" hypothesis (21–24), in which atherosclerosis is considered a healing response of the arterial wall to various injurious stimuli. In order to allow easier understanding of the pathogenesis of the disease and better formulation of therapeutic strategies, we have proposed a pathophysiological classification of vascular injury or damage into three categories representing stages of increasing severity (25). *Type I injury* consists of functional alterations of endothelial cells without substantial morphological changes; *type II injury* is characterized by endothelial denudation; and *type III injury* is characterized by endothelial denudation with damage to the intima and even the media.

Hypercholesterolemia and other risk factors somehow lead to endothelial dysfunction, but not necessarily to endothelial denudation (type I injury). Dysfunctional endothelium is characterized by increased permeability to lipoproteins and the expression of specific adhesive glycoproteins on its surface. Numerous studies have suggested that modified (oxidized) low-density lipoprotein (oxLDL) is a key component in endothelial injury (26–29). Endothelial cells themselves have the ability to oxidize lipoproteins as they are transported into the artery wall (24,30). In vitro, smooth muscle cells and macrophages have also shown the ability to oxidize native LDL (31–37). Malondialdehyde released from platelets also has the ability to modify LDL (38), although probably not in sufficient quantity to have clinical relevance. Curiously, nitric oxide (NO), a potent vasodilator released by endothelium that can prevent platelet aggregation and leukocyte adhesion (39,40), has also been shown capable of promoting oxidation of LDL and all of the deleterious consequences of its formation (41,42).

Interestingly, despite exposure of different areas of the endothelial surface to the same lipoprotein concentration, spontaneous atherosclerotic lesions only develop in certain locations. Therefore, in addition to systemic factors, local factors must play a role in determining the location and possibly the rate of atherosclerosis progression (43). It is believed that variations in local hemodynamics would explain this fact, and regions of low shear stress with complex secondary flow patterns, standing recirculation zones, and oscillatory hemodynamic changes occurring mostly at bending points and bifurcations seem to be the preferred sites of atherogenesis (43–45). Several hypotheses have been advanced to explain the mechanism by which low vessel wall shear stress might promote the development of atherosclerotic lesions. It has been shown that endothelial cells undergo morphological alterations in response to change in the degree and orientation of shear stress; elongated endothelial cells are located in regions of high shear stress, whereas polygonal endothelial cells are located in low-shear-stress regions (46,47). These alterations may be responsible for changes in endothelial cell permeability to atherogenic lipoprotein particles (44). It has been also hypothesized that, in regions of low shear stress, a reduction or stagnation in the velocity of blood flow permits increased uptake of atherogenic particles as a result of increased residence time or prolonged contact with the endothelium, which would allow concentration-induced endocytosis (44,48).

Besides atherogenic lipoproteins and local mechanical forces, other forms of injury such as those that can be induced by circulating vasoactive amines, immunocomplexes, infection, and chemical irritants in tobacco smoke (25) may potentiate type I chronic endothelial injury favoring accumulation of lipids into the intima. In addition, because of the presence of T lymphocytes together with macrophages in early lesions, it has been suggested that there may be an immunological reaction during the process of atherogenesis and that this may be related to specific antigens (23,49–51), but no such an antigen has been identified, although autoantibodies have been found to oxLDL (24).

In summary, atherosclerosis is recognized to be an immune/inflammatory response of the intima to injury. It is also being increasingly realized that the injury may be initiated by lipid in certain parts of the arterial tree where there are disturbances in the pattern of blood flow. Moreover, the presence of dysfunctional endothelium in very early stages of atherosclerosis has been shown in a clinical study (52) in which high-resolution ultrasound performed in femoral and brachial arteries of symptom-free children and young adults with risk factors for atherosclerosis demonstrated impairment in endothelium-dependent vasodilatation even before anatomical evidence of plaque formation.

Lipid Accumulation in the Arterial Wall

Clinical and experimental studies have firmly established that high levels of LDL are associated with atherosclerosis (53–55). Lipids deposited in the atherosclerotic lesions are mostly derived from plasma LDL. The internalization and intravascular accumulation of cholesterol and its esters depends on two mechanisms: one is active and dependent on specific receptors located in the wall membrane, and the other is passive and receptor independent. Endothelial type I injury increases the passive one through the direct interaction of plasma LDL with glycosaminoglycans in the vessel wall (55).

A specific LDL receptor in the liver is responsible for the metabolism and elimination of plasma cholesterol, and another specific LDL receptor in the adrenal cortex and ovary is responsible for the synthesis of hormones (56). The number of these LDL receptors seems to be, in part, genetically determined, and this would help to explain the fact that a positive family history may be a risk factor for coronary artery disease (55). Among various receptors expressed by cultured macrophages, the B-VLDL (very-low-density lipoprotein) and the acetyl-LDL or scavenger receptor are poorly regulated by the intracellular cholesterol levels in contrast to the classic LDL receptor (57). Monocytes/Macrophages take up LDL via the LDL receptor at a very low rate without formation of cholesterol deposits, but chemically modified (oxidized) LDL is taken up much more rapidly via the acetyl-LDL or scavenger receptor (58). Since the macrophage receptors for modified LDL are not regulated, excessive intracellular accumulation of this lipoprotein may lead to destruction of the cell with subsequent release of oxLDL and free radicals, which can cause cytotoxicity and further damage to the endothelium (type II injury).

Monocyte/Macrophage Recruitment into the Arterial Wall

Macrophages are present throughout all stages of plaque pathogenesis, from the initial events in atherogenesis (increased vascular permeability, increased monocyte adherence, and intimal recruitment) to the established disease (monocyte/macrophage derived foam cells and formation of the lipid core). Macrophages may act as an antigen-presenting cell to T lymphocytes, as a scavenger cell to remove noxious materials, and as a source of growth-regulatory molecules and cytokines.

It is likely that the focal accumulation of monocyte/macrophages in atherogenesis involves the local expression of specific adhesive glycoproteins on the endothelial surface and the generation of chemotactic factors by altered endothelium, its adherent leukocytes, and possibly underlying smooth muscle cells. The presence of oxLDL may play an initial role in monocyte recruitment by inducing the expression of adhesive cell-surface glycoproteins in the endothelium, the most important being VCAM-1 (athero-ELAM) (24). Several specific molecules may be relevant in attracting monocytes to the subendothelial space, such as a specific chemotactic protein (monocyte chemotactic protein-1 [MCP-1]) synthesized by vascular cells (59–62), colony-stimulating factors (CSFs) (63), and transforming growth factor β (TGF-β) (64). In more advanced stages, during which there is significant connective tissue production and tissue necrosis, peptide fragments from fibrin, fibronectin, elastin, collagen degradation (65–67), and thrombin (68) may be the predominant monocyte chemoattractants elaborated.

Relationship Between Adaptive Intimal Hyperplasia and Atherosclerosis

When atherosclerosis develops, advanced lesions form first in some regions with adaptive intimal thickening. In humans, the topographic distribution of eccentric intimal thickening and of advanced atherosclerotic lesions is similar in the aorta and major distributing arteries, including coronaries (69–71). These intimal regions coincident with locations at which advanced atherosclerotic lesions develop earlier have been called the atherosclerotic-prone areas of arteries. The question of whether eccentric intimal thickening should be considered atherosclerotic has been much debated. The available evidence indicates that specific mechanical stress, present in locations of the arterial tree with adaptive intimal thickening, causes the thickening whether or not high concentrations of atherogenic lipoproteins are present. When atherogenic lipoproteins exceed certain critical levels, the same mechanical forces may enhance lipoprotein deposition in the same regions, leading to transformation into atheromatous lesions. However, in severely hypercholesterolemic humans and in several species of animals subjected to severe hypercholesterolemia, nearly all regions of the aorta and many arteries may ultimately be the sites of advanced lesions. Thus, advanced lesions may not be confined to regions with adaptive intimal thickening. To answer this question, in 1992 the Committee on Vascular Lesions of the American Heart Association (72) stated that regions of intimal thickening simply mark locations where, under the influence of atherogenic stimuli, lesions form earlier and more rapidly than elsewhere.

PROGRESSION OF ATHEROSCLEROSIS

The progression of early atherosclerotic lesions to clinically manifest, enlarging atherosclerotic plaques, such as those that cause exertional angina, is often more rapid in persons with coronary risk factors and in lesions with certain topographic characteristics (73,74). Diabetes, proximal or midvessel lesion location, elevated cholesterol, interval from infarction, and complex lesion morphology were identified as predictors of angiographic progression in The Coronary Artery Surgery Study (CASS) (74). In some plaques progression is slow and probably a continuation of the complex biological process initiated by chronic type I and II endothelial injury. However, in most growing lesions progression is probably rapid, suggesting recurrent minor fissures of the most fatty or atheromatous plaques (type III injury), with subsequent thrombus formation and fibrotic organization (73).

Slow Step-by-Step Atherosclerosis Progression

Thus far we have set the pathobiological stage for lesion initiation, emphasizing that in lesion-prone sites characterized by dysfunctional endothelium, the combination of increased LDL uptake and enhanced monocyte recruitment are pivotal initiating events for atherosclerosis.

Morphological Features Involved in Slow Step-by-Step Progression

The morphological studies by Stary have provided many answers about plaque evolution (69,75) (Fig. 1). Detailed studies were made of the arteries in young subjects, ranging in age from infancy to more than 30 years, who died suddenly from noncardiac causes. The cases were divided into cohorts by age. It was

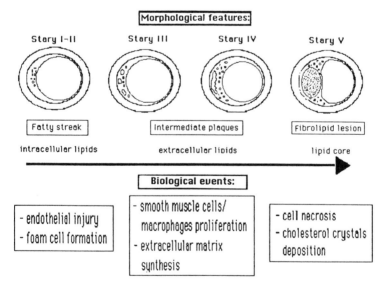

Figure 1 Integrated morphological and biological events in slow step-by-step atherosclerosis progression. See text for details.

therefore possible to see the age at which lesions of any particular type first appeared and the sequence of plaque progression could be inferred. In the first three decades, the composition of the lesions is predominantly lipid and relatively predictable. From the fourth decade on, the composition of advanced lesions becomes unpredictable because some lesions continue to increase by mechanisms other than lipid deposition (75).

The first observation made by Stary was that in all human infants focal thickening of the intima develops due to smooth muscle cell proliferation. Initial atherosclerotic lesions (Stary type I) are only microscopically and chemically perceivable, consisting of isolated groups of macrophages filled with lipid droplets. Macrophages without lipid droplets in this type of lesion are twice the number present in intima normally. Such minimal alterations were found in the coronary arteries of 45% of infants in the first 8 months of life, located in areas of adaptive eccentric intimal thickening where usually more advanced lesions in older children and adults develop (75). Type II lesions or fatty streaks are composed of more lipid-laden cells than initial lesions. Macroscopically, they appear as fatty dots or streaks barely raised above the intimal surface. Each lesion is made up of one or more layers of lipid-filled foam cells within the intima. While most of the electron-microscopically visible lipids are within cells, some smaller lipid particles are extracellular (75). There is usually no evident smooth muscle cell proliferation in fatty streaks, and the endothelium is probably intact (76). There is some evidence that such streaks are precursors of larger plaques. Pathological studies show that transitional forms exist between fatty streaks and advanced plaques and that advanced plaques develop in the sites where fatty streaks are more frequent. However, it is clear that not all fatty streaks progress (77).

The cohort studies by Stary show that progression beyond the fatty streak stage is associated with a sequence of changes starting with the appearance of extracellular lipid, which begins to form a lesion that is more elevated (Stary type III). Extracellular lipids accumulate deep in the elastin-rich part of the intima, just below the layers of foam cells, and replace intercellular matrix proteoglycans and fibers, driving smooth muscle cells apart. Multiple scattered pools of extracellular lipid may progress to a massive, confluent accumulation of extracellular lipid— the lipid core characteristic of Stary type IV lesions. It seems that the formation of the core of an atheroma precedes the smooth muscle cell proliferation and collagen deposition that may eventually thicken the region above the lipid core, forming the fibrotic cap of the lesion. In Stary type IV lesions, most of the tissue above the lipid core represents the upper (proteoglycan) part of the preexisting adaptive thickening. In Stary type V lesions (fibroatheroma) or fibrolipid lesions, smooth muscle cells migrate into and proliferate within the plaque, forming a layer over the luminal side of the lipid core (69,78). More and more collagen is produced, and plaque size increases. The process culminates in what is known as a raised fibrolipid or advanced plaque. In the aorta, such plaques may be a centimeter or more in length. Raised fibrolipid plaques have a very characteristic microanatomy, with a core of extracellular lipid separated from the media by smooth muscle cells and insulated from the lumen by a thick cap of collagen-rich fibrous tissue containing smooth muscle cells. Surrounding the lipid core are lipid-filled foam cells. The lipid-rich core is avascular and usually almost totally acellular, consisting of pultaceous debris with abundant cholesterol crystals (9–12). Part of the core lipid is thought to be derived from the death of lipid-containing macrophage foam cells and the release of their intracytoplasmatic content. Some of the lipid core bordering foam cells appear disintegrated (dead), and cell constituents may be found extracellularly within the core, indicating that cell necrosis has contributed to its growth (79).

Lesions having visible thrombotic deposits and/or marked hemorrhage, in addition to lipid and collagen, are Stary type VI lesions (complicated fibroatheroma or complicated lesions). The term calcific lesion (Stary type VII lesion) may be applied to some advanced, largely mineralized lesions, particularly after the fourth decade. Calcium deposits replace the accumulated remnants of dead cells and extracellular lipid. Finally, some atherosclerotic lesions, possibly more often in arteries of lower extremities, may consist entirely or almost entirely of scar collagen. Lipid may have regressed, or it may never have been in the lesion. The term fibrotic lesion or Stary type VIII lesion is best reserved for lesions with this morphology.

The aorta and coronary arteries of adults show all stages of plaque development, implying that new lesions are generated throughout adult life. Even in a single individual there are major plaque-to-plaque variations in the relative proportions of extracellular lipid and collagen present (1–3). The work by Stary indicates that a plaque can take from 10 to 15 years to fully develop.

Biological Mechanisms Involved in Slow Step-by-Step Progression
The progression of atherosclerotic lesions from one morphological type to another is believed to be dependent on the balance between positive and negative feedback loops generated by its cellular components (Fig. 2). Through a network of cellular interactions, the release of one molecule (growth factors, cytokines, and other chemicals) can lead to expression of a second molecule in a target cell that can then either stimulate its neighbors in a paracrine way or stimulate itself in an autocrine way. The underlying biological events that follow endothelial injury and monocyte recruitment during step-by-step progression include foam cell formation, cellular proliferation, synthesis of connective tissue matrix, and formation of the lipid core. The cells involved in these events are (1) endothelial cells, (2) platelets, (3) macrophages, and (4) smooth muscle cells.

Endothelial Cells. It has been shown that endothelial cells possess growth-promoting activity for cultured smooth muscle cells and fibroblast (80) caused by the secretion of at least two majors mitogens, one of which is closely related to or

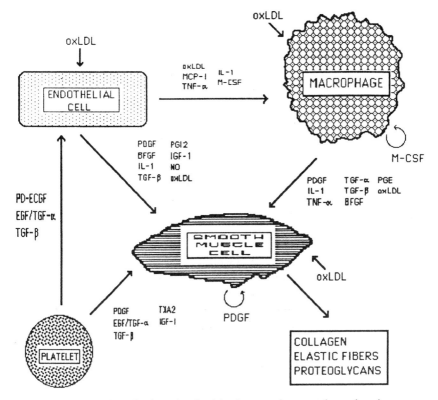

Figure 2 Biological mechanisms involved in slow step-by-step atherosclerosis progression. The principal mediators in cellular interactions between endothelial cells, platelets, macrophages, and smooth muscle cells are shown in this diagram. BFGE, Basic fibroblast growth factor; EGF, endothelial growth factor; IGF-1, insulin-like growth factor-1; IL-1, interleukin-1; MCP-1, monocyte chemotactic protein-1; M-CSF, monocyte colony-stimulating factor; NO, nitric oxide; oxLDL, oxidized low-density lipoprotein; PD-ECGF, platelet-derived endothelial cell growth factor; PDGF, platelet-derived growth factor; PGE, prostaglandin E; PGI2, prostacyclin; TGF-α, transforming growth factor α; TGF-β, transforming growth factor β; TNF-α, tumor necrosis factor α; and TxA$_2$, thromboxane A$_2$. See text for details. (Adapted from Ref. 24.)

identical to platelet-derived growth factor (PDGF). Experiments have suggested that dying endothelial cells produce in less than 3 days as much growth factor as healthy cells produce over a 19-day period (81). These data support a modified version of the response-to-injury hypothesis, in which focal lethal damage to the endothelium could lead to high local mitogenic activity that can attract and stimulate neighboring smooth muscle cells. On the other hand, endothelial cells with intact function may play a central role in maintaining the low proliferative state of smooth muscle cells under normal vascular conditions (82,83).

Platelets.   After type II or III injury, platelets attached to the vascular surface can release PDGF and other mitogens necessary for smooth muscle cell growth. However, the role of platelets in the early phases of atherogenesis (type I injury) is controversial. Scanning electron micrographs have shown small platelet microthrombi located at sites where endothelial dysfunction may have occurred (24). We have demonstrated that extensive, proliferative, atheromatous lesions developed in normal swine on a normal or mildly hypercholesterolemic diet, whereas less proliferative lesions developed in swine homozygous for von Willebrand's disease and thus defective in platelet-vessel wall interaction (84–86). However, several other experimental atherosclerosis models suggest proliferation of smooth muscle cells and intimal thickening without obvious platelet participation (87–89).

Although platelets have been previously suggested to serve as a source of lipid in fatty streaks (90), only recently has this observation been confirmed in experiments showing that platelets could support foam cell formation in cultured aortic smooth muscle cells (91). In the absence of any other source of free cholesterol, platelets aggregated with collagen could produce lipid-rich particles that are avidly taken up by monocyte/macrophage in tissue culture when provided with free fatty acids (oleate), when the cholesterol is esterified, and when foam cell formation is induced (92).

Macrophages.   The subendothelial transmigration of blood monocytes leads to the accumulation of these cells in the intima. Once in the intima, the monocyte is exposed to a milieu of modified lipoproteins, cytokines, chemoattractants, and growth factors, all of which can cause its activation and differentiation into a macrophage. When activated, the macrophage expresses several biologically active molecules such as cytokines, growth factors, and free radicals, which in turn modulate the activity of surrounding cells (93). Some of these secretory products may serve as chemoattractants for further recruitment of monocytes (setting up an amplification system), as well as for vascular smooth muscle cells from the tunica media. Others may stimulate growth of smooth muscle cells. It was recently reported (94) that at least part of the macrophage-derived mitogenic activity is caused by production of a growth factor similar, if not identical, to PDGF. Interleukin-1 (IL-1) and tumor necrosis factor α (TNF-α) can also result in secondary PDGF gene expression by smooth muscle cells or endothelium (24). Free radicals generated by macrophages not only act on the cellular components of the lesion but can also further modify the LDL present in the intima.

Besides being a secretory cell, the macrophage is also an active phagocyte. The scavenger receptor implicated in the ingestion of modified LDL (53,95) is expressed on its surface. Several groups have shown the transformation of the macrophage into a foam cell after lipid ingestion (21–23,96–98). The corpulent foam cell tends to stay in the lesion area, possibly because its motility is inhibited by a chemotactic factor (lysophosphatidylcholine) (99,100). Evidence from studies in hypercholesterolemic swine indicates that occasionally the lipid-laden foam cells do reenter the circulation (97). Foam cells from early lesions may return to the circulation without endothelial damage, whereas foam cells from advanced lesions, being more lipid-laden and larger, may damage the endothelium during their egress (97). Finally, macrophage-derived foam cells may die, releasing their lipid content into the extracellular compartment, and may also release proteases leading to the digestion of extracellular matrix, favoring in this way the formation of a core of atheromatous soft material in the lesion.

Until recently it was thought that the macrophage was a terminally differentiated cell incapable of proliferative activity. Though several decades ago foam cells were shown to incorporate ^{3}H-thymidine into DNA, they were thought to originate from modified smooth muscle cells (101,102). The development of more specific markers and monoclonal antibodies has allowed the reexamination of the identity of these cells. In one study of human coronary atherosclerotic plaques, 27% of proliferating cell nuclear antigen (PCNA)–positive cells were macrophages and 15% were smooth muscle cells (103). Therefore, macrophages continue to proliferate in the intima, contributing to the growth of the lesion. However, it is unclear whether monocytes have entered into the S phase in the marrow, in the plasma, or after entering the artery wall. Both human and rabbit lesions contain immuno-histochemically identifiable monocyte-colony stimulating factor (M-CSF), as well as increased M-CSF mRNA (104).

Smooth Muscle Cells. Proliferation of smooth muscle cells has been established as a key event in the evolution of atherosclerosis. At least two different phenotypes have been described for smooth muscle cells based on the distribution of myosin filaments and the formation of large amounts of secretory protein apparatus, such as rough endoplasmic reticulum and Golgi (105–107). Contractile phenotype cells respond to agents that induce either vasoconstriction or vasodilatation, such as endothelin, catecholamines, angiotensin-II, prostacyclin, neuropeptides, leukotrienes, and NO (108–115). In contrast, synthetic phenotype cells are capable of expressing genes for a number of growth-regulatory molecule and cytokine receptors (116,117); they can respond to those mediators by proliferation and synthesis of extracellular matrix (24). Besides endothelial cells, platelets, and macrophages, one study has shown that smooth muscle cells isolated from atheroma could also secrete mitogenic proteins, some of which resemble PDGF (118). This capacity to produce endogenous, potentially self-stimulating growth factors may help to explain how replication of smooth muscle cells can begin early in atherogenesis even while the endothelial barrier remains morphologically intact, although functionally it may not be.

Extracellular matrix synthethized by smooth muscle cells is made up of collagen, elastin, and proteoglycans. Collagen formation is the major contributor to the growth of atherosclerotic plaques (collagen type I is the primary component of the plaque). Proteoglycans and elastin appear to be important in the binding of extracellular lipids. Dermatan sulfate is the major glycosaminoglycan present in atherosclerotic plaques (23).

Both chemical and mechanical factors stimulate smooth muscle cells to make connective tissue. One of the most ubiquitous and potent chemical factors potentially involved in matrix formation is TGF-β (64,119). Not only is TGF-β a potent inhibitor of mitogenesis, it is also the most potent stimulator of connective tissue formation known. It can induce gene expression for fibronectin, collagen types I, III, IV, and V, as well as numerous glycosaminoglycan and elastic fiber proteins. In most circumstances, TGF-β is present in a latent form that requires exposure to acidic conditions or specific enzymes, such as plasmin or trypsin-like proteases, to release active TGF-β (64,119). TGF-β seems to inhibit cell proliferation by suppression of transcription of the proto-oncogene c-*myc*, as well as by the increased deposition of connective tissue around the cells (64,120,121). Mechanical factors, such as the chronic pulsatile distension in the arterial wall, also favor the synthesis of collagen by smooth muscle cells (122) and, thus, the slow progression of the atherosclerotic plaque.

Rapid Thrombus-Mediated Atherosclerosis Progression

The processes of lipid accumulation, cell proliferation, and extracellular matrix synthesis might be expected to be linear with time. However, angiographic studies show that the progression of coronary artery disease in humans is neither linear nor predictable (123,124). New high-grade lesions often appear in segments of artery that were normal at previous angiographic examination (125). Two thirds of the lesions presumably responsible for acute ischemic events (unstable angina or myocardial infarction) have been previously only mildly to moderately stenotic (125,126). This unpredictable and episodic progression can be explained by the occurrence of plaque rupture with subsequent thrombosis and changes in plaque geometry and local rheology leading to intermittent plaque growth (1,25).

A fibrolipid lesion (Stary type V) surrounded by a thin fibrous capsule can be easily disrupted, leading to type III injury with thrombus formation (127). Plaque rupture occurs when the strain within the fibrous cap exceeds the deformability of its component material (45). Marked oscillations in shear stress, acute changes in coronary pressure or tone, and bending and twisting of an artery during each heart contraction could contribute to plaque rupture (128). For artery walls with heterogeneous composition, like those in atherosclerotic lesions, focal concentrations of circumferential stress may occur that might lead to plaque fissuring (129). Areas with decreased load-bearing capabilities, such as a lipid pool, result in increased stress elsewhere in the vessel wall. The presence of an extracellular lipid pool in a plaque exceeding around 45% of the vessel circumference increases stress at the lateral edge of the plaque, a frequent site of plaque rupture (129). However, plaque rupture may not always occur at the regions of highest stress, suggesting that local variations in plaque material properties also contribute to plaque rupture (130). Thrombi following rupture may be clinically silent, may be lysed or may become organized contributing to the growth of the atherosclerotic plaque (69,76,127,131,132). When thrombi are large and occlusive, they can contribute to acute coronary syndromes such as unstable angina, myocardial infarction, and sudden ischemic death (127,131). Ultimately, they can be partially lysed or become organized into a chronic fibrotic occlusion.

Morphological Features Involved in Rapid Thrombus-Mediated Progression

Pathological studies have revealed that atherosclerotic plaques prone to rupture are commonly composed of a crescentic mass of soft lipids, with abundant cholesterol crystals, separated from the vessel lumen by a fibrous cap (Stary type V lesions) (127) (Fig. 3). Ruptured plaques contain more extracellular lipid and less collagen, smooth muscle cells, and calcium than intact plaques (9). Fissures tend to occur at the margins of plaques where caps are necrotic, very thin, and extensively infiltrated by macrophages (129,133,134). Plaques in which the extracellular lipid core makes up more than 40% of the overall plaque volume are particularly susceptible (135,136). It is likely that many variables determine whether an unstable plaque proceeds rapidly to an occlusive thrombus with the potential for acute infarction or persists at an intermediate stage as a mural nonocclusive thrombus. Local factors such as quantity (fissure size) and quality (plaque composition) of thrombogenic substrate exposed after plaque rupture, as well as rheology of blood flow at the site of rupture, are the major local determinants of thrombotic response after plaque rupture (137). Systemic factors such as the fibrinolytic system, catecholamines, and lipoprotein(a) [Lp(a)] also modulate thrombosis, but it is unknown whether they are more or less influential than substrate thrombogenicity and rheology (73). Lp(a) contains apolipoprotein (a), which has considerable amino acid sequence homology with plasminogen, but, because of a crucial substitution of serine for arginine, Lp(a) has no enzymatic activity (138). Therefore, Lp(a) may compete with serum plasminogen for binding sites and inhibit the activation of plasminogen by t-PA, thus serving as a procoagulant.

Plaque fissuring comes in various shapes and sizes. The tear may be small, measuring 100–200 μm across. Such tears allow blood to enter and expand the plaque but may not result in thrombus formation in the arterial lumen. Plaque expansion produces acute changes in plaque geometry with subsequent alterations in local rheology that may further stimulates lesion progression (43). The healing organization process of intraplaque thrombi may also contribute to subclinical progression of the lesions (135). Indeed, analysis of the coronary tree in patients who died of ischemic heart disease (139,140) showed a morphological appearance

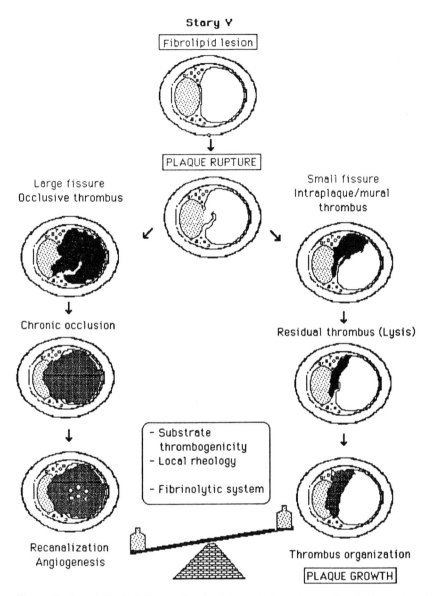

Figure 3 Morphological features involved in rapid thrombus-mediated atherosclerosis progression. See text for details.

consistent with previously healed fissures at different stages of thrombosis and thrombus organization. Furthermore, studies using monoclonal antibodies to identify fibrin, fibrinogen, and their degradation products have also supported the evidence for thrombosis and thrombus incorporation as part of plaque progression (141). In advanced and fibrous plaques, fibrin and fibrin-related products were detected in the intima, in the neointima, and even in the deeper medial layer, particularly around areas with thrombus incorporation (smaller quantities of these fibrin and fibrin-related products were also found in early lesions and in normal arteries). In addition, results of a recently completed 5-year trial of platelet-inhibitor therapy in the angiographic progression of coronary disease (142) indicate that antiplatelet agents are helpful in preventing the angiographic progression of coronary disease, supporting the hypothesis that recurrent thrombotic events are important contributors to the process of atherosclerosis progression.

The tear may be large and the thrombus formed within the lumen may occlude the vessel. It may either be removed by lysis or become replaced in the process of organization by what is the ubiquitous vascular repair response (smooth muscle cell proliferation and collagen production). Of interest, acute occlusive thrombus may be invaded by several neovascular channels and appears partially open at angiography (135).

Biological Mechanisms Involved in Rapid Thrombus-Mediated Progression

Mechanisms underlying plaque rupture, the subsequent healing process, and thrombus-mediated progression of atherosclerotic plaques include biological interactions among macrophages, thrombotic elements, and vascular cells (Fig. 4). Prior to rupture, macrophages may play a key role in determining plaque susceptibility to rupture, whereas, platelets, thrombin, and fibrin(ogen) may play their main role after rupture in promoting lesion progression.

Macrophages. As mentioned earlier, macrophages are involved in the development of fatty streaks and advanced atherosclerotic plaques. Moreover, macrophages may reduce the strength of collagen, elastin, and other structural proteins by releasing proteases (129,143,144). These proteases may lead to erosion of the fibrous cap overlying a lipid pool, increasing the concentration of wall stress while reducing the ability of the cap to resist deformation. Although ruptured plaques contain a higher concentration of macrophages (136,143,145), comparative quantification of protease activity in ruptured and stable plaques has not been performed.

Platelets. As pointed out, the precise role of platelets in the initiation of spontaneous atherosclerosis is currently undefined; however, emerging evidence suggests that platelet aggregation and thrombosis after type II and III injuries have a key role in the progression of the disease. Platelets may induce the migration

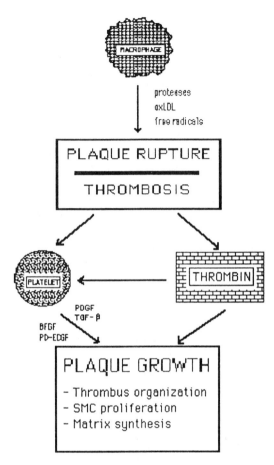

Figure 4 Biological mechanisms involved in rapid-thrombus mediated atherosclerosis progression. BFGF, Basic fibroblast growth factor; PD-ECGF, platelet-derived endothelial cell growth factor; PDGF, platelet-derived growth factor; oxLDL, oxidized low-density lipoprotein; TGF-β, transforming growth factor β. See text for details.

and proliferation of smooth muscle cells by secreting PDGF (23,24,146). Other mitogenic factors such as transforming growth factor β (TGF-β), acidic and basic fibroblast growth factor (FGF), and platelet-derived endothelial cell growth factor (PD-ECGF) may also have a role in thrombosis and the subsequent myofibrotic response of atherosclerosis (23,24,147).

Thrombin. After vascular damage, the enzymatically active thrombin is produced and is incorporated into the thrombus and into the extracellular matrix (148,149). Subsequently, it may be released gradually in active form during spontaneous fibrinolysis or during thrombus organization. Surface-bound fibrin, in particular, may act as a reservoir for enzymatically active thrombin (149). The role that thrombin plays in the progression of atherosclerosis is not yet well understood. Several studies have demonstrated that thrombin is a potent mitogen for mesenchymal-derived cells (150,151) and that it is chemotactic for monocytes (152). In addition, through the activation of platelets and the elaboration of fibrin polymers and fibrinopeptide A and B fragments, thrombin may also bear direct adverse, atherogenic consequences in the vascular milieu.

Fibrin(ogen) and Fibrin(ogen) Degradation Products. Fibrin(ogen) is believed to contribute to atherogenesis through several mechanisms. These include the stimulation of migration and proliferation of smooth muscle cells (153,154). Fibrin(ogen) degradation products released from fibrin(ogen) by the action of plasmin (and thrombin) may also contribute to atherogenesis by their ability to stimulate vascular cell proliferation (153,155–158).

Thus, an interesting and complex set of pathological amplification and inhibition mechanisms exists that sustains the potential adverse effects of prothrombotic determinants in pathogenesis and progression of atherosclerosis.

RELATIONSHIP BETWEEN PLAQUE PROGRESSION AND ACUTE ISCHEMIC SYNDROMES

Acute coronary ischemic syndromes are caused by plaque rupture and coronary thrombi, as confirmed by multiple clinical and autopsy studies (127–131,133). Coronary thrombosis occurs only in arteries containing atherosclerosis, rupture of the surface of an atherosclerotic plaque initiates thrombosis, and clinical consequences of plaque rupture depend on the extent of coronary thrombus formation.

In unstable angina, a relatively small fissuring or disruption of an atherosclerotic plaque may lead to an acute change in plaque structure and a reduction in coronary blood flow, resulting in exacerbation of angina. Transient episodes of thrombotic vessel occlusion at the site of plaque injury may occur, leading to angina at rest (159,160). This thrombus is usually labile and results in temporary vascular occlusion, perhaps lasting only 10–20 minutes (128). In addition, release of vasoactive substances by platelets and vasoconstriction secondary to endothelial vasodilator dysfunction may contribute to a reduction in coronary flow. Alterations in perfusion and myocardial oxygen supply probably account for two thirds of episodes of unstable angina; the rest may be caused by transient increases in myocardial oxygen demand (160).

In non–Q-wave infarction, more severe plaque damage would result in more persistent thrombotic occlusion, perhaps lasting up to 1 hour. About one fourth of patients with non–Q-wave infarction may have an infarct-related vessel occluded for more than 1 hour, but the distal myocardial territory is usually supplied by collaterals (161). ST segment elevation in the electrocardiogram, an early peak in plasma creatine kinase concentration, and a high rate of angiographic patency of the involved vessel in early angiograms support these speculations. Resolution of vasospasm may be also pathogenically important in non–Q-wave infarction (162,163). Therefore, spontaneous thrombolysis, spasm resolution, or presence of collateral circulation are important in preventing the formation of Q-wave infarction by limiting the duration of myocardial ischemia.

In Q-wave infarction, larger plaque fissures may result in the formation of a fixed and persistent thrombus. This leads to an abrupt cessation of myocardial perfusion for more than 1 hour, resulting in transmural necrosis of the involved myocardium. The coronary lesion responsible for the infarction is frequently only mildly to moderately stenotic, which suggests that plaque rupture with super-imposed thrombus rather than the severity of the underlying lesion is the primary determinant of acute occlusion (125,126). It is conceivable that in patients with severe coronary stenosis, well-developed collaterals prevent or reduce the extent of infarction (128). In perhaps one fourth of patients, coronary thrombosis results from superficial intimal injury or blood stasis in areas of high-grade stenosis (132). Additionally, myocardial infarction may be promoted by alterations in hemostasis, such as increased activation of the coagulation system, increased platelet aggregability, increased fibrinogen and factor VII activity, and deficient fibrinolytic mechanisms.

Sudden coronary death probably involves a rapidly progressive coronary lesion in which plaque rupture and resultant thrombosis lead to ischemia and fatal ventricular arrhythmias (131,133). Absence of collateral flow to the myocardium distal to the occlusion or platelet microemboli (164) perhaps contributes to the development of sudden ischemic death.

REGRESSION OF THE ATHEROSCLEROTIC PLAQUE

In approaching the concept of reversibility or arrest of the coronary atherosclerotic process, it is essential to keep in mind that atherosclerosis disease starts at a young age and takes many years to progress into the symptomatic stage. By the time the first symptoms of coronary atherosclerosis appear, the disease is advanced to two- or three-vessel involvement in most patients. These patients must be identified before the first symptoms appear. The development of intravascular ultrasound has allowed the assessment of plaque morphology and composition (165–169). Preliminary data suggest that intravascular ultrasound can be used to differentiate ruptured from stable plaques (170). Ruptured plaques have thinner intimal leading edges and larger intimal sonolucent zones by ultrasound than do stable plaques. However, it remains to be seen whether these plaques' characteristics can be used to identify regions of future rupture. Once vulnerable plaque is identified, approaches may be taken toward retardation in the progression or even reversibility of plaque, reduction in the susceptibility of plaque rupture, and/or prevention of thrombosis after plaque rupture.

Approaches that have been taken toward retardation or even reversibility of atherosclerotic lesions include (1) better control of risk factors, especially by reducing plasma cholesterol levels, (2) enhancement of lipid-removal pathways from the vessel wall, particularly by increasing plasma HDL levels, and (3) reduction of LDL oxidation by using antioxidant agents, such as probucol, which has shown regression of atherosclerosis in animals models to a greater degree than expected by its lipid-lowering action alone (171). Each of these approaches, by acting on the lipid-rich plaque more prone to rupture, might prevent progression and even induce removal of fat and regression of atherosclerotic plaques.

However, very minimal regression of atherosclerotic lesions has been shown in numerous trials, despite substantial reduction in the incidence of acute cardiac events (171–179). It is possible that removal of lipids and an increase in the relative collagen content of plaque together with production of heavily cross-linked collagen may account for increasing plaque stability without significant reduction in plaque size (180). Other approaches, besides hypolipidemic therapy, that may possibly reduce the incidence of plaque rupture include angiotensin-converting enzyme inhibitors and beta-blockers. Recent evidence from trials of angiotensin-converting enzyme inhibitors in patients with ischemic heart disease points to a reduction of 14–28% incidence of ischemic cardiac events (181,182). The mechanism of any possible reduction in the incidence of plaque rupture by angiotensin-converting enzyme inhibitors is uncertain. It may be due to a reduction in arterial wall stress caused by lower blood pressure or reduced levels of neurohumoral activation (183) or their effect on protein synthesis influencing plaque composition in such a way that the likelihood of plaque rupture is reduced. Meta-analysis of secondary prevention trials with beta-blockers has shown a 20% reduction in cardiac mortality, an additional 25% reduction in the incidence of reinfarction, and a 30% reduction in the incidence of sudden death (184). Beta-blockers reduce the circumferential plaque stress by reducing blood pressure and blunting hypertensive pressure surges (185). Beta-blockers may also prevent plaque rupture by increasing the ability of the plaque fibrous cap to withstand stress. Because plaque stiffness increased with heart rate, beta-blockers may increase plaque tensile strength by reducing heart rate (185).

Finally, because platelet thrombus formation appears to be an important factor in the conversion of chronic to acute events and in the progression of the disease after plaque rupture, another promising approach in preventing this process is the use of new strategies in antithrombotic therapy if this does not result in increased risk of bleeding. Direct antithrombins such as hirudin and new strategies such as the combination of warfarin plus aspirin are currently under clinical investigation for the prevention of arterial thrombosis.

REFERENCES

1. Davies MJ. A macro and micro view of coronary vascular insult in ischemic heart disease. Circulation 1990; 82:II38–II46.
2. Bouch DC Montgomery GL. Cardiac lesions in fatal cases of recent myocardial ischaemia from a coronary care unit. Br Heart J 1970; 32:795–803.
3. Hangartner JRW, Charleston AJ, Davies MJ, Thomas AC. Morphological characteristics of clinical significant coronary artery stenosis in stable angina. Br Heart J 1986; 56:501–508.
4. Leary T. Pathology of coronary sclerosis. Am Heart J 1934; 10:328–337.
5. Tracy RE, Kissling, G. E. Age and fibroplasia as preconditions for atheronecrosis in human coronary arteries. Arch Pathol Lab Med 1987; 111:957–963.
6. Tracy RE, Kissling GE. Comparisons of human populations for histologic features of atherosclerosis: a summary of questions and methods for geographic studies. Arch Pathol Lab Med 1988; 112:1056–1065.
7. Corrado D, Thiene G, Penelli N. Sudden death as the first manifestation of coronary artery disease in young people (≤35 years). Eur Heart J 1988; 9:139–144.
8. Angelini A, Thiene G, Frescura C, Baroldi G. Coronary arterial wall and atherosclerosis in youth (1–20 years): a histological study in a norther italian population. Int J Cardiol 1990; 28:361–370.
9. Kragel AH, Reddy SG, Wittes JT, Roberts WC. Morphometric analysis of the composition of atherosclerotic plaques in the four major epicardial coronaries in acute myocardial infarction and sudden coronary death. Circulation 1989; 80:1747–1756.
10. Kragel AH, Reddy SG, Wittes JT, Roberts WC. Morphometric analysis of the composition of coronary arterial plaques in isolated unstable angina pectoris with pain at rest. Am J Cardiol 1990; 66:562–567.
11. Dollar AL, Kragel AH, Fernicola DJ, Wacliwiw MA, Roberts WC. Composition of atherosclerotic plaques in coronary arteries in women <40 years with fatal coronary artery disease and implications for plaque reversibility. Am J Cardiol 1991; 67: 1223–1227.
12. Gertz SD, Malekzadeh S, Dollar AL, Kragel AH, Roberts WC. Composition of atherosclerotic plaques in the four major epicardial coronary arteries in patients ≥90 years of age. Am J Cardiol 1991; 67:1228–1233.
13. Baroldi G, Silver MD, Mariani F, Giuliano G. Correlation of morphological variables in the coronary atherosclerotic plaque with clinical patterns of ischemic heart disease. Am J Cardiovasc Pathol 1988; 2:159–172.
14. Svindland A, Torvik A. Atherosclerotic carotid disease in asymptomatic individuals: a histologic study of 53 cases. Acta Neurol Scand 1988; 78:506–517.
15. Bassiouny HS, Davis H, Massawa N, Gewertz BL, Glagov S, Zarins CK. Critical carotid stenosis: morphologic and chemical similarity between symptomatic and asymptomatic plaques. J Vasc Surg 1989; 9:202–212.
16. Bocan TMA, Guyton JR. Human aortic fibrolipid lesions: progenitor lesions for fibrous plaques, exhibiting early formation of the cholesterol-rich core. Am J Pathol 1985; 120:193–206.
17. Bocan TMA, Schifani TA, Guyton JR. Ultrastructure of the human aortic fibrolipid

lesion: formation of the atherosclerotic lipid-rich core. Am J Pathol 1986; 123: 413–424.

18. Cliff WJ, Heathcote CR, Moss NS, Richenbach DD. The coronary arteries in cases on cardiac and non cardiac sudden death. Am J Pathol 1988; 132:319–329.

19. Rokitansky K. A Manual of Pathological Anatomy. Vol. 4. (Day GE, trans.) London: Sydenham Society, 1852:271–273.

20. Virchow R. Gesammelte Abhandlungen zür Wissenschaftlichen Medicin, Frankfurt: Meidinger u. Sohn Co., 1856:458.

21. Harker LA, Ross R, Glomset JA. The role of endothelial cell injury in platelet response in atherogenesis. Thromb Haemost 1978; 39:312–321.

22. Ross R. Atherosclerosis: a problem of the biology of the arterial wall cells and their interactions with blood components. Atherosclerosis 1981; 1:293–311.

23. Ross R. The pathogenesis of atherosclerosis: an update. N Engl J Med 1986; 314: 488–500.

24. Ross R. The pathogenesis of atherosclerosis: a perspective for the 1990s. Nature 1993; 362:801–809.

25. Ip JH, Fuster, V, Badimon L, Badimon J, Taubman MB, Chesebro JH. Syndromes of accelerated atherosclerosis: role of vascular injury and smooth muscle cell proliferation. J Am Coll Cardiol 1990; 15:1667–1687.

26. Cathcart MK, Morel DW, Chisolm GM. Monocytes and neutrophils oxidize low density lipoprotein making it cytotoxic. J Leuk Biol 1985; 38:341–350.

27. Rosenfeld ME, Palinski W, Yia-Herttuala S, Carew TE. Macrophages, endothelial cells, and lipoprotein oxidation in the pathogenesis of atherosclerosis. Toxicol Pathol 1990; 18:560–561.

28. Steinbrecher UP, Zhang HF, Lougheed M. Role of oxidatively modified LDL in atherosclerosis. Free Radical Biol Med 1990; 9:155–168.

29. Steinberg D. Antioxidants and atherosclerosis. A current assessment (editorial). Circulation 1991; 84:1420–1425.

30. Henriksen T, Mahoney EM, Steinberg D. Interactions of plasma lipoproteins with endothelial cells. Ann NY Acad Sci 1982; 401:102–116.

31. Henriksen T, Mahoney EM, Steinberg D. Enhanced macrophage degradation of low density lipoprotein previously incubated with cultured endothelial cells: recognition by receptors for acetylated low density lipoproteins. Proc Natl Acad Sci USA 1981; 78:6499–6503.

32. Henriksen T, Mahoney EM, Steinberg D. Enhanced macrophage degradation of biologically modified low density lipoprotein. Arteriosclerosis 1983; 3:149–159.

33. Heineke JW, Rosen H, Chait A. Iron and copper promote modification of low density lipoprotein by human arterial smooth muscle cell in culture. J Clin Invest 1984; 74:1890–1894.

34. Morel DW, DiCorleto PE, Chisolm GM. Endothelial and smooth muscle cells alter low density lipoprotein in vitro by free radical oxidation. Arteriosclerosis 1984; 4:357–364.

35. Cathcart MK, Morel DW, Chisolm GM. Monocytes and neutrophils oxidize low density lipoprotein making its cytotoxic. J Leukoc Biol 1985; 38:341–350.

36. Parthasarathy S, Printz DJ, Boyd D, Joy L, Steinberg D. Macrophage oxidation of

LDL generates a modified form recognized by the scavenger receptor. Arterio-sclerosis 1986; 6:505–510.

37. Hiramatsu K, Rosen H, Heinecke JW, Wolfbauer G, Ghait A. Superoxide initiates oxidation of low density lipoprotein by human monocytes. Arteriosclerosis 1987; 7:55–60.

38. Fogelman AM, Schechter I, Seager J, Hokom M, Child JJ, Edward P. Malondialde-hyde alteration of low density lipoprotein leads to cholesteryl-ester accumulation in human monocyte-macrophages. Proc Natl Acad Sci USA 1980; 77:2214–2218.

39. Furchgott RF. Role of endothelium in responses of vascular smooth muscle cell. Cir Res 1983; 53:557–573.

40. Luscher TF. Imbalance of endothelium-derived relaxing and contracting factors. A new concept in hypertension?. Am J Hypertens 1990; 3:317–330.

41. Minor RL, Myers PR, Guerra RJr, Bates JN, Harrison DG. Diet-induced athero-sclerosis increases the release of nitrogen oxides from rabbit aorta. J Clin Invest 1990; 86:2109–2116.

42. Chester AH, O'Neil GS, Moncada S, Tadjkarimi S, Yacoub MH. Low basal and stimulated release of nitric oxide in atherosclerotic epicardial coronary arteries. Lancet 1990; 336:897–900.

43. Gibson CM, Diaz L, Kandarpa K, et al. Relation of vessel wall shear stress to atherosclerosis progression in human coronary arteries. Atherosclerosis and Throm-bosis 1993; 13:310–315.

44. Asakura T, Karino T. Flow patterns and spatial distribution of atherosclerotic lesions in human coronary arteries. Circ. Res 1990; 66:1045–1066.

45. McIsaac AI, Thomas JD, Topol EJ. Toward the quiescent coronary plaque. J Am Coll Cardiol 1993; 22:1228–1241.

46. Levesque MJ, Liepsch D, Moravec S, Nerem RM. Correlation of endothelial cell shape and wall shear stress in a stenosed dog aorta. Arteriosclerosis 1986; 6:220–229.

47. Reidy MA, Bowyer DE. Scanning electron microscopy of arteries: the morphology of aortic endothelium in hemodynamically stressed areas associated with branches. Atherosclerosis 1977; 26:181–194.

48. Glagov S, Zarins C, Giddens DP, Ku DN. Hemodynamics and atherosclerosis: insights and perspectives gained from studies of human arteries. Arch Pathol Lab Med 1988; 112:1018–1031.

49. Munro JM, Cotran RS. The pathogenesis of atherosclerosis: atherogenesis and inflammation. Lab Invest 1988; 58:249–261.

50. Thomas WA, Lee KM, Kim DN. Cell population kinetics in atherogenesis. Cell births and losses in intimal cell mass-derived lesions in the abdominal aorta of swine. Ann NY Acad Sci 1985; 454:305–315.

51. Hansson GK, Jonasson L, Seifert PS, Stemme S. Immune mechanisms in athero-sclerosis. Arteriosclerosis 1989; 9:567–578.

52. Celermajer D, Sorensen K, Gooch V, Spiegelhalter D, Miller O, Sulluvan I, Lloyd J, Deanfield J. Non-invasive detection of endothelial dysfunction in children and adults at risk of atherosclerosis. Lancet 1992; 340:1111–1115.

53. Goldstein JL, Brown MS. The low density lipoprotein pathway and its relation to atherosclerosis. Anna Rev Biochem 1977; 46:987–930.

54. Steinberg D. Lipoprotein and atherosclerosis: a look back and a look ahead. Arteriosclerosis 1983; 3:283–301.
55. Badimon J, Fuster V, Chesebro JH, Badimon L. Coronary atherosclerosis. A multifactorial disease. Circulation 1993; 87:II3–II16.
56. Brown MS, Kovanen PT, Goldstein JL. Regulation of plasma cholesterol by lipoprotein receptors. Science 1976; 212:628–638.
57. Shepherd J, Packard CJ. Lipoprotein receptors and atherosclerosis. Clin Sci 1986; 70:1–6.
58. Goldstein JL, Ho YK, Basu SK, Brown MS. Binding site on macrophages that mediates uptake and degradation of acetylated low density lipoproteins producing massive cholesterol deposition. Proc. Natl. Acad. Sci. USA 1979; 76:333–337.
59. Valente AJ, Rozek MM, Sprague EA, Schwartz CJ. Mechanism in intimal monocyte-macrophage recruitment. A special role for monocyte chemotactic protein-1. Circulation 1992; 86:III20–III25.
60. Valente AJ, Fowler SR, Sprague EA, Kelley JL, Suenram CA, Schwartz CJ. Initial characterization of a peripheral blood mononuclear cell chemoattractant derived from cultured arterial smooth muscle cells. Am J Pathol 1984; 117:409–417.
61. Valente AJ, Graves DT, Vialle-Valentin E, Delgado R, Schwartz CJ. Purification of a monocyte chemotactic factor (SMC-CF) secreted by nonhuman primate vascular smooth muscle cells in culture. Biochemistry 1988; 27:4162–4168.
62. Berliner JA, Territo MC, Almada L, Carter A, Shafonsky E, Fogelman AM. Monocyte chemotactic factor produced by large vessel endothelial cells in vitro. Arteriosclerosis 1986; 6:254–258.
63. Stanley ER. The macrophage colony-stimulating factor, CSF-1. Meth Enzym 1985; 116:564–587.
64. Sporn MB, Roberts AB, Wakefield LM, Crombrugghe B. Some recent advances in the chemistry and biology of transforming growth factor-beta. J Cell Biol 1987; 105:1039–1045.
65. Norris DA, Clark RAF, Swigart LM, Huff JC, Weston WL, Howell SE. Fibronectin fragment(s) are chemotactic for human peripheral blood monocytes. J Immunol 1982; 129:1612–1618.
66. Hunninghake GW, Davidson JM, Rennard S, Szapiel S, Gadek JE, Crystal RG. Elastin fragments attract macrophage precursors to disease sites in pulmonary emphysema. Science 1981; 212:925–927.
67. Postlethwaite AE, Kang AH. Collagen- and collagen peptide-induced chemotaxis of human peripheral blood monocytes. J Exp Med 1976; 143:1299–1307.
68. Bar-Shavit R, Kahn A, Fenton JW, Wilner GD. Chemotactic response of monocytes to thrombin. J Cell Biol 1983; 96:282–285.
69. Stary HC. Evolution and progression of atherosclerotic lesions on coronary arteries of children and young adults. Arteriosclerosis 1989; 9:I19–I32.
70. Wilens SL. The nature of diffuse intimal thickening of arteries. Am J Pathol 1951; 27:825–839.
71. Movat HC, More RH, Haust MD. The diffuse intimal thickening of the human aorta with aging. Am J Pathol 1958; 34:1023–1031.
72. Stary HC, Blankenhorn DH, Chandler AB, et al. A definition of the intima in

humans and of its atherosclerosis-prone areas. A report from the Committee on Vascular Lesions of the Council of Atherosclerosis, American Heart Association. Circulation 1992; 85:391–405.

73. Fuster V, Badimon L, Badimon J, Chesebro JH. The pathogenesis of coronary artery disease and the acute coronary syndromes. N Engl J Med 1992; 326:242–250, 310–318.

74. Alderman EL, Corley SD, Fisher LD, et al. Five-year angiographic follow-up of factors associated with progression of coronary artery disease in the Coronary Artery Surgery Study (CASS). J Am Coll Cardiol 1993; 22:1141–1154.

75. Stary HC. Composition and classification of human atherosclerotic lesions. Virchows Archiv A Pathol Anat 1992; 421:277–290.

76. Davies MJ, Wolf N, Rowles PM, Pepper L. Morphology of the endothelium over atherosclerotic plaques in human coronary arteries. Br Heart J 1988; 60:459–464.

77. Restrepo C, Tracy R. Variations in human aortic fatty streaks among geographic locations. Atherosclerosis 1975; 21:179–193.

78. Falk E. Why do plaques rupture? Circulation 1992; 86:III30–III42.

79. Wilcox JN, Smith KM, Schwartz SM, Gordon D. Localization of tissue factor in the normal vessel wall and in the atherosclerotic plaque. Proc. Natl. Acad. Sci. USA 1989; 86:2839–2843.

80. Gajdusek C, DiCorleto P, Ross R, et al. An endothelial cell-derived growth factor. J Cell Biol 1980; 85:467–472.

81. Fox PL, DiCorleto PE. Regulation of production of a platelet-derived growth factor-like protein by cultured bovine aortic endothelial cells. J Cell Physiol 1984; 11: 298–308.

82. Castellot J Jr, Addonozio ML, Rosenberg R, et al. Cultured endothelial cells produce an heparinlike inhibitor of smooth muscle cell growth. J Cell Biol 1981; 90:372–379.

83. Badimon L, Badimon J, Penny W, Webster MW, Chesebro J, Fuster V. Endothelium and atherosclerosis. J Hypertension 1992; 10:S43–S50.

84. Fuster V, Bowie EJW, Lewis JC, et al. Resistance to atherosclerosis in pigs with von Willebrand's disease: spontaneous and high cholesterol diet-induced arteriosclerosis. J Clin Invest 1978; 61:722–730.

85. Fuster V, Fass DN, Kaye MP, et al. Arteriosclerosis in normal and von Willebrand pigs: long-term prospective study and aortic transplantation study. Cir Res 1982; 51:587–593.

86. Fuster V, Griggs TR. Porcine von Willebrand disease: implications for the pathophysiology of atherosclerosis and thrombosis. Prog Hemost Thromb 1986; 8: 159–183.

87. Aqel Nm, Ball RY, Waldmann H, et al. Identification of macrophages and smooth muscle cells in human atherosclerosis using monoclonal antibodies. J Pathol 19485; 146:197–204.

88. Gown AM, Tsukada T, Ross R. Human atherosclerosis: II. Immunicytochemical analysis of the cellular composition of human atherosclerotic lesions. Am J Pathol 1986; 125:191–207.

89. Davies PF, Ridy MA, Goode TB, et al. Scanning electron microscopy in the

evaluation of endothelial integrity of the fatty lesion in atherosclerosis. Atherosclerosis 1976; 25:125–130.

90. Chandler AB, Hand RA. Phagocytized platelets: a source of lipids in human thrombi and atherosclerotic plaques. Science 1961; 134:946–947.

91. Kruth HS. Platelet-mediated cholesterol accumulation in cultured aortic smooth muscle cells. Science 1985; 227:1243–1245.

92. Mendelsohn ME, Loscalzo J. Role of platelets in cholesteryl ester formation by U-937 cells. J Clin Invest 1988; 81:62–68.

93. Faruqi RM, DiCorleto PE. Mechanisms of monocyte recruitment and accumulation. Br Heart J 1993; 69:S19–S29.

94. Singh JP, Pearson JD, McCroskey MC, et al. Competence inducing mitogenic activities secreted by human macrophage: molecular characterization of a macrophage derived growth factor MDGF (abstract). Circulation 1989; 80:II–452.

95. Brown MS, Goldstein JL. Lipoprotein metabolism in the macrophage. Anna Rev Biochem 1983; 52:223–261.

96. Gerrity RG. The role of monocytes in atherogenesis. I. Transition of blood-borne monocytes into foam cells in fatty lesions. Am J Pathol 1981; 103:181–190.

97. Gerrity RG. The role of the monocyte in atherogenesis. II. Migration of foam cells from atherosclerotic lesions. Am J Pathol 1981; 103:191–200.

98. Klurfeld DM. Identification of foam cells in human atherosclerotic lesions as macrophages using monoclonal antibodies. Arch Pathol Lab Med 1985; 109:445–449.

99. Quinn MT, Parthasarathy S, Steinberg D. Endothelial cell-derived chemotactic activity for mouse peritoneal macrophages and the effects of modified forms of low density lipoproteins. Proc Natl Acad Sci USA 1985; 82:5949–5953.

100. Quinn MT, Parthasarathy S, Steinberg D. Lysiphosphatidylcoline: a chemotactic factor for human monocytes and its potential role in atherogenesis. Proc Natl Acad Sci USA 1988; 85:2805–2809.

101. McMillan GC, Duff GL. Mitotic activity in the aortic lesions of experimental atherosclerosis in rabbits. Arch Pathol 1948; 46:179–182.

102. Benditt EP, Benditt JM. Evidence for a monoclonal origin of human atherosclerotic plaques. Proc Natl Acad Sci USA 1973; 70:1753–1756.

103. Gordon D, Reidy MA, Benditt EP, Schwartz SM. Cell proliferation in human coronary arteries. Proc Natl Acad Sci USA 1990; 87:4600–4604.

104. Rosenfeld ME, Ylä-Herttnala S, Lipton B, Ord V, Steimberg D. Macrophage-colony stimulating factor mRNA and protein in atherosclerosis lesions of rabbits and humans. Am J Path 1992; 140:291–300.

105. Campbell GR, Campbell JH, Manderson JA, Horrigan S, Rennick RE. Arterial smooth muscle cell. A multifunctional mesenchymal cell. Arch Path Lab Med 1988; 112:977–986.

106. Thyberg J, Hedin U, Sjölund M, Palmberg L, Bottger BA. Regulation of differentiated properties and proliferation of arterial smooth muscle cells. Arteriosclerosis 1990; 10:966–990.

107. Nagai R, Kuro-o M, Babij P, Periasamy M. Identification of two types of smooth muscle myosin heavy chain isoforms by cDNA cloning and immunoblot analysis. J Biol Chem 1989; 264:9734–9737.

108. Yanagisawa M, Kurihara H, Kimura S, et al. A novel potent vasoconstrictor peptide produced by vascular endothelial cells. Nature 1988; 332:411–415.

109. Majesky MW, Daemen MJAP, Schwartz SM. Alpha 1-adrenergic stimulation of platelet-derived growth factor A-chain gene expression in rat aorta. J Biol Chem 1990; 265:1082–1088.

110. Taubman ME, Berk BC, Izumo S, Tsuda D, Alexander RW, Nadal-Ginard B. Angiotensin-II induces c-fos mRNA in aortic smooth muscle. Role of Ca2+ metabolization and protein kinase C activation. J Biol Chem 1989; 264:526–530.

111. Owen NE. Effect of prostaglandin E1 on DNA synthesis in vascular smooth muscle cells. Am J Physiol 1986; 250:C584–C588.

112. Morisaki N, Kanzaki T, Motoyama N, Saito Y, Yoshida S. Cell cycle-dependent inhibition of DNA synthesis by prostaglandin I2 in cultured rabbit aortic smooth muscle cells. Atherosclerosis 1988; 71:165–171.

113. Nilsson J, von Euler AM, Dalsgaard CJ. Stimulation of connective tissue cell growth by substance P and substance K. Nature 1985; 315:61–63.

114. Palmberg L, Claesson HE, Thyberg J. Leukotrienes stimulate initiation of DNA synthesis in cultured arterial smooth muscle cells. J Cell Sci 1987; 88:151–159.

115. Moncada S, Higgs EA, eds. Nitric Oxide from L Arginine: A Bioregulatory System. Amsterdam: Excerpta Medica, 1990.

116. Libby P, Warner SJC, Salomon RN, Birinyi LK. Production of platelet-derived growth factor-like mitogen by smooth muscle cells from human atheroma. N Engl J Med 1988; 318:1493–1498.

117. Sjölund M, Hedin U, Sejersen T, Heldin CH, Thyberg J. Arterial smooth muscle cells express platelet-derived growth factor (PDGF) A chain mRNA, secrete a PDGF-like mitogen, and bind exogenous PDGF in a phenotype- and growth state-dependent manner. J Cell Biol 1988; 106:403–413.

118. Libby P, Warner SJC, Salomon RN, et al. Production of platelet-derived growth factor-like mitogen by smooth muscle cells from human atheroma. N Engl J Med 1988; 318:1493–1498.

119. Roberts AB, Sporn MB. In: Sport MB, Roberts AB, eds. In: Handbook of Experimental Pharmacology: Peptide Growth Factors and Their Receptors I. Berlin: Springer, 1990:419–472.

120. Moses HL, Coffey RJr, Leof EB, Lyons RM, Keski-OjaJ. Transforming growth factor beta regulation of cell proliferation. J Cell Physiol 1987; 5:1–7.

121. Moses HL, Yang EY, Pietenpol JA. TGF-beta stimulation and inhibition of cell proliferation: new mechanistic insights. Cell 1990; 63:245–147.

122. Leung DYM, Glagov S, Mathews MB. Cyclic stretching stimulates synthesis of matrix components by arterial smooth muscle cells in vitro. Science 1976; 191: 475–477.

123. Bruschke A, Wijers T, Kolsters W, Landmann J. The anatomic evaluation of coronary artery disease demonstrated by coronary angiography in 256 non operated patients. Circulation 1981; 63:527–540.

124. Rafflenbeul W, Nellensen U, Galvao P, Kreft M, Peters S, Lichtlen P. Progression and regression of coronary artery disease assessed with sequential coronary angiography. Z Kardiol 1984; 73:II33–II40.

125. Ambrose J, Tannenbaum M, Alexopoulos D, et al. Angiographic progression of coronary artery disease and the development of myocardial infarction. J Am Coll Cardiol 1988; 12:56–62.

126. Little WC, Constantinescu M, Applegate RJ, Kutcher MA, Burrows MT, Kahl FR, Sontamore WP. Can coronary angiography predict the site of a subsequent myocardial infarction in patients with mild-to moderate coronary artery disease? Circulation 1988; 78:1157–1166.

127. Richardson PD, Davies MJ, Born GVR. Influence of plaque configuration and stress distribution on fissuring of coronary atherosclerotic plaques. Lancet 1989; 2: 941–944.

128. Fuster V, Badimon L, Cohen M, Ambrose JA, Badimon JJ, Chesebro J. Insights into the pathogenesis of acute ischemic syndromes. Circulation 1988; 77:1213–1220.

129. Richardson PD, Davies MJ, Born GV. Influence of plaque configuration and stress distribution on fissuring of coronary atherosclerotic plaques. Lancet 1989; 2: 941–944.

130. Cheng GC, Loree HM, Kamm RD, Fishbein MC, Lee RT. Distribution of circumferential stress in ruptured and stable atherosclerotic lesions. A structural analysis with histopathological correlation. Circulation 1993; 87:1179–1187.

131. Falk E. Unstable angina with fatal outcome: dynamic coronary thrombosis leading to infarction and/or sudden death: autopsy evidence of recurrent mural thrombosis with peripheral embolization culminating in total vascular occlusion. Circulation 1985; 71:699–708.

132. Davies MJ, Bland MJ, Hangartner WR, Angelini A, Thomas AC. Factors influencing the presence or absence of acute coronary thrombi in sudden ischemic death. Eur Heart J 1989; 10:203–208.

133. Davies MJ, Thomas A. Thrombosis and acute coronary artery lesions in sudden cardiac ischemic death. N Engl J Med 1984; 310:1137–1140.

134. Constantinides P. Causes of thrombosis in human atherosclerosis arteries. Am J Cardiol 1990; 66:37G–40G.

135. Davies MJ, Woolf N. Atherosclerosis: what is it and why does it occur? Br Heart J 1993; 69:S3–S11.

136. Davies MJ, Richardson PD, Woolf N, Katz DR, Mann J. Risk of thrombosis in human atherosclerosis plaques: role of extracellular lipid, macrophages, and smooth muscle cell content. Br Heart J 1993; 69:377–381.

137. Chesebro JH, Ebster MW, Zoldhelyi P, Roche PC, Badimon L, Badimon J. Antithrombotic therapy and progression of coronary artery disease: Antiplatelets versus antithrombins. Circulation 1992; 86:III100–III111.

138. Moliterno DJ, Lange RA, Meidell RS, et al. Relation of plasma lipoprotein (a) to infarct artery patency in survivors of myocardial infarction. Circulation 1993; 88:935–940.

139. Davies MJ. Thrombosis and coronary atherosclerosis. In: Julian DG, Kublen W, Norris RM, et al. eds. Thrombolysis in Cardiovascular Disease. New York: Marcel Dekker, 1989:25–43.

140. Davies MJ, Thomas AC. Plaque fissuring: the cause of acute myocardial infarction, sudden ischemic death and crescendo angina. Br Heart J 1985; 53:363–373.

141. Bini A, Fenoglio JJ, Mesa-Tejada R, et al. Identification and distribution of fibrinogen, fibrin and fibrin(ogen) degradation products in atherosclerosis: Use of monoclonal antibodies. Arteriosclerosis 1989; 9:109–121.

142. Chesebro J, Webster MW, Smith HC, et al. Antiplatelet therapy in coronary artery disease progression: reduced infarction and new lesion formation. Circulation 1989; 80:II266.

143. Woolf N, Davies MJ. Interrelationship between atherosclerosis and thrombosis. In: Fuster V, Verstraete M, eds. Thrombosis in Cardiovascular Disorders. Philadelphia: W.B. Saunders, 1992:41–77.

144. Shah PK, Falk E, Badimon J, Levy G, Fernandez-Ortiz A, Fallon J, Fuster V. Human monocyte-derived macrophages express collagenase and induce collagen breakdown in atherosclerotic fibrous caps: implications for plaque rupture (Abstract). Circulation 1993; 88:1361.

145. Jonasson L, Holm J, Skalli O, Bondjers G, Hansson GK. Regional accumulation of T cells, macrophages and smooth muscle cells in the human atherosclerosis plaque. Arteriosclerosis 1986; 6:131–138.

146. Ferns GAA, Raines EW, Sprugel KH, Motani AS, Reidy MA, Ross R. Inhibition of neointimal smooth muscle accumulation after angioplasty by an antibody to PDGF. Science 1991; 253:1129–1132.

147. Loscalzo J. The relation between atherosclerosis and thrombosis. Circulation 1992; 86:III95–III99.

148. Bar-shavit R, Eldor A, Vlodavsky I. Binding of thrombin to subendothelial extracellular matrix protection and expression of functional properties. J Clin Invest 1989; 84:1096–1104.

149. Weitz JI, Hudoba M, Massel D, Maraganore J, Hirh J. Clot-bound thrombin is protected from inhibition by heparin-antithrombin III but is susceptible to inactivation by antithrombin III-independent inhibitors. J Clin Invest 1990; 86:385–391.

150. Bar-Shavit R, Benezra M, Eldor A, et al. Thrombin immobilized to extracellular matrix is a potent mitogen for vascular smooth muscle cells: nonenzymatic mode of action. Cell Regul 1990; 1:453–463.

151. Berk BC, Taubman MB, Griendling KK, et al. Thrombin stimulated events in cultured vascular smooth muscle cells. Biochem J 1991; 265:17334–17340.

152. Jones A, Geczy CL. Thrombin and factor Xa enhance the production of interleukin-1. Immunology 1990; 71236–241.

153. Ishida T, Tanaka K. Effects of fibrin and fibrinogen degradation products on the growth of rabbit aortic smooth muscle cells in culture. Atherosclerosis 1982; 44:161–174.

154. Naito M, Hayashi T, Kuzuya M, et al. Effects of fibrinogen and fibrin on the migration of vascular smooth muscle cells in vitro. Atherosclerosis 1990; 83:9–14.

155. Senior RM, Skogan WF, Griffith GL, Wilner GD. Effects of fibrinogen derivates upon the inflammatory response. J Clin Invest 1986; 77:1015–1019.

156. Kaplan KL, Mather T, DeMarco L, Solomon S. Effect of fibrin on endothelial cell production of prostacyclin and tissue plasminogen activator. Arteriosclerosis 1989; 9:43–49.

157. Rowland FN, Donovan MJ, Picciano PT, Wilner GK, Kreurzer DL. Fibrin-mediated

vascular injury: identification of fibrin peptides that mediate endothelial cell retraction. Am J Pathol 1984; 117:418–428.

158. Dang LC, Bell WR, Kaiser D, Wong A. Disorganization of cultured vascular endothelial cell monolayers by fibrinogen fragment D. Science 1985; 277:1487–1490.

159. Sherman CT, Litvak F, Grundfest W. Coronary angioscopy in patients with unstable angina. N Engl J Med 1986; 315:913–919.

160. Gotoh H, Minamino T, Hatoh O, et al. The role of intracoronary thrombus in unstable angina: angiographic assessment and thrombolytic therapy during ongoing anginal attacks. Circulation 1988; 77:526–534.

161. DeWood MA, Stifter WF, Simpson CS, et al. Coronary arteriographic findings soon after non-Q-wave myocardial infarction. N Engl J Med 1986; 315:417–423.

162. Gibson RS, Beller GA, Gheorghiade M, et al. The prevalence and clinical significance of residual myocardial ischemia 2 weeks after uncomplicated non-Q-wave infarction: a prospective natural history study. Circulation 1986; 73:1186–1198.

163. Timmis AD, Griffin B, Crick JCP, et al. The effects of early coronary patency on the evolution of myocardial infarction: a prospective angiographic study. Br Heart J 1987; 58:345–351.

164. Fuster V, Stein B, Ambrose JA, Badimon L, Badimon JJ, Chesebro J. Atherosclerosis plaque rupture and thrombosis: evolving concepts. Circulation 1992; III47–III59.

165. Liebson PR, Klein LW. Intravascular ultrasound in coronary atherosclerosis: a new approach to clinical assessment. Am Heart J 1992; 123:1643–1650.

166. Werner GS, Sold G, Buchwald AT, et al. Intravascular ultrasound imaging of human coronary arteries after percutaneous transluminal angioplasty: morphologic and quantitative assessment. Am Heart J 1991; 122:212–220.

167. Honye J, Mahon DJ, Jain A, et al. Morphological effect of coronary balloon angioplasty in vivo assessed by intravascular ultrasound imaging. Circulation 1992; 85:1012–1025.

168. Bartorelli AL, Neville RF, Keren G, et al. In vitro and in vivo intravascular ultrasound imaging. Eur Heart J 1992; 13:102–108.

169. White CW, Wilson RF, Marcus ML. Methods of measuring myocardial blood flow in humans. Prog Cardiovasc Dis 1988; 31:79–94.

170. Nissen SE, Gurley JC, Booth DC, et al. Differences in intravascular ultrasound plaque morphology in stable and unstable patients (abstract). Circulation 1991; 84:II-436.

171. Carew TE, Schwenke DC, Steinberg D. Antiatherogenic effects of probucol unrelated to its hypocholesterolemic effect: evidence that antioxidant in vivo can selectively inhibit low-density lipoprotein degradation in macrophage-rich fatty streaks and slow the progression of atherosclerosis in Watanabe heritable hyperlipidemic rabbit. Natl Acad Sci USA 1987; 84:7725–7729.

172. Brown G, Albers JJ, Fisher LD, et al. Regression of coronary artery disease as a result of intensive lipid lowering therapy in men with high levels of apolipoprotein B. N Engl J Med 1990; 323:1289–98.

173. Osnish D, Brown SE, Scherwitz LW, et al. Can lifestyle changes reverse coronary artery disease? The lifestyle Heart Trial. Lancet 1990; 336:129–133.

174. Blakenhorn DH, Nessim SA, Johnson RL, et al. Beneficial effects of combined colestipol-niacin therapy on coronary atherosclerosis and coronary venous bypass grafts. JAMA 1987; 257:3233–3240.

175. Brensike JF, Levy RI, Kelsey SF, et al. Effects of therapy with cholestyramine on progression of coronary atherosclerosis: results of the NHLBI Type II Coronary Intervention Study. Circulation 1984; 69:313–324.

176. Cashin-Hemphill L, Mack WJ, Pogoda JM, et al. Beneficial effects of colestipol-niacin on coronary atherosclerosis. JAMA 1990; 264:3013–3017.

177. Buchwald H, Varco RI, Matts JP, et al. Effect of partial ileal bypass surgery on mortality and morbidity from coronary heart disease in patients with hyper-lipidemia. N Engl J Med 1990; 323:946–955.

178. Watts GF, Lewis B, Brunt JN, et al. Effects on coronary artery disease of lipid-lowering diet, or diet plus cholestyramine, in the St Thomas' Atherosclerosis Regression Study (STARS). Lancet 1992; 339:563–569.

179. Vogel RS. Comparative clinical consequences of aggressive lipid management, coronary angioplasty and bypass surgery in coronary artery disease. Am J Cardiol 1992; 69:1229–1233.

180. Blakenhorn DH, Kramsch DM. Reversal of atherosis and sclerosis. The two components of atherosclerosis. Circulation 1989; 79:1–7.

181. Pfeffer MA, Braunwald E, Moye LA, et al. Effect of captopril on mortality and morbidity in patients in patients with left ventricular dysfunction after myocardial infarction: results of the Survival and Ventricular Enlargement Trial. N Engl J Med 1992; 327:669–677.

182. The SOLVD investigators. Effect of enalapril on mortality and the development of heart failure in asymptomatic patients with reduced left ventricular ejection fractions. N Engl J Med 1992; 327:685–691.

183. Francis GS. Neurohumoral activity in congestive heart failure. Am J Cardiol 1990; 66:33D–39D.

184. Yusuf S, Peto J, Lewis J, et al. Beta blockade during and after myocardial infarction: an overview of the randomized trials. Prog Cardiovasc Dis 1985; 27:335–371.

185. Frishman WH, Lazar EJ. Reduction of mortality, sudden death and non-fatal reinfarction with beta-adrenergic blockers in survivors of acute myocardial infarction: a new hypothesis regarding the cardioprotective action of beta-adrenergic blockade. Am J Cardiol 1990; 66:66G–70G.

5

Dyslipoproteinemia in Older People

Walter H. Ettinger, Jr., and William R. Hazzard

Bowman Gray School of Medicine, Wake Forest University, Winston-Salem, North Carolina

INTRODUCTION

Coronary heart disease (CHD) is the leading cause of death and disability in the elderly,* and the majority of total CHD deaths occur in older people (1) (Fig. 1). Although coronary heart disease mortality is declining, even among the very old, the prevalence, yearly incidence rates, and costs of CHD in the elderly will increase dramatically over the next 30 years because of the aging of the population (2,3). Therefore, efforts to reduce the premature mortality, disability, and cost associated with CHD in the elderly are an important public health priority.

Dyslipoproteinemia is a modifiable risk factor for CHD in young and middle-aged adults (4). However, the role of lipoprotein lipids as risk factors for CHD in older adults is controversial. Several studies have suggested that the association between lipid concentrations and CHD weakens with age and that there is little

---

*Considerable debate persists over the proper language to describe older persons, debate that has significantly clouded thinking as to hyperlipidemia "in the elderly." Herein we shall refer collectively to those above the age of 65 interchangeably as "older persons" and "the elderly" (eschewing such euphemisms as "senior citizens" or "seniors"). However, as noted below, "the elderly" are notoriously heterogeneous and progressively more so with advancing age, those above 85 (the "oldest old") differing markedly from those 65–75 as a group as well as among themselves. Hence generalizations regarding "the elderly" are risky at best and, as stressed below, individualization of assessment and treatment regimens is the watchword in the appropriate approach to management of dyslipoproteinemia in the older patient.

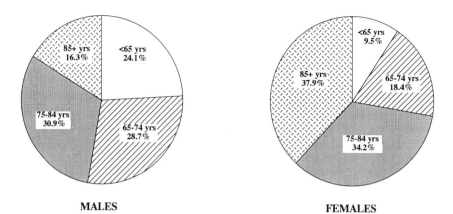

MALES   **FEMALES**

Figure 1   Percentage of ischemic heart disease deaths by age for males (left) and females (right) in 1988. The total number of deaths in males and females were 264,505 and 245,087, respectively. (From Vital Statistics of the United States, 1988.)

potential benefit, and perhaps even adverse effects, from screening and treating older adults for dyslipoproteinemia (5,6). Data such as these have led some groups, such as the Canadian Task Force on Prevention, to recommend that screening and treatment of hyperlipidemia not be done in persons over age 60 (7). In contrast, other reports suggest that lipoprotein levels remain a significant risk factor for CHD in older people (8,9), and that, indeed, treatment of dyslipoproteinemia in older people may have a greater impact on coronary heart disease mortality than in younger people because the population-attributable risk from dyslipoproteinemia is greater in the older age group (10,11). However, there is no direct evidence that treating dyslipoproteinemia reduces the risk for CHD in older people. This patchwork of reports has led not only to widely varying opinions as to the importance of dyslipoproteinemia in the elderly and especially its treatment, but especially to confusion in the minds of practitioners and their patients (and generally no great enthusiasm in either group to treat aggressively, especially with drugs). This lack of consensus and clarity most likely reflects the interacting confounding effects of aging, age-related disease, and the approaching upper limit of the human lifespan in advanced old age.

    In an effort to disentangle this seeming conundrum, in this chapter we shall (1) describe the interactions of age and lipoprotein metabolism and the distribution and correlates of lipoprotein lipids in older people, (2) review the evidence linking lipoprotein levels with CHD in older people, (3) describe the important effects of CHD and other diseases and disease processes common in the elderly on lipids and mortality in this age group, and (4) discuss considerations surrounding treatment of dyslipoproteinemia in older people.

AGING AND LIPOPROTEIN METABOLISM

Aging is a continuous process that begins at conception and ends at death. The human lifespan is characterized by phases of growth, development, and gradual decline in physiological reserve capacity, and terminal loss of homeostatic control, the rates of which vary widely by age and among individuals. The periods of most rapid change in the absence of clear-cut disease occur in utero, the neonatal–early childhood period, adolescence, and (for women) the menopause. In those of middle and advanced age, changes attributable to aging tend to be subtle until their limited homeostatic reserve is disclosed during periods of stress, notably that caused by disease. As detailed below, this aging-disease interaction tends to confound interpretation of data related to lipoprotein levels and attendant CHD in the elderly.

These phases and periods of transition are reflected in average population lipid levels across the human lifespan. Cord blood lipids (the only well-characterized reflection of lipid levels in utero) are typically low (total cholesterol concentrations averaging 60 mg/dl, equally distributed between LDL and HDL). These rise rapidly during infancy, stabilizing at two to three times cord blood levels by age 2 and remaining equivalent between the sexes until adolescence (Fig. 2). Thereafter in Western societies population average total and LDL cholesterol levels rise gradually yet progressively with age between puberty and the menopause in both genders, albeit faster and to a greater degree in males than females, generally in parallel with adult-acquired adiposity. Across the menopause average LDL levels rise rapidly in women, with LDL cholesterol levels in postmenopausal women (absent hormone-replacement therapy) exceeding those in men of comparable age throughout the remainder of the lifespan. Thus it would appear that endogenous estrogens exert a powerful hypocholesterolemic, specifically hypobetalipoproteinemic effect (and their waning secretion is responsible for the higher LDL cholesterol levels postmenopausally). In contrast, HDL cholesterol levels decline progressively in males across puberty (in parallel with Tanner developmental stage) and remain lower in men than women throughout the remainder of the lifespan. This is presumably attributable to the hypoalphalipoproteinemic effect of androgens. Interestingly, loss of most endogenous estrogens across the menopause is not clearly associated with declining HDL cholesterol levels in women [although prospective studies may detect a subtle decline (12)], nor do HDL levels rise in aging men [albeit a suggestive increase in average concentrations was seen in Lipid Research Clinics population data (Fig. 2) (13)]. Thus endogenous estrogens appear to account for the lower LDL cholesterol concentrations in premenopausal women than men of comparable age, while androgens appear to account for lower HDL levels in males at all ages beyond puberty. The additive result of these hormonal effects is higher average total cholesterol levels in women than men at all ages beyond the menopause (since both LDL and HDL concentrations are higher).

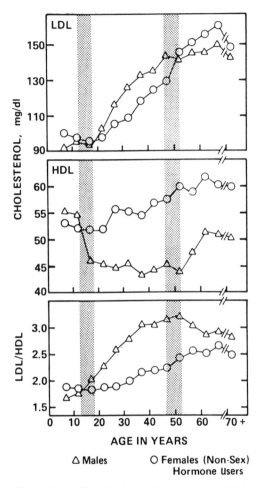

Figure 2   Median North American population high-density lipoprotein (HDL) cholesterol, low-density lipoprotein (LDL) cholesterol, and the ratio between the two versus age in white subjects. (From Ref. 13.)

Hence the prevalence of cholesterol concentrations exceeding recommended levels (200, 240 mg/dl etc.) is higher in older women than in men (Fig. 3), and more women therefore seek medical advice for management of those levels than do men, despite the fact that their LDL/HDL ratios are still lower than in men and their age-specific CHD risk also remains substantially lower. This confusion was only partially corrected in the 1993 ATP II NCEP guidelines (14) in the specific recommendations related to menopausal status and subtraction of a risk factor for

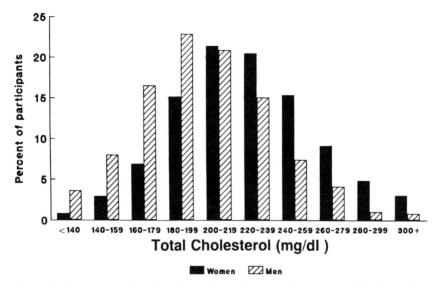

Figure 3 Comparison of total cholesterol levels in men and women participating in the Cardiovascular Health Study. (From Ref. 21.)

an HDL level exceeding 60 mg/dl, a common finding in older women (and especially in those taking estrogens). In this regard exogenous sex hormones, particularly when administered orally such that portal vein concentrations clearly exceed physiological levels, not only exaggerate the effects of comparable endogenous steroids but also exert additional influences: oral estrogens not only depress LDL levels but also increase HDL, while oral androgens (notably anabolic steroids) both depress HDL and increase LDL concentrations.

These developmental and midlife alterations in population total plasma and lipoprotein cholesterol levels attributable to physiological alterations in sex hormones and adiposity are largely complete by age 60, beyond which adiposity remains generally stable in both genders, as do sex hormones until advanced old age or the advent of disease. Thus, effects of "aging" upon lipoprotein metabolism are principally those during infancy, adolescence, and early to mid-middle age, and no clear-cut alteration in lipoprotein metabolism occurs beyond middle age in the absence of changes in those factors—notably diet, body composition, caloric expenditure, drugs, or disease—that may impact upon lipid levels at any age. Thus there is no rationale for age-specific approaches to lipid altering per se other than a changing relationship between lipid levels and CHD risk with advancing age or possible age-related alterations in risks or cost-benefit analysis of such therapy (both addressed below).

Several cohort studies have described lipoprotein lipid levels in populations of older people (9,13,15–18). It is difficult to compare and contrast the lipid levels among studies because of differences in sampling procedures, recruitment rates, age, race, medication use, health status, and measurement techniques. Nevertheless, several consistent findings emerge from these studies. As noted above, lipid levels are different in older women and men, with women having higher average concentrations of total cholesterol, LDL cholesterol, and HDL cholesterol than men at all ages beyond the era of the menopause. In contrast, triglyceride concentrations are similar in older men and women. However, population average lipid values change with age within the heterogeneous group of older people. In men, and to a lesser extent in women, triglyceride, cholesterol, and LDL cholesterol concentrations decline at advanced ages. There is a slight increase in HDL concentrations with age in males, but HDL cholesterol does not vary by age in women.

Recent evidence suggests that, just as in younger persons, population average cholesterol levels are decreasing in older people and total lipid profiles may be improving, presumably due to secular changes in diet, as well as recognition and treatment of dyslipoproteinemia (18–20). Nevertheless, the number of older persons eligible for referral and treatment of dyslipoproteinemia using the National Cholesterol Education Program Guidelines is substantial (19,21). For example, using data from the NHANES III, Sempos et al. (19) have estimated that approximately 50% of men and women over the age of 65 would require dietary therapy for dyslipoproteinemia by the Adult Treatment Panel II Guidelines (14). Furthermore, assuming a 10% reduction in LDL cholesterol from diet, they calculated that between 16 and 18% of people over the age of 65 require pharmacological therapy for hyperlipidemia, including a substantial number of individuals over 80 years old. As pointed out by Garber et al. (5), the annual direct costs of such therapy would be in the billions of dollars.

DETERMINANTS OF LIPID LEVELS IN OLDER PEOPLE

A number of biological and social factors are strongly associated with lipid levels in older people. For example, persons of higher income and education tend to have more favorable lipid profiles (higher HDL cholesterol and lower LDL cholesterol and triglyceride) than those of lower socioeconomic status (16,18). Measures of obesity (and especially central body fat distribution) and glucose intolerance are all strongly associated with lower HDL cholesterol and higher triglycerides and, in some cases, higher LDL cholesterol levels (16,18,22,23). Similarly, people who are physically active tend to have higher HDL cholesterol levels (24).

Race is also an important determinant of lipid levels in older persons. In the Cardiovascular Health Study (CHS), African American men had lower triglyceride and higher HDL cholesterol levels than whites. In contrast, older

African American women had lower triglycerides, total cholesterol, and LDL cholesterol than white women of comparable age (18).

As in younger persons, use of certain medications influences lipid levels in older people. Women who use estrogen have significantly higher HDL cholesterol and lower total and LDL cholesterol than nonusers (25,26). Both men and women who use beta-blockers or diuretics tend to have higher triglycerides and lower HDL cholesterol levels (18,27). Corticosteroids appear to raise levels of VLDL, LDL, and HDL cholesterol (27). While the effects of these medications in older persons are generally similar to those in younger people, the prevalence and hence the population-attributable effects of such medication usage are greater in the elderly.

Health status also appears to be an important determinant of lipid levels in older people. Most familiar are the secondary dyslipoproteinemias caused or aggravated by such disorders as diabetes mellitus and hypothyroidism, diseases that progressively increase in prevalence and incidence with advancing age. Thus, most older diabetic patients (nearly all of whom have type II diabetes), including those who are under pharmacological treatment, tend to have higher triglycerides and lower HDL cholesterol levels than nondiabetic persons. Of specific note, however, are those elderly diabetics who have *low* total cholesterol levels, reflecting depressed LDL levels as well, the added significance of which is discussed below (26,28). Older persons with mild renal impairment also have elevated triglyceride and depressed HDL cholesterol levels (18). Thus the population-attributable burden of secondary dyslipoproteinemias is greatest among the older segment and will continue to grow as the proportion of elderly persons grows and their median age increases for the next half century. Perhaps the only factor aggravating dyslipoproteinemias at earlier ages that actually declines beyond middle age is relative body weight (though adiposity may continue to increase well beyond peak body weight).

This decline in body mass may be a key pathogenetic feature causing population cholesterol and triglyceride levels to decline in advanced old age, reflecting the advent of acquired hypocholesterolemia as an ominous prognostic survival index. Illness and debility regardless of age are commonly associated with lower total cholesterol, LDL cholesterol, VLDL cholesterol, and HDL cholesterol levels (28–30). Acquired hypocholesterolemia is a common finding in a wide variety of acute and chronic conditions that are characterized by inflammation, activation of the immune system, and the acute-phase response (31). A substantial and rapid fall in serum cholesterol is associated with infections, surgery, myocardial infarction, and trauma (31–33). Not surprisingly the incidence of such declines is greatest among the elderly because of their vulnerability to such health problems. In one study, nearly 10% of subjects over the age of 65 years had a fall in serum cholesterol to less than 120 mg/dl while hospitalized, a decline associated with increased risk of complications, prolonged hospitalization, added cost, and increased mortality (32). At the population level this effect of health status is suggested by epidemiological studies showing that older persons with hypocholesterolemia are more likely to have poor health status, higher levels of physical disability, and greater medical comorbidity than subjects with average or high cholesterol levels (29,30).

Finally, it has been reported that older subjects have an especially large variation in lipid measurements when repeated over time compared to younger subjects (34). Changes in body weight, health status, and medication use may explain some of this variation. Furthermore, the effect of health status makes the clinical interpretation of lipid values as well as the interpretation of associations between risk factors and subsequent disease in epidemiological studies more difficult in older people.

ASSOCIATION OF LIPOPROTEIN LIPIDS WITH CHD IN OLDER PEOPLE

The importance of CHD as a cause of death and disability in older people and the high prevalence of dyslipoproteinemia in older persons lead to two important questions:

1.   Is dyslipoproteinemia associated with coronary heart disease in older people?
2.   Does treatment of dyslipoproteinemia decrease morbidity and mortality from coronary heart disease in older people?

Most but not all studies that have examined the association between lipoprotein lipids and incident CHD have shown a statistically significant, though often weak, positive association in older cohorts (9,11,35,36). In a meta-analysis of data from 25 different populations, Manolio et al. (37) showed that total cholesterol, LDL cholesterol, and triglycerides retained a positive relationship with subsequent CHD in older men. Similar relationships were found in women, though the pooled association of total cholesterol and LDL cholesterol with incident heart disease was marginal. In women, HDL cholesterol was inversely associated with the risk of CHD, while, unlike in younger persons, the relationship of HDL cholesterol and CHD was not statistically significant in men. However, though the *relative* risk of developing CHD is lower in older people with hypercholesterolemia compared with those of normal lipid status, the *attributable* risk increases, and therefore the absolute number of cases of CHD attributable to dyslipoproteinemia increases in older people (10,11). This apparent paradox frames the practitioner's dilemma in advising his elderly patient as to whether and how to treat her or his dyslipoproteinemia.

In the aggregate, the available data suggest that there is an association of lipoprotein lipids and CHD in older people. However, there are several notable exceptions to this generalization that may complicate the issue. Perhaps the most important issue is the relationship of cholesterol and *total* mortality in older people. Several studies have shown an inverse relationship between cholesterol and *total* mortality in older people, with the lowest mortality occurring in persons whose cholesterol values are well above those considered to confer increased risk of heart disease (6,38,39). In one study of truly elderly women (nuns in a French nursing home of mean age 82 ± 8.6 years), the cholesterol level associated with the highest chance of survival over the ensuing 5 years was 7.0 mmol/liter (260 mg/dl), and survival rates decreased in graded, progressive fashion with lower levels of cholesterol (38). Although not specified, the prevalence and, especially, incidence of serious disease in this dependent subject group were no doubt high, and acquired hypocholesterolemia was probably common. However, in another study (39) of 840 hypertensive subjects ≥60 years of age who were participants in a clinical trial and thus generally of good functional status, baseline serum cholesterol was also inversely related to all-cause mortality after adjustment for treatment modality, age, blood pressure, body weight, and hemoglobin. All factors being equal, the model showed that a 2.3 mmol (89 mg)/dl higher serum cholesterol at randomization in this study group was associated with a 1-year *increase* in survival. While these studies clearly do not prove a causal relationship between higher cholesterol and increased survival, they do introduce considerable uncertainty as to the net effects of aggressive hypocholesterolemic therapy in such persons, and hence they make one question the value of cholesterol-lowering therapy, particularly in advanced old age, when there is no direct evidence that doing so improves clinical outcomes.

A second important issue in interpreting epidemiological data is whether age modifies the lipoprotein-CHD relationship. In other words, does the positive association of CHD with cholesterol and other lipids diminish in old age? In support of a negative age effect, nearly all studies show that the strength of the association between lipid levels and heart disease, as measured by relative risk, decreases with age. Furthermore, Kronmal et al. (6), in an analysis of the Framingham data, showed a potent interaction between age and lipid levels on CHD risk, such that the relationships between total cholesterol, HDL cholesterol, and LDL cholesterol and CHD disappeared by age 70 and, in fact, became negative for total cholesterol and LDL cholesterol beyond that age. Additional support for a weakening association of cholesterol and CHD in old age comes from Jacobsen et al. (40), who showed in an angiographic study that the strong positive association of cholesterol with coronary artery disease in younger men disappeared in men over age 70. In contrast, three other studies that have looked at the strength of the relationship between total cholesterol and heart disease in persons over 75 have found a positive association; however, the number of very old people in those studies was small, and the confidence intervals around the point estimates of risk were large (10,41,42). Thus, whether or not there is an association of lipid levels and the risk for CHD in the very old remains unclear.

There are several potential explanations for discrepant results among these epidemiological studies. The decrease in the strength of the association of lipid levels and CHD with age is, in part, due to the selective premature mortality of persons with dyslipoproteinemia who are susceptible to the effects of high cholesterol, the so-called harvest effect (43). Selective premature mortality also may affect the association of dyslipoproteinemia with CHD more than other CHD risk factors such as hypertension and glucose intolerance. Dyslipoproteinemia usually is manifest by midlife, with few new cases of dyslipoproteinemia occurring at older ages (44). In contrast, the incidence of hypertension and glucose intolerance continues to rise into advanced old age (45,46). Thus, though selective mortality of persons susceptible to the effect of hypertension and disorders of glucose metabolism may occur in midlife, new cases continues to develop with increasing age in previously unaffected but susceptible persons, maintaining the strength of the association of these risk factors in the population.

A second issue that may affect the association between lipids and CHD in older people is health status. As noted above, poor health status or illness often results in a lowering of serum cholesterol and lipid subfractions (17,29,30). The incidence of chronic and acute diseases increases with age. Thus lipid levels measured in older people may not reflect long-term exposure. If persons with lower cholesterol due to health factors subsequently develop CHD related to their previously high cholesterol levels, the lifetime risk of the association between cholesterol and CHD will be underestimated at the time of its clinical manifestation. Changes in lipid levels due to disease also may explain why associations of lipids and CHD among persons who had cholesterol measured in the distant past are stronger than those whose lipid measurements are more proximate to the event.

In addition to influencing cholesterol levels, health status appears to affect the direction of the association of cholesterol with CHD. Perhaps most informative in this regard was the analysis of the NHANES I data by Harris et al. (47,48), who have showed the effect of health status on the relationship between total cholesterol and incident CHD (Fig. 4). When subjects were stratified either by activity level (46) or prior weight loss (47), different patterns of the lipid–heart disease association were seen. Among older persons who had not lost weight or who reported a high physical activity level, there was a graded *positive* response between total cholesterol and CHD. In contrast, among persons who had lost weight or were inactive, the relationship between cholesterol and coronary heart disease was *negative*. When these data were combined, no overall association was found between cholesterol and incident CHD. Therefore, the positive association between lipoprotein lipids and heart disease may be maintained in older people who are vigorous and otherwise healthy. However, the association may not be recognized if the population under study is heterogeneous with respect to health status and even become negative if substantial numbers of frail and failing subjects are included.

Finally, there may be interactions between other CHD risk factors and lipid

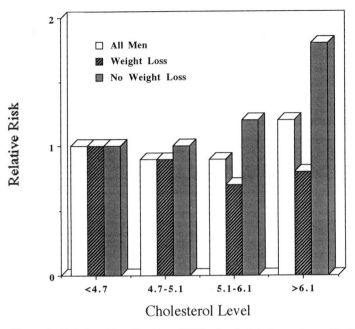

Cholesterol Level

Figure 4 Relative risks of incident CHD by level of cholesterol, stratified by weight loss at baseline in men 65–74 years of age. Weight loss was estimated by subtracting baseline weight from maximum lifetime weight by self-report and dichotomized as <10% (no weight loss) or ≥10% (weight loss). (Adapted from Ref. 48.)

values. Of particular interest are older patients with diabetes, who, in addition to those with secondary hyperlipoproteinemia, include an increased proportion with abnormally low lipid levels. This presumably reflects poor health status, since these are also the subset with poor prognosis for survival. In like fashion changes in cholesterol levels are associated with change in weight, so that persons who have a history of weight loss tend to have lower cholesterol values than those who weigh the same but have not had a change in weight. Thus, potential interaction of risk factors must be examined in analyses of association with CHD.

TO TREAT OR NOT TO TREAT?

The answer to this question can be brief: In 1994 we have no clear answer. There is no direct evidence that treatment of dyslipoproteinemia reduces morbidity or mortality from CHD or affects total mortality in older people. While a small number of clinical trials have included persons who are above the age of 65 years, few reports have sufficiently segregated their responses or outcomes to permit examination of the role of age per se, and generally the number of older participants (especially >75 years) has been small. Hence interpretation of these data is difficult. As pointed out by the Expert Panel on Detection, Evaluation and Treatment of High Blood Cholesterol in Adults (NCEP-ATP II) (14), the evidence for treatment of dyslipoproteinemia in older people must be extrapolated from trials showing reduction of CHD risk in middle-aged people, an extrapolation that may or may not be justified given the generally uncertain and inconsistent relationship between cholesterol status and CHD risk in the elderly noted above.

Since there is no direct evidence that treatment is beneficial to older people, it may be worthwhile to review the indirect evidence that may suggest benefit. The first argument in favor of treatment is that CHD is the leading cause of death and physical disability in older people (see Fig. 1) and that because the rate of cardiac events is higher than in younger people, treatment of dyslipoproteinemia will have a much greater effect in the older population as a whole because the total number of potentially preventable events would be higher. However, many deaths due to CHD occur in persons over the age of 80, especially in women, and one wonders whether such deaths, that occur near the end of the human lifespan, are truly "preventable." Nonetheless, at the level of the individual patient, every experienced geriatric practitioner has witnessed the transformation of an independent free-living elderly patient into a dependent, frail, and failing person following a stroke or disabling myocardial infarction. To prevent such a tragic terminal cascade through effective preventive intervention carries a powerful premium for patients, their families, and physicians alike.

Second, as noted above, the same forces that modulate lipoprotein lipid levels in younger adults appear to be at least equally as effective in the elderly. While not exhaustively or systematically studied, the similarities demonstrated suggest that old age–specific approaches to lipid alteration will be the exception rather than the rule. Nevertheless, it will be important to confirm this general hypothesis through systematic study of specific lipid interventions in an older cohort of volunteer subjects, including those in robust as well as marginal health, those on various medications and nonpharmacological regimens for treatment of coexisting diseases, and those in various institutional or free-living circumstances.

In the meantime, dietary guidelines would seem the same as those for younger individuals at comparable risk for cardiovascular disease and generally according to AHA Step 1 guidelines (per NCEP-ATP II) (14) appropriate for coronary disease prevention in the population at large. However, common sense demands attention to individual preferences (often dictated by lifelong eating habits), coexisting conditions or diseases that may alter access to or assimilation of given foods (e.g., the edentulous state, gastrointestinal disorders, depression, etc.), and precarious nutritional balance (highly prevalent especially in those of marginal or deteriorating health status). Thus, in the frail elderly even Step 1 dietary intervention may carry some risk or at least not confer sufficient benefit to warrant its aggressive prescription and follow-up.

As to other aspects of lifestyle, here the most independent, vigorous, active lifestyle (both physically and socially) should be encouraged, and the expected alterations in lipid levels should be seen given adequate compliance (e.g., HDL levels should rise with physical activity as well as modest alcohol ingestion and discontinuation of cigarette smoking).

In a similar vein pharmacological lipid-altering therapy should be expected to produce responses at least equivalent to those in younger individuals. This has proven to be the case for all drugs carefully examined to date. Those specifically studied include lovastatin (49), simvastatin (50), and nicotinic acid (51).

Clearly the best and most specifically studied agent has been the HMG CoA reductase inhibitor lovastatin. This was examined in the Cholesterol Reduction in Seniors Program (CRISP) trial, designed as a pilot study of hypolipidemic therapy in older subjects, including those with prevalent cardiovascular disease (49). This study was a randomized, double-masked clinical trial with placebo and 20- and 40-mg lovastatin treatment arms. Participants averaged approximately 71 years of age, and African Americans (21%) and women (71%) were well represented. CHD was present in 12–20%, and approximately one third were hypertensive. All had LDL levels between 160 and 220 mg/dl on repeated occasions. As in studies of lovastatin in younger subjects, gratifying declines in LDL concentration were seen, with no significant drug-associated side effects: in the 20- and 40-mg lovastatin groups, total cholesterol fell 17 and 20%, respectively, LDL cholesterol 24 and 28%, respectively, and triglycerides 4.4 and 9.9%, respectively, while HDL cholesterol rose 7 and 9%, respectively. Notable was the marginal additional increase in lipid response to the 40-mg dose, suggesting that for most older dyslipoproteinemic patients the smaller dose in a once-a-day regimen should suffice to reach target LDL levels (though more aggressive hypolipidemic therapy may be required to reach the 100 mg/dl LDL cholesterol goal in secondary prevention). Unfortunately the CRISP pilot study did not proceed to a full-scale trial of the effect of cholesterol reduction or the central question of this treatise: whether or not such treatment prevents CHD, either primarily or secondarily, in the elderly. Lacking such central information, we are left with clinical judgment largely uninformed by scientific evidence in formulating individual management strategies for dyslipoproteinemia in elderly patients.

Thus, apart from side effects that may prove particularly troublesome for older subjects (e.g., flushing, dyspepsia, or aggravation of glucose intolerance with nicotinic acid or exaggeration of a tendency toward constipation with bile acid–binding resins), lack of responsiveness or unacceptable side effects should not generally inhibit aggressive hypolipidemic therapy even among the very old.

A similar rationale for hypolipidemic therapy in the elderly as in the nonelderly is supported by the fact that, from the standpoint of pathology, atherosclerotic CHD appears to be the same in older people as it is in younger people. Although serum cholesterol is more closely related to the development of early, lipid-rich atherosclerotic lesions such as fatty streaks than it is to fibrous plaques, angiographic studies have shown that advanced coronary atherosclerosis responds to cholesterol-lowering treatment, suggesting that even complex lesions are sensitive to changes in levels of circulating lipoproteins.

Finally, treating dyslipoproteinemia in the elderly may be relatively cost-effective since much of the treatment would be for secondary prevention of CHD and in people at high risk for CHD because of the presence of multiple other risk factors. For example, in the CHS study approximately 25% of the population in that study had prevalent CHD, 50% had hypertension, and 25% had diabetes (18).

The arguments in favor of treatment notwithstanding, there are several lines of evidence to support a conservative approach to treatment of hyperlipidemia, especially among the very old ($\geq$75 years). First, as noted above, the epidemiological evidence linking dyslipoproteinemia to incident CHD is conflicting and certainly less compelling than in younger people. Several studies suggest that the relationship between lipid levels and incident CHD declines in old age, and one study suggests that the relationship between total cholesterol and LDL cholesterol in CHD mortality is inverse. Furthermore, several studies suggest that total mortality is negatively related to cholesterol level. While these studies do not prove a causal link between higher cholesterol and longevity, they do not support the hypothesis that treatment of dyslipoproteinemia would favorably affect clinical outcomes.

A second consideration of great concern to elderly patients is drug cost. Until a change in health care policy permits reimbursement for drugs purchased, retirees on fixed incomes will be reluctant to accept advice for such treatment when their purchase must compete with expenditures for items of more immediate priority such as food or housing.

Finally, the first therapeutic dictum is, of course, "above all, do no harm." Even with the reassuringly low incidence of significant side effects of hypolipidemic drugs in elderly persons thus far studied, given that the numbers of adverse drug effects in older persons in general is exponentially related to the number of drugs taken (52) and that many older persons with dyslipoproteinemia will have co-morbid conditions under pharmacological therapy, addition of any drug to the regimen of an older patient carries special risk, at the least, of confusion and poor compliance. Therefore careful, individualized estimates of the risk $\times$ cost/benefit ratio will be essential in the decision of whether or not to treat a given elderly dyslipoproteinemia patient. This includes considered attention to the time-to-effect of the hypolipidemic intervention (up to 2 years may be required before lipid lowering may be expected to attenuate atherogenesis in primary prevention, while regression in secondary prevention seems to begin immediately). And, of course, expected duration of benefit as well as expected longevity and its quality must factor into the therapeutic equation.

In this latter regard, even assuming that there is a relationship between dyslipoproteinemia and CHD and that the treatment is effective, Grover et al. (53) have shown that the effect on life expectancy and morbidity is small in older people. The maximum average expected years of life saved by treating high-risk individuals over age 65 with severe hypercholesterolemia is 0.46 year in women and 0.54 year in men. Similarly, the maximum years free of nonfatal CHD from treatment of dyslipoproteinemia in high-risk older individuals with severe dyslipoproteinemia is 0.71 year in women and 0.53 year in men. This is in contrast to younger and middle-aged persons, who have substantially greater expected longevity benefit from treatment.

Clearly, CHD is a growing public health problem among the elderly. However, in addition to preventing premature mortality, a primary goal of treatment must be preventing disability from the disease. Recent evidence suggests a continued decline in mortality from CHD in the elderly. However, several lines of evidence suggest that mortality is decreasing faster than morbidity and, therefore, the number of older persons with prevalent CHD is increasing. This is particularly troublesome since many such people with nonfatal CHD progress to congestive heart failure (3), a severely disabling condition that is already the most common DRG diagnosis for hospitalization among Medicare beneficiaries. Therefore, it is important to demonstrate that treating dyslipoproteinemia will decrease the incidence of CHD as well as reduce mortality. A corollary is that if premature death from CHD is prevented in older people, they will survive into even more advanced old age and thus be more susceptible to chronic disabling diseases such as osteoporosis, osteoarthritis, and Alzheimer's disease. Therefore it is important to examine the net effects of preventing CHD on total disability as well as quality of life.

In summary, there is no direct evidence that treating dyslipoproteinemia will have a positive effect on health status in older individuals. While there are several strong arguments for treating dyslipoproteinemia in older people, the epidemiological evidence supporting the relationship between dyslipoproteinemia and CHD is mixed and the net effect of preventing CHD deaths in older people is unknown. However, central to unraveling this puzzle seems to be accurate assessment of whether a given elderly dyslipidemic patient is fundamentally well (and hence may benefit from hypolipidemic therapy to sustain his or her vigor) or frail and possibly failing, in which case attention to other issues of immediate urgency may be more appropriate.

So what is the poor doctor to do while we await clear evidence from clinical trials on this issue? Quite simply, be a doctor (which, as always, involves making a decision with an imperfect information base and living with uncertainty)!

ACKNOWLEDGMENT

This work was supported in part by NIH grants P60 AG10484 and K07 AG00421.

REFERENCES

1. Bild D, Fitzpatrick A, Fried L, Wong N, Haan M, Lyles M, Bovill E, Polak J, Schulz R. Age-related trends in cardiovascular morbidity and physical functioning in the elderly: The Cardiovascular Health Study. J Am Geriatr Soc 1993; 41:1047–1056.
2. Weinstein M, Coxson P, Williams L, Pass T, Stason W, Goldman L. Forecasting coronary heart disease incidence, mortality, and cost: the coronary heart disease policy model. Am J Public Health 1987; 77:1417–1426.
3. Bouneux L, Barendredregt JJ, Meeter K, Bonsel GJ, van der Maas PJ. Estimating clinical morbidity due to ischemic heart disease and congestive heart failure: the future rise of heart failure. Am J Public Health 1994; 84:20–28.
4. LaRosa J. Cholesterol lowering, low cholesterol, and mortality. Am J Cardiol 1993; 72:776–785.
5. Garber A, Littenberg B, Sox H, Wagner J, Gluck M. Costs and health consequences of cholesterol screening for asymptomatic older Americans. Arch Intern Med 1991; 151:1089–1095.
6. Kronmal R, Cain K, Ye Z, Omenn G. Total serum cholesterol levels and mortality risk as a function of age. Arch Intern Med 1993; 153:1065–1073.
7. Canadian Consensus Conference on Cholesterol: Final Report. The Canadian Consensus Conference on the prevention of heart and vascular disease by altering serum cholesterol and lipoprotein risk factors. Can Med Assoc J 1988; 139(suppl):1–8.
8. Cohen D, Mindell J. Serum cholesterol and older people. Br J Hosp Med 1991; 46: 323–325.
9. Castelli W. Risk factors in the elderly: a view from Framingham. Am J Geriatr Cardiol 1993; 2:8–19.
10. Rubin S, Sidney S, Black D, Browner W, Hulley S, Cummings S. High blood cholesterol in elderly men and the excess risk for coronary heart disease. Ann Intern Med 1990; 113:916–920.
11. Benfante R, Reed D. Is elevated serum cholesterol level a risk factor for coronary heart disease in the elderly? JAMA 1990; 263:393–396.
12. Matthews KA, Meilahan E, Kuller LH, et al. Menopause and risk factors for coronary heart disease. N Engl J Med 1989; 321:641–646.
13. The Lipid Research Clinics Program Epidemiology Committee. Plasma lipid distributions in selected North American populations. Circulation 1979; 60:427–439.
14. Expert Panel on Detection, Evaluation and Treatment of High Blood Cholesterol in Adults. Summary of the second report of the National Cholesterol Education Program (NCEP). JAMA 1993; 269:3015–3023.
15. Mittlemark M, Luepker R, Slater J, Pirie P, Murray D, Rastam L, Blackburn H. Total blood cholesterol levels in older adult participants in community-wide screening programs (abstract). J Am Geriatr Soc 1991; 39:A7.
16. Curb J, Reed D, Yano K, Kautz J, Albers J. Plasma lipids and lipoproteins in elderly Japanese-American men. J Am Geriatr Soc 1986; 34:773–780.
17. Wallace R, Colsher P. Blood lipid distributions in older persons prevalence; and correlates of hyperlipidemia. Ann Epidemiol 1992; 2:15–21.
18. Ettinger W, Wahl P, Kuller L, Bush T, Tracy R, Manolio T, Borhani N, Wong N,

O'Leary D. Lipoprotein lipids in older people; results from the Cardiovascular Health Study. Circulation 1992; 86:858–869.

19. Sempos C, Cleeman J, Carroll M, Johnson C, Bachorik P, Gordon D, Burt V, Briefel R, Brown C, Lippel K, Rifkind B. Prevalence of high blood cholesterol among US adults. JAMA 1993; 269(23):3009–3014.

20. Johnson C, Rifkind B, Sempos C, Carroll M, Bachorik P, Briefel R, Gordon D, Burt V, Brown C, Lippel K, Cleeman J. Declining serum total cholesterol levels among US adults. JAMA 1993; 269:3002–3008.

21. Manolio T, Furberg C, Wahl P, Tracy R, Borhani N, Gardin J, Fried L, O'Leary D, Kuller L. Eligibility for cholesterol referral in community-dwelling older adults; the Cardiovascular Health Study. Ann Intern Med 1992; 116:641–649.

22. Haffner S, Stern M, Hazuda H, Mitchell B, Patterson J. Cardiovascular risk factors in confirmed prediabetic individuals. JAMA 1990; 263:2893–2898.

23. Haarbo J, Hassager C, Riis B, Christiansen C. Relation of body fat distribution to serum lipids and lipoproteins in elderly women. Atherosclerosis 1989; 80:57–62.

24. Schwartz R, Cain K, Shuman W, Larson V, Stratton J, Beard J, Kahn S, Cerqueira M, Abrass I. Effect of intensive endurance training on lipoprotein profiles in young and older men. Metabolism 1992; 41:649–654.

25. Wahl P, Walde C, Knopp R, Hoover J, Wallace R, Heiss G, Rifkind B. Effect of estrogen/progestin potency on lipid/lipoprotein cholesterol. N Engl J Med 1983; 308:862–876.

26. Manolio, T, et al. Associations of postmenopausal estrogen use with cardiovascular disease and its risk factors in older women. Circulation 1993; 88:2163–2171.

27. Henkin Y, Como J, Oberman A. Secondary dyslipidemia. Inadvertent effects of drugs in clinical practice. JAMA 1992; 267:961–968.

28. Franzblau A, Criqui M. Characteristics of persons with marked hypocholesterolemia: a population-based study. J Chronic Dis 1984; 37:387–395.

29. Manolio T, Ettinger W, Tracy R, Kuller L, Borhani N, Lynch J, Fried L. Epidemiology of low cholesterol levels in older adults; The Cardiovascular Health Study. Circulation 1993; 87:728–737.

30. Ives D, Bonino P, Traven N, Kuller L. Morbidity and mortality in rural community-dwelling elderly with low total serum cholesterol. J Gerontol Med Sci 1993; 48: M103–M107.

31. Ettinger W, Harris T. Causes of hypocholesterolemia. Coronary Artery Dis 1993; 4: 854–859.

32. Noel M, Smith T, Ettinger W. Characteristics and outcomes of hospitalized older patients who develop hypocholesterolemia. J Am Geriatr Soc 1991; 39:455–461.

33. Sammalkorpi K, Valtonen V, Kerttula Y, Nikkila E, Taskinen M. Changes in serum lipoprotein pattern induced by infections. Metabolism 1988; 35:859–865.

34. Frishman W, Lock Ooi W, Derman M, Eder H, Gidez L, Ben-Zeev D, Zimetbaum P, Heiman M, Aronson M. Serum lipids and lipoproteins in advanced age-intraindividual changes. Ann Epidemiol 1992; 2:43–50.

35. Barrett-Connor E, Suarez L, Khaw K, Criqui M, Wingard D. Ischemic heart disease risk factors after age 50. J Chronic Dis 1984; 37:903–908.

36. Bass K, Newschaffer C, Klag M, Bush T. Plasma lipoprotein levels as predictors of cardiovascular death in women. Arch Intern Med 1993; 153:2209–2216.

37. Manolio T, Pearson T, Wenger N, Barrett-Connor E, Payne G, Harlan W. Cholesterol and heart disease in older persons in women—review of an NHLBI workshop. Ann Epidemiol 1992; 2:161–176.

38. Forette B, Tortrat D, Wolmark Y. Cholesterol as risk factor for mortality in elderly women. Lancet 1989; 1:868–870.

39. Staessen J, Amery A, Birkenhager W, Bulpitt C, Clement D, de Leeuw P, Deruyttere M, Schaepdryver A, Dollery C, Fagard R, Fletcher A, Forette F, Forte J, Henry J, Koistinen A, Leonetti G, Nissinen A, O'Brien E, O'Malley K, Pelemans W, Petrie J, Strasser T, Terzoli L, Thijs L, Tuomilehto J, Webster J, Williams B. Is a high serum cholesterol level associated with longer survival in elderly hypertensives? J Hypertension 1990; 8:755–761.

40. Jacobsen S, Freedman D, Hoffman R, Gruchow H, Anderson A, Barboriak J. Cholesterol and coronary artery disease: age as an effect modifier. J Clin Epidemiol 1992; 45:1053–1059.

41. Sorkin J, Andres R, Muller D, Baldwin H, Fleg J. Cholesterol as a risk factor for coronary heart disease in elderly men—the Baltimore longitudinal study of aging. Ann Epidemiol 1992; 2:59–67.

42. Zimetbaum P, Frishman W, Lock Ooi W, Derman M, Aronson M, Gidez L, Eder H. Plasma lipids and lipoproteins and the incidence of cardiovascular disease in the very elderly; the Bronx aging study. Arteriosclerosis Thromb 1992; 12:416–423.

43. Anderson K, Castelli W, Levy D. Cholesterol and mortality; 30 years of follow up from the Framingham Study. JAMA 1987; 257:2176–2180.

44. Hazzard W. Aging and atherosclerosis: teasing out the contributions of time, secondary aging and primary aging. Clin Geriatr Med 1985; 1:251–285.

45. Savage P, Wahl P, Tracy R, Borhani N, Ettinger W. Association of abnormal glucose tolerance with coronary heart disease in older men and women: The Cardiovascular Health Study (CHS). (abstract) Circulation 1991; 84(suppl II):II-548.

46. Applegate WB, Ruton GH. Advances in the management of hypertension in older persons. J Am Geriatr Soc 1992; 40:1164–1174.

47. Harris T, Makuc D, Kleinman J, Gillum R, Curb J, Schatzkin A, Feldman J. Is the serum cholesterol-coronary heart disease relationship modified by activity level in older persons? J Am Geriatr Soc 1991; 39:747–754.

48. Harris T, Kleinman J, Makuc D, Gillum R, Feldman J. Is weight loss a modifier of the cholesterol-heart disease relationship in older persons? Data from the NHANES I epidemiologic follow up study. Ann Epidemol 1992; 2:35–41.

49. LaRosa JC, Applegate WB, Crouse JR III, Hunninghake DB, Grimm R, Knopp R, Eckfeldt JH, Davis CE, Gordon DJ. Cholesterol lowering in the elderly: results of the cholesterol reduction in seniors program (CRISP) pilot study. Arch Int Med 1994; 154:529–539.

50. Antonicelli R, Onorato G, Pagelli P, Peirazzoil L, Paciaroni E. Simvastatin in the treatment of hypercholesterolemia in elderly patients. Clin Ther 1990; 12:165–71.

51. Keenan JM, Bae CY, Fontaine PL, et al. Treatment of hypercholesterolemia: compari-

son of younger versus older patients using wax-matrix sustained release niacin. J Am Geriatric Soc 1991; 40:12–18.

52. Vestal R. Clinical pharmacology. In: Hazzard WR, et al., eds. Principles of Geriatric Medicine and Gerontology. 2d ed. New York: McGraw-Hill, 1990.

53. Grover S, Abrahamowicz M, Joseph L, Brewer C, Coupal L, Suissa S. The benefits of treating hyperlipidemia to prevent coronary heart disease. Estimating changes in life expectancy and morbidity. JAMA 1992; 267:816–822.

6

Cholesterol Lowering in Women

John C. LaRosa

Tulane University Medical Center, New Orleans, Louisiana

GENDER AND NORMAL LIPOPROTEINS: MEAN LIPOPROTEINS IN MALES AND FEMALES

Partly because of differences in endogenous, gonadal hormone levels, lipoprotein levels are not identical in males and females of the same age. Androgens lower high-density lipoprotein (HDL) cholesterol and triglyceride levels and raise low-density lipoprotein (LDL) cholesterol levels (1). Estrogens, on the other hand, raise HDL and triglyceride levels and lower LDL levels.

As a result, from puberty on, females have higher HDL levels than males (Fig. 1). Even the estrogen deficiency of postmenopause, however, does not equalize HDL levels in men and women (2). While a longitudinal study of a cohort of women progressing through menopause does demonstrate a small fall in HDL after menopause, it is not of the degree that might be expected (3).

LDL differences between men and women are more complex. Until about age 55 (i.e., until the postmenopause), women have lower LDL levels than men. After age 55, the relationship is reversed. From then on, women have higher levels than men of the same age. This postmenopausal rise in LDL is confirmed in longitudinal follow-up. These differences in later life may also be related to the higher coronary mortality in younger, hypercholesterolemic men, leaving only men with lower cholesterol levels in older age groups.

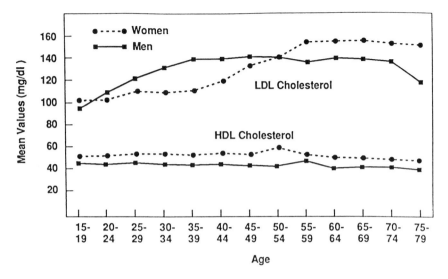

Figure 1 Age trends in lipoprotein-cholesterol fractions. (From Ref. 2.)

LIPOPROTEINS AND THE FEMALE LIFE CYCLE

Other phases of the female life cycle, in addition to menopause, may affect circulating lipoprotein levels. The decline in HDL at puberty in males (probably the result of rising circulating androgen levels), is not seen in females. This is likely due to rising estrogen levels. As a result, HDL levels are higher in females than in males from early adolescence on (4).

During the monthly menstrual cycle, LDL levels decline early and remain low even through the progesterone-dominated second half of the cycle. HDL levels remain unchanged (5). Thus, estrogen appears to have the dominant effect throughout the cycle. This is compatible with studies of exogenously micronized progesterone demonstrating no estrogen-opposing effect with regard to lipoprotein levels (6). On the other hand, the lack of variability in HDL levels during the cycle, particularly in the estrogen-dominant first phase, is unexplained.

During pregnancy, both LDL and HDL levels rise, perhaps as a result of increasing levels of endogenous estrogen and progesterone. While LDL levels remain elevated until several weeks after delivery (7), HDL levels fall at about the 24th week of pregnancy, so that they are almost at pregravid levels by the time of delivery. Like HDL, triglyceride levels rise during pregnancy but fall more rapidly after delivery than do LDL levels (8).

It might be expected that because the sustained increase in LDL is potentially atherogenic, multiple pregnancies would predict an increased risk of coronary heart disease (CHD) in later life. Indeed, recent evidence suggests that such an association does exist, the relative risk of CHD increasing 50% with six or more pregnancies (9).

LIPOPROTEINS AS RISK FACTORS FOR ATHEROSCLEROSIS IN WOMEN

Coronary risk rises with increases in LDL levels and falls with increases in HDL levels in women as well as men (10). There are, however, some quantitative differences. HDL is a more potent predictor of risk in women than in men, implying that HDL may function differently in women (11).

LDL, for reasons that are unclear, does not predict risk in women as strongly as in men (10). Studies in female primates demonstrate that the presence of circulating estrogen interferes with LDL uptake in the arterial wall (12). This may be related to estrogen's antioxidant properties (13). At any given level, then, LDL is probably less atherogenic in women than in men.

Because HDL comprises a greater fraction of total cholesterol in women and is more protective and because LDL is potent, total cholesterol is a less reliable predictor of CHD risk in women (14). It is erroneous, however, to conclude from this that lipoproteins are unimportant as risk factors in women. As in men, total cholesterol, in long-term follow-up, predicts both CHD and total mortality risk in women (15).

In younger women, as in men, triglyceride levels are not a statistically independent predictor of coronary risk when HDL and LDL are considered in multivariant analysis (16). This is not the case in older, postmenopausal women, however, in whom triglycerides are an independent predictor of coronary risk (17). This may not be a direct effect of triglycerides themselves. High triglycerides are associated with the appearance of a small, dense, more easily oxidized form of LDL (18). This form of LDL is more prominent in older women (19).

Lipoprotein(a) [Lp(a)] is a relatively new lipoprotein risk factor. Lp(a) is a complex molecule, comprised of one LDL molecule, including a full complement of lipids and one molecule of apoB-100, linked by a disulfide bond to a large protein of variable size, called apo(a). Apo(a) is about 80% homologous to plasminogen. Thus, Lp(a) is a molecule with potential effects on both the lipoprotein and clotting systems (20).

Little is known about the physiological function and metabolism of Lp(a). Apo(a) is synthesized in the liver. Lp(a) can be bound by both LDL receptors and plasminogen receptors (21).

Lp(a) appears to be, in cross-sectional studies, a cardiovascular risk factor in both men and women (22,23). Lp(a) levels are higher in women than in men throughout most of the lifespan, particularly in the postmenopausal period (23).

Most dietary factors or drugs that affect other lipoproteins do not influence

Lp(a) levels (24). *Trans* fatty acids, however, raise Lp(a) levels in both men and women (25). Only one commonly used lipid-lowering agent, niacin, has been shown to lower Lp(a) levels (26). In addition, both estrogen and tamoxifen have been shown to lower Lp(a) levels (27,28). The significance of these findings is uncertain, however, since there is as yet no evidence that changing Lp(a) levels has any effect on changing the risk of CHD.

ENVIRONMENTAL MODIFIERS OF LIPOPROTEINS IN WOMEN

Diet

Because dietary recommendations assume that men and women will respond in an equivalent fashion, there is no difference in recommendations for women and men. This may not be appropriate. Small studies, mostly in premenopausal women, suggest that women experience less lowering of *both* LDL and HDL on a low-cholesterol, low-saturated-fat diet compared to men of the same age (29–31). While these studies involved small numbers of subjects, a large study of 2000 *postmenopausal* women (compared to men of comparable age) demonstrated a greater decline in LDL and triglyceride levels in men than in women and a greater decline in HDL levels in women than in men when all were put on a very low-fat diet (32). Thus, while low-saturated-fat diets are not harmful in women, their effect is less dramatic than in men of the same age.

Weight and Body Fat Distribution

Like those involving dietary compositional change, studies of weight loss suggest that weight loss in women may not necessarily be associated with favorable changes in circulating lipoproteins. Indeed, in one study, women actually lowered HDL with the same degree of weight loss that, in the male subjects in the study, was associated with an increase in HDL levels (33).

It is tempting to speculate that these differences in diet and weight loss responses between genders may be related to the differences in body fat distribution. In men, fat accumulates in the "android" pattern, around the trunk, increasing the waist-to-hip ratio. In most women, however, body fat is more likely to accumulate in the buttocks and thighs, the "gynecoid" pattern. Truncal fat in both men and women is correlated with increases in LDL, decreases in HDL, and an increased risk of coronary disease and total mortality (34,35).

It is likely that in men these adverse changes in CHD risk are in part mediated by increased insulin resistance and hyperinsulinemia. Relationships between truncal fat and these factors are not as well demonstrated in women (36), even though women with a tendency to distribute fat in the male pattern have higher levels of circulating testosterone (37).

The levels of other endogenous estrogens may also be of importance. Adipose tissue in postmenopausal women can serve as a source of estrogen, synthesizing it from androstenedione (38). The benefits of weight reduction in older women, then, may be offset by the loss of estrogen-producing adipose tissue. As a result, expected increases in HDL and declines in LDL may be less prominent than in men. The exact relationships between total body fat, body fat distribution, gender, and endogenous hormones remain to be delineated.

Exercise

Like low-fat diets and weight reduction, exercise, too, appears to be more effective in raising HDL and lowering LDL in men than in women (39). In addition, premenopausal women respond to exercise with lipoprotein changes that are smaller than older, postmenopausal women (40). The relationships between exercise, changes in total body fat and body fat distribution, and changes in endogenous hormone levels are incompletely understood. Taken together with previously noted differences in diet and weight, however, they imply that endogenous estrogen, particularly in younger women, may act as a physiological buffer, tending to maintain lipoprotein levels in the face of changes in diet and body fat. This is not a reason to avoid such hygienic interventions but rather to temper expectations of their effects, particularly in premenopausal women.

Diabetes

Until they reach their sixties and seventies, nondiabetic women have lower rates of CHD than nondiabetic men (Fig. 2) (41). Diabetic men and women, however, have virtually the same rates of coronary disease (Fig. 3). One explanation for this profoundly adverse effect of diabetes in women is that diabetic women have greater and more adverse changes in their lipoprotein levels than do diabetic men. In some studies LDL levels are higher in diabetic women than diabetic men (42). Other studies, however, demonstrate that diabetic women continue to have more favorable lipoprotein levels than diabetic men (43). In any case, the magnitude of the differences is insufficient to explain such a profound effect of diabetes in women. Changes in circulating lipoprotein levels may be part of the explanation for the equalization of coronary risk in diabetic men and women, but they are unlikely to be the whole explanation.

EXOGENOUS GONADAL HORMONES AS CHD RISK FACTORS

The effect of exogenous gonadal hormones in the form of oral contraceptives (OCs) or hormone-replacement therapy (HRT) is dependent on their relative potency as androgens or estrogens. Even though modern, low-dose OCs have somewhat adverse effects on lipoproteins—raising LDL and lowering HDL levels

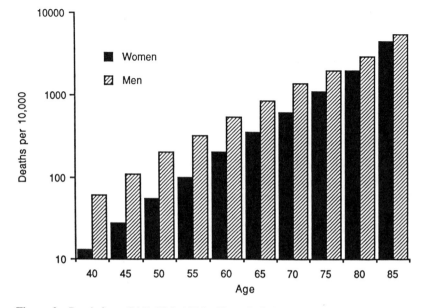

Figure 2   Death from CAD (U.S. 1986). (From Ref. 41.)

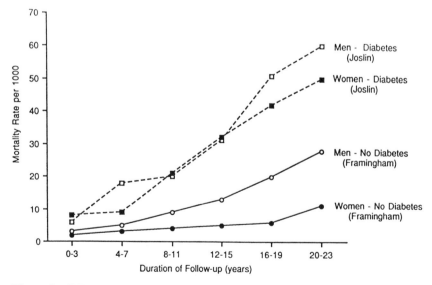

Figure 3   CHD mortality and diabetes. (From Ref. 43.)

(44)—current or past use of OCs, even after 15 years of follow-up, is associated with no increase in cardiovascular disease risk (45).

The situation is somewhat different with postmenopausal HRT. In numerous cohort studies, unopposed estrogen has been associated with dramatically lowered risk of both CHD and all-cause mortality (46). Unopposed estrogen has favorable effects on lipoproteins—lowering LDL and raising HDL. When progestins are added, these effects are somewhat inhibited (47). Even so, in one cohort study, estrogen alone or combined with progestin was associated with a lower risk of nonfatal myocardial infarction (48). How much of these beneficial effects is related to favorable lipoprotein changes is uncertain.

TREATMENT OF LIPOPROTEIN DISORDERS IN WOMEN: EVIDENCE OF BENEFIT

Since lipoprotein metabolism is not identical in women and men, it is not unreasonable to ask whether cholesterol lowering is of the same importance in preventing CHD in women. Coronary disease is the most common cause of death and disability in women as well as men, although coronary disease is generally clinically manifest in women about 5–10 years later than in men (Fig. 2). The issue is often raised that there is insufficient clinical trial data to justify cholesterol-lowering interventions in women, most cholesterol-lowering studies having been done in men. Nevertheless, what data exist indicate that cholesterol lowering is important in both women and men. In the Newcastle (49), Edinburgh (50), and Finnish (51) Mental Hospitals studies, all done in Europe in the late 1960s and early 1970s, both men and women were included and the gender-specific data analyzed separately. A meta-analysis of these three studies (52) demonstrated that equivalent cholesterol lowering in women and men was associated with equivalent reductions in coronary death rates.

In addition, aggressive LDL cholesterol lowering and HDL increasing in severely hypercholesterolemic men and women (53) has been shown to inhibit progression and induce regression of coronary disease in both sexes.

CURRENT GUIDELINES FOR TREATMENT OF LIPOPROTEIN DISORDERS IN WOMEN

Guidelines for the detection, evaluation, and treatment of hypercholesterolemia and related disorders have been developed by the National Cholesterol Education Program (NCEP) (54). These guidelines recommend that LDL cholesterol be the primary target of intervention but that the total and HDL cholesterol be the primary screening parameters. Table 1 lists NCEP definitions of desirable, border-line, and high total and LDL cholesterol levels. These were not considered to be different in women than in men. HDL, however, is not considered a primary target

Table 1   NCEP Cutpoints for Total and LDL
Cholesterol Levels

Level	Total cholesterol	LDL cholesterol
Desirable	<200 mg/dl	<130 mg/dl
Borderline	200–239 mg/dl	130–159 mg/dl
High	>240 mg/dl	>160 mg/dl

Who should receive drug therapy?

   Those with LDL > 190 mg/dl after trial of diet
   Those with LDL > 160 mg/dl after trial of diet,
      if CHD is present or in the presence of two
      other risk factors (women) or one other risk
      factor (men)
   Those with LDL > 130 mg/dl if CHD is present.

for intervention. Individuals with low HDL cholesterol levels (defined in the guidelines as less than 35 mg/dl) are treated as having an additional risk factor.

Triglycerides are not addressed in depth in these guidelines. When triglyceride levels are >200 mg/dl, however, they are recommended as targets for therapeutic intervention, using hygienic means, including diet, weight loss, and exercise. Drugs to lower triglycerides, however, are suggested only if other lipoprotein abnormalities are present, in other words, if LDL cholesterol levels are also high, if HDL cholesterol is low, if triglycerides are very high (>1000 mg/dl), or if there is a personal or family history of CHD or other manifestations of atherosclerosis.

Selection of both male and female subjects for evaluation beyond total cholesterol and HDL cholesterol measurement is limited to those with HDL cholesterol <35 mg/dl, to those with total cholesterol >240 mg/dl, to those whose screening cholesterols are in the 200–239 mg/dl range and who also have at least two other risk factors, or to all cases with established coronary disease. Similarly, subjects with LDL levels >190 mg/dl or between 160 and 190 mg/dl in the presence of other risk factors or >130 mg/dl in the presence of CHD are candidates for drug therapy.

Other risk factors leading to full lipoprotein evaluation include age (males >45, females >55), sex, cigarette smoking, hypertension, diabetes, HDL cholesterol <35 mg/dl, and a family history of coronary disease. However, a high HDL cholesterol (≥60 mg/dl) is regarded as a negative risk factor.

The major difference in applying these guidelines to men and women is in the recognition that men develop CHD at an earlier age and therefore should be considered candidates for more intense interventions at age >45 compared to women, in whom risk increases at age >55. For example, a 46-year-old man with a total cholesterol in the 200–239 mg/dl range who also smokes cigarettes should, according to the guidelines, have an LDL cholesterol measurement. A 46-year-old woman who is a smoker, however, would require one other risk factor before triggering an LDL cholesterol measurement. In similar fashion, a 46-year-old man with an LDL cholesterol between 160 and 190 mg/dl who was a smoker, would become a candidate for drug therapy, while a 46-year-old woman with the same LDL cholesterol would require another risk factor before drugs were considered.

In this way and by designating an HDL cholesterol >60 mg/dl as a negative risk factor, the guidelines attempted to account for the fact that the risk of coronary disease is lower in women than in men through most of the lifespan and that, therefore, drug intervention, in particular, should be more limited in women than in men.

Coronary disease is the most common form of death and disability in women as well as in men. No algorithm, however, can substitute for physician judgment. For example, diabetic women have the *same* risks of CHD as diabetic men. As a result, it may be appropriate to initiate drug therapy in a diabetic woman under age 35.

As noted above, there is reason to expect a lesser response to diet in women than in men. There is no reason, however, to expect that drug therapy will be less effective in women than in men. Indeed, in the regression study involving hypercholesterolemic men and women noted above (53), LDL decreases and HDL increases were virtually identical in women compared to men.

A unique aspect of lipoprotein management in women, however, may be estrogen-replacement therapy (ERT). ERT itself is viewed in the NCEP guidelines as an intermediate therapy, between diet and hypolipidemic drugs, for hypercholesterolemic, postmenopausal women, since estrogen tends to lower LDL and raise HDL levels. These recommendations are based on prospective population studies (46). Clinical trials clearly demonstrating the benefits (and perhaps the drawbacks) of ERT for coronary morbidity and mortality and indicating the most beneficial hormone regimen(s) have not yet been done.

SUMMARY

The female life cycle has important effects on lipoprotein. As girls pass through puberty, they do not experience the same fall in HDL as boys. During pregnancy, LDL, HDL, and triglyceride levels all rise, although only LDL levels stay elevated until well after delivery. In the postmenopause, LDL levels rise sharply while HDL levels decline modestly. All of these effects probably reflect, at least in part, the impact of endogenous, gonadal hormones.

Hormones may also affect lipoprotein responses to other lifestyle changes. Diet composition, weight loss, and exercise all appear to be less effective in producing favorable alterations in circulating lipoprotein levels in women than in men. Pharmacological therapy, on the other hand, is equally effective in both.

Lipoprotein disorders occur in women as well as in men. Moreover, they result in an increase in CHD risk in both sexes. There are differences, however. HDL is a better predictor of risk in women than in men. Triglyceride appears to be a better independent predictor of coronary risk in postmenopausal women than in men of the same age. LDL, in other words, appears to be a less important risk factor in women, perhaps because estrogen itself protects the arterial wall against LDL deposition. Clinical trial evidence of the benefit of cholesterol lowering in women is scarce. Available evidence, however, indicates that alterations in circulating lipoproteins are beneficial in preventing CHD in women as well as men.

Exogenous gonadal hormones in both OCs and postmenopausal HRT are potentially important in modifying lipoprotein levels and, therefore, coronary risk, in women. Available cohort studies indicate that OC use has little effect in coronary risk, whereas HRT lowers such risk.

Current guidelines for detection and treatment of lipoprotein abnormalities set higher thresholds for lipoprotein evaluations in hypercholesterolemic women and higher thresholds for drug treatment for women who are found to have elevations of LDL. They do not, however, exclude women from cholesterol detection, evaluation, or treatment. For women as well as men, treatment of hyper-cholesterolemia is an important step in lowering CHD risk.

REFERENCES

1. Seed M. Sex hormones, lipoproteins, and cardiovascular risk. Atherosclerosis 1991; 90:1–7.
2. Kannel WB. Nutrition and the occurrence and prevention of cardiovascular disease in the elderly. Nutr Rev 1984; 46:68–78.
3. Matthews KA, Meilahn E, Kuller LH, Kelsey SF, Caggiula AW, Wing RR. Menopause and risk factors for coronary heart disease. N Engl J Med 1989; 321:641–646.
4. Rifkind BM, Segal P. Lipid Resource Clinics reference values for hyperlipidemia and hypolipidemia. JAMA 1983; 250:1869–1872.
5. Kim HJ, Kalkhoff RK. Changes in lipoprotein composition during the menstrual cycle. Metabolism 1979; 28:663–668.
6. Ottosson UB, Johansson BG, von Schoulz B. Subfractions of high-density lipoprotein cholesterol during estrogen replacement therapy: a comparison between progestogens and natural progesterone. Am J Obstet Gynecol 1985; 151:746–750.
7. Fahraeus L, Larsson-Cohn U, Walletin L. Plasma lipoproteins including high density lipoprotein subfractions during normal pregnancy. Obstet Gynecol 1985; 66: 468–472.
8. Desoye G, Schweditsch MO, Pfeiffer KP, Zechner R, Kostner GM. Correlations of hormones with lipid and lipoprotein levels during normal pregnancy and postpartum. J Clin Endocrinol Metab 1987; 64:704–712.
9. Ness RB, Harris T, Cobb J, et al. Number of pregnancies and the subsequent risk of cardiovascular disease. N Engl J Med 1993; 328:1528–1533.
10. Eaker ED, Castelli WP. Coronary heart disease and its risk factors among women in

the Framingham Study. In: Eaker ED, Packard B, Wenger N, et al., eds. Coronary Heart Disease in Women. New York: Haymarket Doyma, 1987:122–130.

11. Gordon DJ, Probstfield JL, Garrison RF, et al. High-density lipoprotein cholesterol and cardiovascular disease: four prospective studies. Circulation 1989; 79:8–15.

12. Wagner JD, Clarkson TB, St. Clair RW, Schwenke DC, Adams MR. Estrogen replacement therapy (ERT) and coronary artery (CA) atherogenesis in surgically postmenopausal cynomologus monkeys (abstr). Circulation 1989; 80(suppl II):331.

13. Rifici VA, Khachadurian AK. The inhibition of low-density lipoprotein oxidation by 17-β estradiol. Metabolism 1992; 41:1110–1114.

14. Jacobs D, Blackburn H, Higgins M, et al. Report of the conference on low blood cholesterol: Mortality associations. Circulation 1992; 86:1046–1060.

15. Knapp RG, Sutherland SE, Keil JE, Rust PF, Lackland DT. A comparison of the effects of cholesterol on CHD mortality in black and white women: twenty-eight years of follow-up in the Charleston Heart Study. J Clin Epidemiol 1992; 45:1119–1129.

16. Austin MA. Epidemiologic associations between hypertriglyceridemia and coronary heart disease. Semin Thromb Hemost 1988; 14:137–142.

17. Castelli WP. The triglyceride issue: a view from Framingham. Am Heart J 1986; 112:432–437.

18. de Graaf J, Hak-Lemmers HLM, Hectors MPC, et al. Enhanced susceptibility to in vitro oxidation of the dense low density lipoprotein subfraction in healthy subjects. Arteriosclerosis 1991; 11:298–306.

19. Campos H, McNamara JR, Wilson PWF, Ordovas JM, Schaefer EJ. Differences in low density lipoprotein subfractions and apolipoproteins in premenopausal and postmenopausal women. J Clin Endocrinol Metab 1988; 67:30–35.

20. Scanu AM, Fless GM. Lipoprotein(a): heterogeneity and biological relevance. J Clin Invest 1990; 85:1709–1715.

21. Utermann G. The mysteries of lipoprotein(a). Science 1989; 246:904–910.

22. Dahlen GH. Incidence of Lp(a) lipoprotein among populations. In: Scanu AM, ed. Lipoprotein(a). New York: Academic Press, 1990:151–173.

23. Sandkamp M, Assmann G. Lipoprotein(a) in PROCAM participants and young myocardial infarction survivors. In: Scanu AM, ed. Lipoprotein(a). New York: Academic Press, 1990:205–209.

24. Brewer HB. Effectiveness of diet and drugs in the treatment of patients with elevated Lp(a) levels. In: Scanu AM, ed. Lipoprotein(a). New York: Academic Press, 1990:211–220.

25. Mensink RP, Zock PL, Katan MB, Hornstra G. Effect of dietary cis and trans fatty acids on serum lipoprotein(a) levels in humans. J Lipid Res 1992; 33:1493–1501.

26. Carlson LA, Hamsten A, Asplund A. Pronounced lowering of serum levels of lipoprotein(a) in hyperlipidaemic subjects treated with nicotinic acid. J Int Med 1989; 226:271–276.

27. Kim CJ, Jang HC, Min YK. Hormone replacement therapy lowers the plasma concentration of lipoprotein(a) in postmenopausal women (abstr). Circulation 1992; 86(suppl I):866.

28. Shewmon DA, Stock JL, Heiniluoma KM, Arpano M, Ukena TR, Weale VW.

Tamoxifen lowers Lp(a) in males with heart disease (abstr). Circulation 1992; 86(suppl I):338.

29. Mensink RP, Katan MB. Effect of monounsaturated fatty acids versus complex carbohydrates on high-density lipoproteins in healthy men and women. Lancet 1987; 8525:122–125.

30. Masarei JRL, Rouse IL, Lynch WJ, Robertson K, Vandongen R, Beilin LJ. Effects of a lacto-ovo vegetarian diet on serum concentration of cholesterol, triglyceride, HDL-C, HDL_2-C, HDL_3-C, apoprotein-B, and Lp(a). Am J Clin Nutr 1984; 40:468–478.

31. Ernst N, Bowen P, Fisher M, Schaefer EJ, Levy RI. Changes in plasma lipids and lipoproteins after a modified fat diet. Lancet 1980; 1:111–113.

32. Barnard RF. Effects of life-style modification on serum lipids. Arch Intern Med 1991; 151:1389–1394.

33. Brownell KD. Differential changes in plasma high-density lipoprotein-cholesterol levels in obese men and women during weight reduction. Arch Intern Med 1981; 141: 1142–1146.

34. Soler JT, Folsom AR, Kushi LH, Prineas RJ, Seal US. Association of body fat distribution with plasma lipids, lipoproteins, apolipoproteins AI and B in post-menopausal women. J Clin Epidemiol 1988; 41:1075–1081.

35. Lapidus L, Bengtsson C, Larsson B, Pennert K, Rybo E, Sjostrom L. Distribution of adipose tissue and risk of cardiovascular disease and death: a 12 year follow up of participants in the population study of women in Gothenburg, Sweden. Br Med J 1984; 289:1257–1261.

36. Donahue RP, Orchard TJ, Becker DJ, Kuller LH, Drash AL. Physical activity, insulin sensitivity, and the lipoprotein profile in young adults: The Beaver County study. Am J Epidemiol 1988; 127:95–103.

37. Hauner H, Ditschuneit HH, Pal SB, Moncayo R, Pfeiffer EF. Fat distribution, endocrine and metabolic profile in obese women with and without hirsutism. Metabolism 1988; 37:281–286.

38. Grodin JM, Siiteri PK, MacDonald PC. Source of estrogen production in post-menopausal women. J Clin Endocrinol Metab 1973; 36:207–214.

39. Lokey EA, Tran ZV. Effects of exercise training on serum lipid and lipoprotein concentrations in women: a meta-analysis. Int J Sports Med 1989; 10:424–429.

40. Hartung GH, Moore CE, Mitchell R, Kappus CM. Relationship of menopausal status and exercise levels to HDL cholesterol in women. Exp Aging Res 1984; 10:13–18.

41. Bush TL. The epidemiology of cardiovascular disease in postmenopausal women. Ann NY Acad Sci 1990; 592:263–271.

42. Ikeda T, Terasawa H, Ishimura M, et al. Sex differences in plasma cholesterol and apolipoprotein B levels in non-obese type 2 diabetic subjects. Diabete Metab 1992; 18:465–467.

43. Krolewski AS, Warram JH, Valsania P, Martin BC, Laffel LMB, Christlieb AR. Evolving natural history of coronary artery disease in diabetes mellitus. Am J Med 1991; 90(2A):56S–61S.

44. Notelovitz M, Feldman EB, Gillespy M, Gudat J. Lipid and lipoprotein changes in women taking low-dose, triphasic oral contraceptives: a controlled, comparative, 12-month clinical trial. Am J Obstet Gynecol 1989; 160:1269–1280.

45. Stampfer MJ, Willett WC, Colditz GA, Speizer FE, Hennekens CH. A prospective study of past use of oral contraceptive agents and risk of cardiovascular disease. N Engl J Med 1988; 319:1313–1317.

46. Henderson BE, Paganini-Hill A, Ross RK. Risk factors for coronary artery disease. Arch Intern Med 1991; 151:75–78.

47. Lobo RA. Effects of hormonal replacement on lipids and lipoproteins in postmenopausal women. J Clin Endocrinol Metab 1991; 73:925–930.

48. Falkeborn M, Persson I, Adami H-O, et al. The risk of acute myocardial infarction after oestrogen and oestrogen-progestogen replacement. Br J Obstet Gynaecol 1992; 99:821–828

49. Group of Physicians of the Newcastle upon Tyne Region. Trial of clofibrate in the treatment of ischaemic heart disease. Br Med J 1971; 4:767–775.

50. Research Committee of the Scottish Society of Physicians. Ischaemic heart disease: a secondary prevention trial using clofibrate. Br Med J 1971; 4:775–784.

51. Miettinen M, Karvonen MJ, Turpeinen O, Elosuo R, Paavilainen E. Effect of cholesterol-lowering diet on mortality from coronary heart-disease and other causes. Lancet 1972; 2:835–838.

52. Rossouw JF. International trials (abstr). Cholesterol and Heart Disease in Older Persons and in Women. National Heart, Lung, and Blood Institute, National Institutes of Health, Bethesda, MD, June 18–19, 1990. (Unpublished.)

53. Kane JP, Malloy MJ, Ports TA, Phillips NR, Diehl JC, Havel RJ. Regression of coronary atherosclerosis during treatment of familial hypercholesterolemia with combined drug regimens. JAMA 1990; 264:3007–3012.

54. Summary of the second report of the National Cholesterol Education Program (NCEP) expert panel on detection, evaluation, and treatment of high blood cholesterol in adults (Adult Treatment Panel II). JAMA 1993; 269:3015–3023.

7

An Approach to Cholesterol Levels in Children and Adolescents

Ronald M. Lauer

University of Iowa, Iowa City, Iowa

RATIONALE FOR CHOLESTEROL LOWERING IN CHILDHOOD

There is now compelling evidence that the atherosclerotic process has its origins in childhood and progresses slowly into adult life, at which time it leads to morbidity and mortality. While an important decline in mortality from coronary heart disease (CHD) has occurred during the past two decades, it remains a major cause of morbidity and mortality. Each year about 1.25 million Americans suffer a myocardial infarction, about 300,000 coronary artery bypass operations are carried out, and more than 500,000 deaths from CHD occur. About 20% of hospital discharges for acute CHD occurs in patients younger than 55 years.

A large amount of evidence supports a direct relationship between the level of blood total cholesterol and low-density lipoprotein (LDL) cholesterol levels and the rate of CHD in adults (1–5). In both males and females, within-population studies have shown an association between cholesterol levels and CHD rates (4,6). Population studies comparing populations have shown that countries whose subjects consume diets rich in saturated fats and cholesterol have high levels of blood cholesterol and high rates of CHD (7,8). In addition, studies of migrating populations whose country of origin have low LDL cholesterol levels and low rates of CHD show that within a generation of immigration, both cholesterol levels and CHD rates rise to resemble those of the new country of residence (9,10).

In adult populations clinical trials have demonstrated that lowering LDL cholesterol levels in men with high levels decreases the incidence of CHD. This has been demonstrated both by dietary intervention and with drugs (11–17). In addition, in adults with angiographic evidence of atherosclerotic vascular disease, cholesterol-lowering therapy has been shown to slow the progression and to result in regression of coronary atherosclerosis (18–23).

While diverse studies in adult populations have confirmed a direct relationship of LDL cholesterol levels to CHD and lowering LDL cholesterol levels has an ameliorative effect upon morbidity and mortality, no such studies have been carried out beginning with populations first observed in their childhood years. Thus the significance of cholesterol levels in childhood must be inferred from less direct evidence.

Pathology of Atherosclerosis in Youth

Fatty streaks may occur in the coronary arteries during the second decade of life, and fibrous plaques, the lesions that narrow the arteries, begin to appear in some young persons in the United States as early as the second decade. Fibrous plaques are found in many subjects older than 20 (24). Children with familial hyper-lipidemia (FH) have marked elevations of total and LDL cholesterol and also develop coronary artery disease early in life, thus demonstrating that young arteries are not resistant to the atherogenic effect of high serum cholesterol levels. Indeed myocardial infarction has occurred in subjects with FH before the age of 10 years (25,26).

Relationship of Serum Cholesterol Levels to Arterial Lesions in the Young

Because clinically evident CHD does not occur in children or adolescents (except in the rare homozygous FH subject), the relationship of serum cholesterol levels to clinical disease cannot be easily evaluated in childhood. Serum lipid levels have been related to the extent of arterial lesions of atherosclerosis in children who have died traumatic deaths. Aortic fatty streaks were related to antemortem levels of both total and LDL cholesterol and negatively correlated with the ratio of high-density lipoprotein (HDL) to LDL cholesterol (27–29). A multicenter study, Pathobiological Determinants of Atherosclerosis in Youth (PDAY), has evaluated vascular tissues obtained from young subjects and related them to postmortem blood measurements. Intimal surface involvement with atherosclerotic lesions in both the aorta and right coronary arteries were positively associated with serum very low-density lipoprotein (VLDL) and LDL cholesterol levels and negatively associated with HDL cholesterol levels. Serum thiocyanate levels, a measure of cigarette smoking, was also strongly associated with the prevalence of raised lesions, particularly in the aorta (30). These observations from both premortem and postmortem blood samples indicate the importance of serum lipoprotein cholesterol concentrations in the early stages of atherosclerosis in children and adolescents.

Familial Aggregation of Elevated Lipid and Lipoprotein Levels and Their Relationship to Atherosclerotic Disease

The familial aggregation of lipid, lipoprotein, and apoprotein levels and CHD is another important line of evidence linking the atherosclerotic process to cholesterol levels in childhood. CHD occurs more frequently in the adult members of families in which children's cholesterol levels are elevated (31–35). Among children with high cholesterol levels (>200 mg/dl, approximately the 95th percentile), the prevalence of CHD in adult relatives was significantly increased compared with school children who have normal levels of cholesterol (32,33). In addition, the first-degree relatives of children with high levels of total and LDL cholesterol have higher levels of total and LDL cholesterol than the first-degree relatives of children whose cholesterol levels are not elevated (36). Several studies have shown familial aggregation of total LDL and HDL cholesterol levels in children and parents (36,37). In addition, studies have shown that among school-age children screened because they have a positive family history of premature CHD in a parent or grandparent, 30–50% have some form of dyslipoproteinemia (31,35,38). Thus, children of families in which premature CHD occurs in adult members are at an elevated risk of having abnormal lipids.

International Studies

Total blood cholesterol levels in children vary geographically. In adults the major nutritional determinant of differences in serum cholesterol levels between countries appears to be the proportion of saturated fat in the diet (39–41). Table 1 shows that a similar relationship exists in childhood.

In countries such as the Philippines, Italy, and Ghana, saturated fat constitutes about 10% or less of dietary calories and the serum cholesterol of boys 7–9 years of age is on average less than 160 mg/dl (42–44). In boys from countries such as the United States, the Netherlands, and Finland, the saturated fat intake varies between 13.5 and 17.7% of caloric intake and mean serum cholesterol levels in similar age boys exceed 160 mg/dl (Table 2).

Although blood cholesterol levels are lowest in countries in which nutrition is not optimal and growth is delayed, normal growth occurs in many industrialized countries such as Portugal, Israel, and Italy. Higher serum cholesterol levels in children are associated with higher levels in middle-aged adults in the same country and, with some exceptions, with higher CHD mortality rates in the adult population (42,44–52).

Table 1 Mean Dietary Saturated Fat and Cholesterol Intake and Serum Total Cholesterol in Boys Ages 7–9 in Six Countries

	Dietary intake		Serum total cholesterol (mg/dl)
Country	Saturated fat (% of energy)	Cholesterol (mg/1000 calories)	
Philipines	9.3	97	147
Italy	10.4	159	159
Ghana	10.5	48	128
United States	13.5	151	167
Netherlands	15.1	142	174
Finland	17.7	157	190

Source: Refs. 42–44.

Tracking of Cholesterol Levels

Tracking is a term used to describe consistency of rank order over time. When applied to cholesterol levels in childhood it describes whether cholesterol levels in childhood are good indicators of later childhood and adult levels of cholesterol. Several studies have shown that childhood rank order of cholesterol is maintained over time, although not as well as rank order of height and weight (53–57). Thus, children whose cholesterol levels are observed to be high generally have higher levels as adults; however, many have levels that are not as high as would have been predicted by their childhood levels.

Several studies have related childhood cholesterol levels to later young adult levels (57,58). One study (58) examined data on children 5–18 years of age who were observed to have cholesterol levels greater than the 90th percentile at a single measurement. At 20–30 years of age, 43% of these individuals were observed to have levels greater than the 90th percentile (about four times the percentage expected), 62% greater than the 75th percentile (about two to three times the percentage expected), and 81% greater than the 50th percentile (about one and a half times the percentage expected) (58). Of children whose cholesterol levels were greater than the 90th percentile on two occasions, 75% had higher than desirable levels ($\geq$200 mg/dl), and 25% had desirable levels ($<$200 mg/dl) at age 20–25 years (59). Because 200 mg/dl is approximately the 75th percentile for adults in their twenties, this percentage of individuals with levels at or above 200 mg/dl is about three times the percentage expected for the general population.

According to the criteria established by the U.S. National Cholesterol Education Program (NCEP) Adult Treatment Panel, adults (20 years of age and older) require individual intervention if they have high-risk LDL cholesterol ($\geq$160 mg/dl). Adults with borderline high-risk LDL cholesterol (130–159 mg/dl) may require

Table 2 Serum Total Cholesterol in Boys and in Middle-Age Men and CHD Mortality Rates in Middle-Age Men in Several Industrialized Countries

Country	Serum total cholesterol (mg/dl)[a] Boys	Men	CHD mortality per 100,000 men ages 45–54
Portugal	149	203	71
Israel	155	204	119
Italy	159	200	91
Hungary	159	203	276[b]
United States	167	217	170
Netherlands	171	221	134
Poland	176	192	218[b]
Finland	190	240	264

[a]Cholesterol measurements were standardized to the Abell-Kendall reference method.
[b]High CHD mortality rate in Hungary and Poland are part of an observed pattern of generally high CHD mortality rates in eastern Europe; these may be related to factors other than elevated cholesterol levels.
Sources: Refs. 42, 44–52.

individual intervention if they also have CHD or two or more other CHD risk factors (60). In the tracking study cited above, it is important to note that many children with cholesterol levels greater than the 90th percentile on two successive examinations did not qualify for intervention when they reached their twenties (59). This was because either their adult LDL cholesterol levels were below 130 mg/dl or, despite having LDL cholesterol levels of 130–159 mg/dl, they had fewer than two other CHD risk factors. This indicates that although children with high cholesterol levels have a greater risk of having elevated adult cholesterol levels than the general population, quite a few of these children will have adult levels that do not require intervention.

Other Risk Factors for Coronary Heart Disease

The risk of elevated blood cholesterol levels in adults is compounded by the presence of other risk factors that independently influence the occurrence of CHD and, moreover, tend to aggregate in individuals. Even in children and adolescents, obesity is associated with increased total cholesterol and triglyceride levels and blood pressure, and cigarette smoking is correlated with higher VLDL and LDL and with lower HDL (31,61–63). Family history of cardiovascular disease is associated with high levels of cholesterol in childhood (32–35). Children and adolescents identified as being potentially at risk on the basis of a single risk factor, such as elevated blood cholesterol or high blood pressure, may well have other risk factors such as obesity or smoking that should be addressed.

INTERVENTION STRATEGIES IN CHILDHOOD AND ADOLESCENCE

The Population Approach

Because children in the United States have higher blood cholesterol levels and higher intakes of saturated fats and cholesterol than their counterparts in many other countries and American adults have higher blood cholesterol levels and higher rates of CHD morbidity and mortality than adults in these other countries, a number of advisory groups (64–69) have recommended changes in American children's nutritional habits. In general these various groups are in concert. The Expert Panel on Blood Cholesterol Levels in Children and Adolescents of the National Cholesterol Education Program (70) recommended the following nutritional intake averaged over a period of several days for children beyond the age of 2–3 years:

Saturated fatty acids = less than 10% of total calories
Total fat = no more than 30% of calories
Dietary cholesterol = no more than 300 mg/day

In following these recommended levels, nutritional adequacy should be achieved by eating a wide variety of foods. Energy (calories) should be adequate to support growth and development and to reach or maintain desirable body weight. The reason that infants were excluded from these recommendations relates to their need for more fat in their diets to provide greater caloric density so that sufficient calories can be consumed in small volumes.

The Individual Approach

This approach focuses upon children who appear to be destined to become adults with high blood cholesterol and increased risk for CHD. The Expert Panel on Cholesterol Levels in Children and Adolescents of the NCEP indicated that an LDL cholesterol level of 130 mg/dl or higher (95th percentile), when associated with a parental history of atherosclerotic cardiovascular disease or parental history of hypercholesterolemia, is sufficiently elevated to warrant further evaluation, treatment, and follow-up. This targeted approach was selected because hypercholesterolemia in children whose parents or grandparents have premature CHD are at greater risk because of an increased familial frequency of CHD (31, 33,35,71).

Screening Recommendations

The following selective screening recommendations have been made by the NCEP for children and adolescents (70):

- Screen children and adolescents whose parents or grandparents, at 55 years of age or less, underwent diagnostic coronary arteriography and were found to have coronary atherosclerosis. This includes parents and grandparents who have undergone balloon angioplasty or coronary artery bypass surgery.
- Screen children and adolescents whose parents or grandparents, at 55 years of age or less, suffered a documented myocardial infarction, angina pectoris, peripheral vascular disease, cerebrovascular disease, or sudden cardiac death.
- Screen the offspring of a parent who has been found to have high blood cholesterol (240 mg/dl or higher).
- For children and adolescents whose parental or grandparental history is unobtainable, particularly those with other risk factors, physicians or other health care providers may choose to measure cholesterol levels in order to identify those in need of individual nutritional and medical advice.

A classification of total and LDL cholesterol levels in children and adolescents from families with hypercholesterolemia or premature CHD is shown in Table 3.

What Should Be Measured When

No treatment recommendations have been made for children under the age of 2 years, and thus it has been recommended that blood samples be obtained beyond the age of 2 years. For children who have a parent with high blood cholesterol (>240 mg/dl), the initial screening test should be a measurement of total blood cholesterol levels. This measurement is convenient (it does not require that the subject be fasting), and it is less expensive than a lipoprotein profile. For children with a positive family history of cardiovascular disease the initial screening recommended is a lipoprotein profile because many of these children have

Table 3   Classification of Total and LDL Cholesterol Levels in Children and Adolescents from Families with Hypercholesterolemia or Premature Cardiovascular Disease

Category	Total cholesterol (mg/dl)[a]	LDL cholesterol (mg/dl)[a]
Acceptable	<170	<110
Borderline	170–199	110–129
High	≥200	≥130

[a]To convert cholesterol values in mg/dl to mmol/liter, multiply by 0.02586.

dyslipidemia that requires characterization. Figures 1 and 2 show the algorithms for the screening of children identified in these two ways.

Treatment

   Dietary Intervention.   Table 4 presents the total and LDL cholesterol levels for dietary intervention in children and adolescents recommended by the NCEP Expert Panel (70). The Step 1 diet calls for an average intake of saturated fat of less than 10% of total calories, total fat less than 30% of calories, and cholesterol less

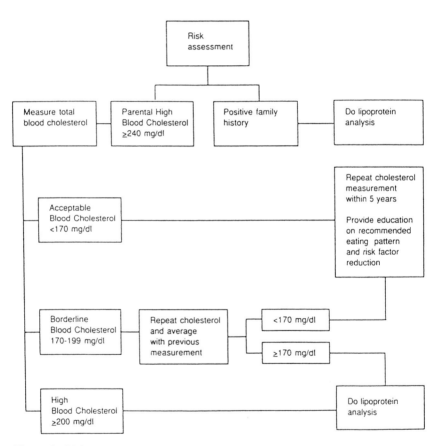

Figure 1   Risk assessment. Positive family history is defined as a history of premature (before age 55 years) cardiovascular disease in a parent or grandparent. (From the National Cholesterol Education Program. Report of the Expert Panel on Blood Cholesterol Levels in Children and Adolescents. U.S. Department of Health and Welfare, PHS, NIH, NHLBI, NCEP. NIH Publication No. 91-2732, September 1991.)

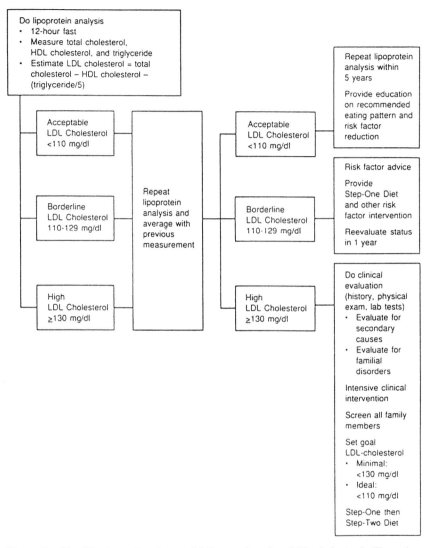

Figure 2 Classification, education, and follow-up based on LDL cholesterol. (From the National Cholesterol Education Program. Report of the Expert Panel on Blood Cholesterol Levels in Children and Adolescents. U.S. Department of Health and Welfare, PHS, NIH, NHLBI, NCEP. NIH Publication No. 91-2732, September 1991.)

Table 4 Cutpoints of Total and LDL Cholesterol for Dietary Intervention in Children and Adolescents with a Family History of Hypercholesterolemia of Premature Cardiovascular Disease

Category	Total cholesterol (mg/dl)	LDL cholesterol (mg/dl)	Dietary intervention
Acceptable	<170	<110	Recommended population eating pattern
Borderline	170–199	110–129	Step 1 diet prescribed, other risk factor intervention
High	⩾200	⩾130	Step 1 diet prescribed, then Step 2 diet if necessary

than 300 mg/day. This diet should be provided in a medical setting by a health professional experienced in nutritional counseling. After 3 months on the Step 1 diet, if the minimal goals of therapy are not achieved, a Step 2 diet should be prescribed. This calls for saturated fat to account less than 7% of calories and cholesterol less than 200 mg per day. The Step 2 diet requires careful planning by a qualified health professional to ensure adequate energy as well as macronutrient and micronutrient intake.

Drug Therapy. Drug therapy should be considered for children over the age of 10 years who have had an adequate trial of diet therapy for 6–12 months. Diet therapy should be continued when drug therapy is instituted. Drug therapy should be considered if (1) LDL cholesterol remains >190 mg/dl (>99th percentile) or (2) LDL cholesterol remains >160 mg/dl (>95th percentile) and the there is a positive family history of premature CVD or the subject has two or more other CVD risk factors (Table 5).

Table 5 Other Risk Factors that May Contribute to Earlier Onset of Coronary Heart Disease

Family history of premature CHD, cerebrovascular, or occlusive peripheral vascular disease (definite onset before the age of 55 years in a sibling, parent, or sibling or a parent)
Cigarette smoking
Elevated blood pressure
Low HDL cholesterol concentration (<35 mg/dl)
Severe obesity (⩾95th percentile weight for height)[a]
Diabetes mellitus
Physical inactivity

[a]This corresponds to ⩾30% overweight.

Table 6 Initial Dosage Schedule for Treatment of
Familial Hypercholesterolemic Children and Adolescents
with a Bile Acid Sequestrant[a]

Daily doses of bile acid sequestrant[b]	Cholesterol levels after diet (mg/dl)	
	Total cholesterol	LDL cholesterol
1	<245	<195
2	245–300	195–235
3	301–345	236–280
4	>345	

[a]These are generally recommended doses and may require adjustment based on the patient's response.
[b]One dose is the equivalent of 9-g packet of cholestyramine (containing 4 g cholestyramine and 5 g filler), one bar of cholestyramine, or 5 g of colestipol.
Source: Ref. 64.

The only drugs that have been recommended for routine use in children and adolescents are bile acid sequestrants because of their apparent safety. Because bile acid sequestrants are not absorbed by the intestine, they lack systemic toxicity and thus are potentially safe in children. The dosage schedule for bile acid sequestrants is shown in Table 6 (72).

REFERENCES

1. Pooling Project Research Group. Relationship of blood pressure, serum cholesterol, smoking habit, relative weight and ECG abnormalities to incidence of major coronary events: final report of the Pooling Project. Am Heart Assoc Monograph 1978;60.
2. Gordon T, Kannel WB, Castelli WP, Dawber TR. Lipoproteins, cardiovascular disease, and death: the Framingham study. Arch Intern Med 1981; 141:1128–1131.
3. Castelli WP, Garrison RJ, Wilson PWF, Abbott RD, Kalousdian S, Kannel WB. Incidence of coronary heart disease and lipoprotein cholesterol levels: the Framingham Study. JAMA 1986; 256:2835–2838.
4. Stamler J, Wentworth D, Neaton JD. Is the relationship between serum cholesterol and risk of premature death from coronary heart disease continuous and graded? Findings in 356,222 primary screenees of the Multiple Risk Factor Intervention Trial (MRFIT). JAMA 1986; 256:2823–2828.
5. Anderson KM, Castelli WP, Levy D. Cholesterol and mortality: 30 years of follow-up from the Framingham Study. JAMA 1987; 257:2176–2180.
6. Jacobs D, Blackburn H, Higgins M, et al. Report of the conference on low blood cholesterol: mortality associations. Circulation 1992; 86:1046–1060.

7. Keys A, Menotti A, Aravanis C, et al. The seven countries study: 2,289 deaths in 15 years. Prev Med 1984; 13:141–154.
8. People's Republic of China-United States Cardiovascular and Cardiopulmonary Epidemiology Research Group. An epidemiological study of cardiovascular and cardiopulmonary disease risk factors in four populations in the People's Republic of China: baseline report from the P.R.C.-U.S.A. Collaborative Study. Circulation 1992; 85:1083–1096.
9. Kagan A, Harris BR, Winkelstein W Jr, et al. Epidemiological studies of coronary heart disease and stroke in Japanese men living in Japan, Hawaii and California: demographic, physical, dietary and biochemical characteristics. J Chron Dis 1974; 27:345–364.
10. Toor M, Katchalsky A, Agmon J, Allalouf D. Atherosclerosis and related factors in immigrants to Israel. Circulation 1960; 22:265–279.
11. Lipid Research Clinics Program. The Lipid Research Clinics Coronary Primary Prevention Trial Results; I: reduction in incidence of coronary heart disease. JAMA 1984; 251:351–364.
12. Lipid Research Clinics Program. The Lipid Research Clinics Coronary Primary Prevention Trial Results; II: the relationship of reduction in incidence of coronary heart disease to cholesterol lowering. JAMA 1984; 251:365–374.
13. Frick MH, Elo MO, Haapa K, et al. Helsinki Heart Study: primary-prevention trial with gemfibrozil in middle-aged men with dyslipidemia. Safety of treatment, changes in risk factors, and incidence of coronary heart disease. N Engl J Med 1987; 317:1237–1245.
14. Manninen V, Tenkanen L, Koskinen P, et al. Joint effects of serum triglyceride and LDL cholesterol and HDL cholesterol concentrations on coronary heart disease risk in the Helsinki Heart Study: implications for treatment. Circulation 1992; 85:37–45.
15. Mann JI, Marr JW. Coronary heart disease prevention: trials of diets to control hyperlipidemia. In: Miller NE, Lewis B, eds. Lipoproteins, Atherosclerosis and Coronary Heart Disease. Amsterdam: Elsevier/North-Holland Biomedical Press, 1981: 197–210.
16. Holm I. An analysis of randomized trials evaluating the effect of cholesterol reduction on total mortality and coronary heart disease incidence. Circulation 1990; 82:1916–1924.
17. Rossouw JE. Clinical trial of lipid-lowering drugs. In: Rifkind BM, ed. Drug Treatment of Hyperlipidemia. New York: Marcel Dekker, 1991:67–88.
18. Buchwald H, Varco RL, Matts JP, et al. Effect of partial ileal bypass surgery on mortality and morbidity from coronary heart disease in patients with hypercholesterolemia: report of the Program on the Surgical Control of Hyperlipidemias (POSCH). N Engl J Med 1990; 323:946–955.
19. Blankenhorn DH, Nessim SA, Johnson RL, Sanmarco ME, Azen SP, Cashin-Hemphill L. Beneficial effects of combined colestipol-niacin therapy on coronary atherosclerosis and coronary venous bypass grafts. JAMA 1987; 257:3233–3240.
20. Brown G, Albers JJ, Fisher LD, et al. Regression of coronary artery disease as a result of intensive lipid-lowering therapy in men with high levels of apolipoprotein B. N Engl J Med 1990; 323:1289–1298.

21. Kane JP, Malloy MJ, Ports TA, Phillips NR, Diehl JC, Havel RJ. Regression of coronary atherosclerosis during treatment of familial hypercholesterolemia with combined drug regimens. JAMA 1990; 264:3007–3012.
22. Ornish D, Brown SE, Scherwitz LW, et al. Can lifestyle changes reverse coronary heart disease: the Lifestyle Heart Trial. Lancet 1990; 336:129–133.
23. Watts GF, Lewis B, Brunt JNH, et al. Effects on coronary artery disease of lipid-lowering diet, or diet plus cholestyramine, in the St. Thomas' Atherosclerosis Regression Study (STARS). Lancet 1992; 339:563–569.
24. Strong JP, McGill HC Jr. The pediatric aspects of atherosclerosis. J Atheroscler Res 1969; 9:251–265.
25. Mabuchi H, Koizumi J, Shimizu M, Takeda R, Hokuriku FH-CHD Study Group. Development of coronary heart disease in familial hypercholesterolemia. Circulation 1989; 79:225–232.
26. Sprecher DL, Schaefer EJ, Kent KM, et al. Cardiovascular features of homozygous familial hypercholesterolemia: analysis of 16 patients. Am J Cardiol 1984; 54:20–30.
27. Newman WP III, Freedom DS, Voors AW, et al. Relation of serum lipoprotein levels and systolic blood pressure to early atherosclerosis. The Bogalusa Heart Study. N Engl J Med 1986; 314:138–144.
28. Freedman DS, Newman WP III, Tracy RE, et al. Black-white differences in aortic fatty streaks in adolescence and early childhood: The Bogalusa Heart Study. Circulation 1988; 77:856–864.
29. Newman WP III, Wattigney W, Berenson GS. Autopsy studies in U.S. children and adolescents. Relationship of risk factors to atherosclerotic lesions. Ann NY Acad Sci 1991; 623:16–25.
30. PDAY Research Group. Relationship of atherosclerosis in young men to serum lipoprotein cholesterol concentrations and smoking. A preliminary report from the Pathobiological Determinants of Atherosclerosis in Youth (PDAY) Research Group. JAMA 1990; 264:3018–3024.
31. Hennekens CH, Jesse MJ, Klein BE, Gourley JE, Blumenthal S. Cholesterol among children of men with myocardial infarction. Pediatrics 1976; 58:211–217.
32. Schrott HG, Clarke WR, Abrahams P, Wiebe DA, Lauer RM. Coronary artery disease mortality in relatives of hypertriglyceridemic school children: The Muscatine Study. Circulation 1982; 65:300–305.
33. Moll PP, Sing CF, Weidman WH, et al. Total cholesterol and lipoproteins in school children: prediction of coronary heart disease in adult relatives. Circulation 1983; 67:127–134.
34. Freedman DS, Srinivasan SR, Shear CL, Franklin FA, Webber LS, Berenson GS. The relation of apolipoproteins A-1 and B in children to parental myocardial infarction. N Engl J Med 1986; 315:721–726.
35. Lee J, Lauer RM, Clarke WR. Lipoproteins in the progeny of young men with coronary artery disease: children with increased risk. Pediatrics 1986; 78:330–337.
36. Morrison JA, Namboodiri K, Green P, Martin J, Glueck CJ. Familial aggregation of lipids and lipoproteins and early identification of dyslipoproteinemia. The Collaborative Lipid Research Clinics Family Study. JAMA 1983; 250:1860–1868.
37. Beaty TH, Self SG, Chase GA, Kwiterovich PO. Assessment of variance component

models on pedigrees using cholesterol, low-density, and high-density lipoprotein measurements. Am J Med Genet 1983; 16:117–129.

38. Glueck CJ, Fallat RW, Tsang R, Buncher CR. Hyperlipidemia in progeny of parents with myocardial infarction before age 50. AJDC 1974; 127:70–75.

39. Blackburn H, Berenson GS, Christakis GS, et al. Conference on the health effects of blood lipids: optional distributions for populations [Workshop report: Epidemiological section]. Prev Med 1979; 8:612–678.

40. Keys A. Coronary heart disease in seven countries. Circulation 1970; 41(suppl I):I1–I211.

41. Knuiman JT, West CE, Katan MB, Hautvast JGAJ. Total cholesterol and high density lipoprotein cholesterol levels in populations differing in fat and carbohydrate intake. Arteriosclerosis 1987; 7:612–619.

42. Knuiman JT, Westenbrink S, van der Heyden L, et al. Determinants of total and high density lipoprotein cholesterol in boys from Finland, the Netherlands, Italy, the Philippines and Ghana with special reference to diet. Hum Nutr Clin Nutr 1983; 37C: 237–254.

43. National Center for Health Statistics, Carroll MD, Abraham S, Dresser CM. Dietary intake source data: United States 1976–80. Vital and Health Statistics. Series 11, No. 231. Hyattsville, MD: U.S. Department of Health and Human Services, Public Health Service, National Center for Health Statistics, March 1983. DHHS Publication (PHS) 83-1681.

44. Total Serum Cholesterol Levels in Children 4–17 Years: United States 1971–74. Data from the National Health Survey. Hyattsville, MD: U.S. Department of Health, Education, and Welfare, 1978. DHEW Publication (PHS) 78-1655.

45. Knuiman JT, Hermus RJJ, Hautvast JGAJ. Serum total and high density lipoprotein (HDL) cholesterol concentrations in rural and urban boys from 16 countries. Atherosclerosis 1980; 36:529–537.

46. Halfon S-T, Rifkind BM, Harlap S, et al. Plasma lipids and lipoproteins in adult Jew of different origins: The Jerusalem Lipid Research Clinic Prevalence Study. Isr J Med Sci 1982; 18:1113–1120.

47. World Health Organization. World Health Statistics Annual, 1986. Geneva: World Health Organization, 1986.

48. World Health Organization. World Health Statistics Annual, 1987. Geneva: World Health Organization, 1987.

49. World Health Organization. World Health Statistics Annual, 1988. Geneva: World Health Organization, 1988.

50. Halfon S-T, Eisenberg S, Tamir D, Stein Y. Risk factors for coronary heart disease among Jerusalem school children: preliminary findings. Prev Med 1983; 12: 421–429.

51. NCHS-NHLBI Collaborative Lipid Group. Trends in serum cholesterol levels among US adults aged 20 to 74 years: data from the National Health and Nutrition Examination Surveys, 1960 to 1980. JAMA 1987; 257:937–942.

52. Knuiman JT, West CE, Burema J. Serum Total and high density lipoprotein cholesterol concentrations and body mass index in adult men from 13 countries. Am J Epidemiol 1982; 116:631–642.

53. Clark DA, Allen MF, Wilson FH. Longitudinal study of serum lipids. 12-year report. Am J Clin Nutr 1967; 20:743–752.

54. Clarke WR, Schrott HG, Leaverton PE, Connor WE, Lauer RM. Tracking of blood lipids and blood pressures in school age children: The Muscatine Study. Circulation 1978; 58:626–634.

55. Frerichs RR, Webber LS, Voors AW, Srinivasan SR, Berenson GS. Cardiovascular disease risk factor variables in children at two successive years: The Bogalusa Heart Study. J Chron Dis 1979; 32:251–262.

56. Laskarzewski PM, Morrison JA, deGroot I, et al. Lipid and lipoprotein tracking in 108 children over a four-year period. Pediatrics 1979; 64:584–591.

57. Freedom DS, Shear CL, Srinivasan SR, Webber LS, Berenson GS. Tracking of serum lipids and lipoproteins in children over an 8-year period: The Bogalusa Heart Study. Prev Med 1985; 14:203–216.

58. Lauer RM, Lee J, Clarke WR. Factors affecting the relationship between childhood and adult cholesterol levels: The Muscatine Study. Pediatrics 1988; 82:309–318.

59. Lauer RM, Clarke WR. Use of cholesterol measurements in childhood for the prediction of adult hypercholesterolemia. The Muscatine Study. JAMA 1990; 264: 3034–3038.

60. National Cholesterol Education Program. Report of the Expert Panel on Detection, Evaluation, and Treatment of High Blood Cholesterol in Adults. Bethesda, MD: U.S. Department of Health and Human Services, Public Health Service. National Institutes of Health, National Heart, Lung, and Blood Institute, January 1989. NIH Publication 89-2925.

61. Lauer RM, Connor WE, Leaverton PE, Reiter MA, Clarke WR. Coronary heart disease risk factors in school children: The Muscatine Study. J Pediatr 1975; 86: 697–706.

62. Glueck CJ, Heiss G, Morrison JA, Khoury P, Moore M. Alcohol intake, cigarette smoking and plasma lipids and lipoproteins in 12- to 19-year-old children. The Collaborative Lipid Research Clinics Prevalence Study. Circulation 1981; 64(suppl III):III-48–III-56.

63. Craig WY, Palomaki GE, Johnson AM, Haddow JE. Cigarette smoking-associated changes in blood lipid and lipoprotein levels in the 8- to 19-year-old age group: a meta-analysis. Pediatrics 1990; 85:155–158.

64. National Cholesterol Education Program. Report of the Expert Panel on Population Strategies for Blood Cholesterol Reduction. Bethesda, MD: U.S. Department of Health and Human Services, Public Health Service, National Institutes of Health, National Heart, Lung, and Blood Institute, November 1990; NIH Publication 90-3046.

65. U.S. Department of Agriculture/Department of Health and Human Services. Nutrition and Your Health: Dietary Guidelines for Americans, 3rd ed. Washington, DC: U.S. Government Printing Office, Home and Garden Bulletin No. 232, 1990.

66. Weidman W, Kwiterovich P Jr, Jesse MJ, Nugent E. Diet in the healthy child. Task Force Committee of the Nutrition Committee and the Cardiovascular Disease in the Young Council of the American Heart Association. Circulation 1983; 67: 1411A–14A.

67. Department of Health and Human Services. Public Health Service. The Surgeon General's Report on Nutrition and Health. Summary and Recommendations. Washington, DC: Public Health Service, DHHS Publication No. (PHS) 88-50211, 1988.
68. National Cancer Institute. Diet, Nutrition & Cancer Prevention: The Good News. Bethesda, MD: Department of Health and Human Services, Public Health Service, National Institutes of Health, NIH Pub. No. 87-2878, December 1986.
69. National Research Council. Diet and Health: Implications for Reducing Chronic Disease Risk. Washington, DC: National Academy Press, 1989.
70. National Cholesterol Education Program. Report of the Expert Panel on Blood Cholesterol Levels in Children and Adolescents. Pediatrics 1992; 89(3):525–574.
71. Schrott HG, Clarke WR, Wiebe DA, Connor WE, Lauer RM. Increased coronary mortality in relatives of hypercholesterolemic school children: The Muscatine Study. Circulation 1979; 59:320–326.
72. Farah JR, Kwiterovich PO Jr, Neill CA. Dose effect relation of cholestyramine in children and young adults with familial hypercholesterolemia. Lancet 1977; I:59–63.

8

Population Strategy

Thomas A. Pearson

Mary Imogene Bassett Research Institute, Cooperstown, and Columbia University, New York, New York

INTRODUCTION

Atherosclerotic cardiovascular disease in general and coronary heart disease (CHD) in particular vary markedly from country to country. Initially, northern Europe, North America, and Australasia were affected, with more recent epidemics of CHD in eastern Europe and China and evidence of rapidly increasing CHD rates in the other parts of the developing world (1). The rates among countries differ over a sixfold range (e.g., Japan, the lowest, versus Northern Ireland and Scotland, the highest) even among industrialized nations (2). Thus, rather than a disease simply of modernization, CHD can aptly be classified as a disease characterizing Western culture. Hypercholesterolemia, in addition to having a genetic component, is also a consequence of that Western culture and serves as a link between CHD and those habits and customs that define that culture.

This chapter will examine one strategy to reduce disease on the national level—the population strategy. Whereas the high-risk strategy and community-intervention strategy are described in other chapters, this discussion will focus on interventions at the national level, including evidence for feasibility of this approach, the components of such a national strategy, and barriers to its implementation.

DEFINITION OF THE POPULATION STRATEGY

If hypercholesterolemia is a characteristic of Western culture, then the implication is that virtually the entire population may have cholesterol levels above the optimum. The population strategy, using public education, taxation, regulation, and other persuasive techniques, seeks to move the distribution of serum cholesterol levels to the left for the entire population (Fig. 1) (3). While the change shown in Figure 1 describes a 10% decrease in serum cholesterol level, it is much less dramatic than the differences between certain populations, such as the Japanese and the Finns (Fig. 2) (4). This emphasizes one of the characteristics of the population strategy, namely, the slow and gradual nature of changes in a culture that eventually may lead, over many decades, to the marked differences seen in Figure 2.

A complex but more realistic scenario is the simultaneous occurrence of the population strategy along with the high-risk strategy (Fig. 3). This would result in a much greater reduction in subjects with high-risk serum cholesterol levels, both through the population shift to the left and the removal of those at the right end of the distribution. Moreover, this combination of strategies would result in fewer subjects in the middle part of the cholesterol distribution. While it is likely that both strategies coexist at least to some extent in the United States and many Western countries, little is known about the interaction between population and

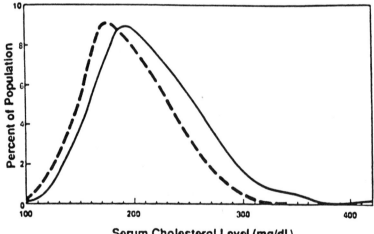

Figure 1   Distribution of serum cholesterol in the U.S. population aged 20–74 years for 1976–1980 showing the expected shift in population distribution if population-based recommendations resulted in a 10% reduction in blood cholesterol levels. Dashed line shows effect of recommendations. (From Ref. 2.)

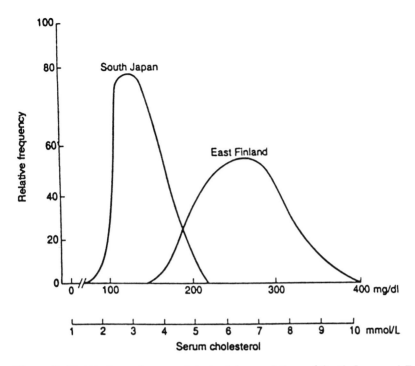

Figure 2 Distributions of serum cholesterol in populations of South Japan and East Finland, showing cultural differences in serum cholesterol distribution. (From Ref. 4.)

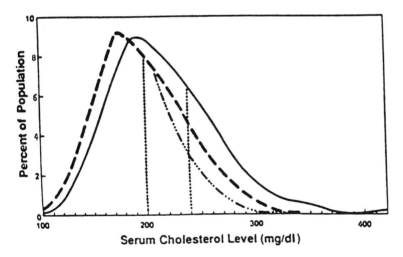

Figure 3 Distribution of serum cholesterol in the U.S. population aged 20–74 years for 1976–1980 showing combined effects of recommendations of the Adult Treatment Panel (high risk approach, dotted-dashed line) and the Population Panel (population strategy, dashed line) of the National Cholesterol Education Program. (From Ref. 2.)

high-risk strategies. One suggestion is that they should be additive; however, it is possible that the combination of strategies is synergistic, resulting in a greater overall effect on the population's cholesterol levels than the sum of each of the two approaches.

THE RATIONALE FOR THE POPULATION STRATEGY

Serum Cholesterol Elevations as a Result of Culture

While a genetic component certainly exists in the determination of an individual's level of serum cholesterol, marked differences in the distribution of serum cholesterol exist between populations. Perhaps the most extreme contrast is that between Japan and Finland (Fig. 2) (3). The high rate of CHD in Finland may be due in part to the proportion of the population with serum cholesterol levels above 5 mmol/liter, just as the low rate of CHD in Japan is due to the small proportion of the Japanese population above that level. (It should be noted that levels of cigarette smoking and hypertension are about equally high in both countries.) In this context, the high-risk strategy loses its meaning, in that the Japanese at highest risk are equal to the Finns at lowest risk. Thus, while the serum cholesterol levels seem to distinguish between these two cultures, the population strategy seeks to alter those characteristics in the Finns that cause their cholesterol levels to be high as a group.

Evidence for the Ability to Change the Rate of CHD in a Culture

The impact of populationwide changes in a culture, especially changes in diet, are apparent following mass disruption in the usual lifestyles of a society through war or natural disaster. A prime example of this is the abrupt reduction in coronary events in Sweden during World War II (Fig. 4) (5). Many countries in northern Europe experienced abrupt reductions in coronary mortality due to deprivations caused by the destruction of their economies, nutritional infrastructures, and health care systems. Neutral Sweden, on the other hand, had only the disruption of sea lanes, depriving it of foreign sources of foodstuffs, particularly cooking fats such as coconut oil (6). An abrupt reduction in coronary events was observed in Sweden, with an equally abrupt restoration of the high coronary rates after 1945. Thus, populationwide changes in dietary patterns obviously can lead to large and rather rapid changes in coronary mortality rates for entire cultures.

Less drastic evidence of a nation's dietary pattern strongly influencing its coronary disease rate is found in cross-sectional and longitudinal analyses of national nutritional data. Nutritional data from men and women of 36 countries showed a strong, direct association between the average annual CHD mortality rate from 1984 to 1987 and the food disappearance data from the Food and Agricultural Organization (1979–1981) (7,8). The correlation coefficients are

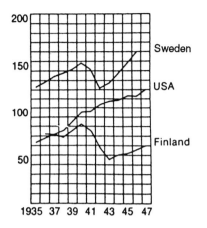

Figure 4 Death rates due to atherosclerotic heart disease during World War II in Sweden, Finland, and the United States. (From Ref. 5.)

particularly large between CHD and animal fat consumption (R = 0.46) and inverse between CHD and vegetable fat consumption (R = −0.40) and cereal consumption (R = −0.38). Moreover, trends in CHD mortality between 1969 and 1975 in 20 countries also correlate with trends in consumption of macronutrients, such as dietary cholesterol, saturated fat, as well as meat, poultry, and egg consumption (9). These data suggest that the alterations of national levels of consumption of saturated fat and cholesterol would in fact lead to changes in mortality rates due to coronary disease.

The Population Strategy Versus the High-Risk Approach

A second part of the rationale for the population approach is derived from the distribution of serum cholesterol levels in Western populations, such as the United States (Fig. 1). While a distinct excess of CHD cases occurs in persons with high cholesterol levels (greater than 240 mg/dl or 6.21 mmol/liter), a much larger proportion of the population have borderline high levels (200–239 mg/dl, 5.17–6.18 mmol/liter). Even though the rate of CHD is lower in this group, approximately 60% of CHD deaths occur in this midportion of the cholesterol distribution (3). While the high-risk approach seeks to identify and treat those persons with cholesterol levels greater than 240 mg/dl, it accomplishes little in terms of reducing the total number of persons with cholesterol levels of 200–239 mg/dl. The population approach, on the other hand, which involves a shift of the population to the left, will allow a reduction in the large number of persons at moderate risk.

The Population Strategy Versus the Community Intervention Approach

Finally, there is a need for a national program, in addition to local community interventions, to reduce dietary cholesterol and saturated fat. While community-based approaches show considerable promise in the mobilization of local populations to effect changes in awareness and behaviors, a program of national scope is also essential. Many influences on eating behavior do not originate at the local level. Mass media, makers and distributors of foodstuffs, retail supermarket chains, and legislation, such as that regarding food labeling, all operate on a national basis. These form powerful determinants of national eating patterns (see below). The sociology of eating within a nation should provide a backdrop for interventions at the community level. Thus, a nationwide program is an important way to support local community interventions, all leading to populationwide changes in nutrition and serum cholesterol levels.

DIETARY CHANGE AS THE CORNERSTONE OF THE POPULATION STRATEGY

Goals for Populationwide Changes in Diet and Serum Cholesterol

The link between dietary saturated fat and/or cholesterol and serum cholesterol has been established by animal studies, cross-sectional studies of individuals, clinical trials, and international correlations. Additional data suggest that dietary fats and cholesterol are related to CHD even after the adjustment for serum cholesterol levels (10). This information has been used by a variety of U.S. organizations, including the National Cholesterol Education Program, the American Heart Association, the Surgeon General, the American Academy of Pediatrics, and the National Research Council to endorse the reduction of dietary fat, saturated fat, and cholesterol as a means to reduce a variety of chronic diseases (3). In general, most (but not all) organizations suggest that populationwide goals for adults should be diets containing less than 10% of calories from saturated fat, less than 30% of calories of fat, and less than 300 mg per day of dietary cholesterol (Table 1). Most organizations also emphasize the achievement and maintenance of ideal body weight. Thus, in the United States at least, there seems to be widespread consensus among governmental and health-related organizations about the reduction of serum cholesterol through a mass modification of the American diet. These goals for fat and saturated fat have been incorporated into the national Health Promotion and Disease Prevention Objectives for the United States, known as Healthy People 2000 (11), which recommends that 30% of calories as total fat and 10% of calories as saturated fat be the maximum average fat intakes for American adults by the year 2000.

The objectives of Healthy People 2000 also seek to reduce the mean blood

Table 1 Current Dietary Recommendations for Consumption of Fat, Saturated Fat, Other Fats, and Cholesterol Compared to Current Consumption

Nutrient	Recommended intake	Estimated current U.S. intake
Total fat	<30% of calories	36–37% of calories
Saturated fat	<10% of calories	13.2% of calories
Other fats	Up to 20% of calories	21–23% of calories
Cholesterol	<300 mg/day	435 mg/day (men)
		304 mg/day (women)

Source: Ref. 3.

cholesterol level for adults to no more than 200 mg/dl (5.17 mmol/liter) by the year 2000. The optimal goal for a population mean serum cholesterol level has been the object of extensive debate and is beyond the scope of the discussion here (12). It should be noted, however, that groups with mean serum cholesterol levels of 180 mg/dl (4.65 mmol/liter) or less have still lower CHD rates than those with serum cholesterol levels of 180–199 mg/dl (4.65–5.17 mmol/liter). This has been the basis for the suggestion that the eventual goal for serum cholesterol levels in the United States is 180 mg/dl (4.65 mmol/liter) or less—similar to those in Japan, which enjoys a low rate of coronary disease. In addition, the Healthy People 2000 objectives seek to reduce the prevalence of adults with serum cholesterol levels greater than 240 mg/dl to less than 20% through both dietary and pharmacological interventions.

Determinants of Populationwide Food Consumption

Numerous factors influence an individual's eating pattern (Fig. 5) (13). The pleasure derived from consuming a food may be affected by taste, culture, and cost. Consumption may be related to the consumer's concern for health, knowledge of the food's nutritional value, and general level of nutrition education. Food availability is related to cost, difficulties in purchasing certain foods in restaurants or preparing them at home, advertising, the food distribution system, and the manufacturer. Strategies to alter food consumption generally target one or more of these determinants of eating behavior. It must be emphasized that the determinants interact in complex ways, including the cancellation or magnification of a strategy's effect through effects of conditions elsewhere in the food chain. For example, early efforts to promote low-fat, low-cholesterol foods were frequently hampered by poor taste, high cost, little advertising, low production, and poor distribution. Thus, strategies must target multiple steps in the food chain, a slow process in order to get all links in the chain producing positive results.

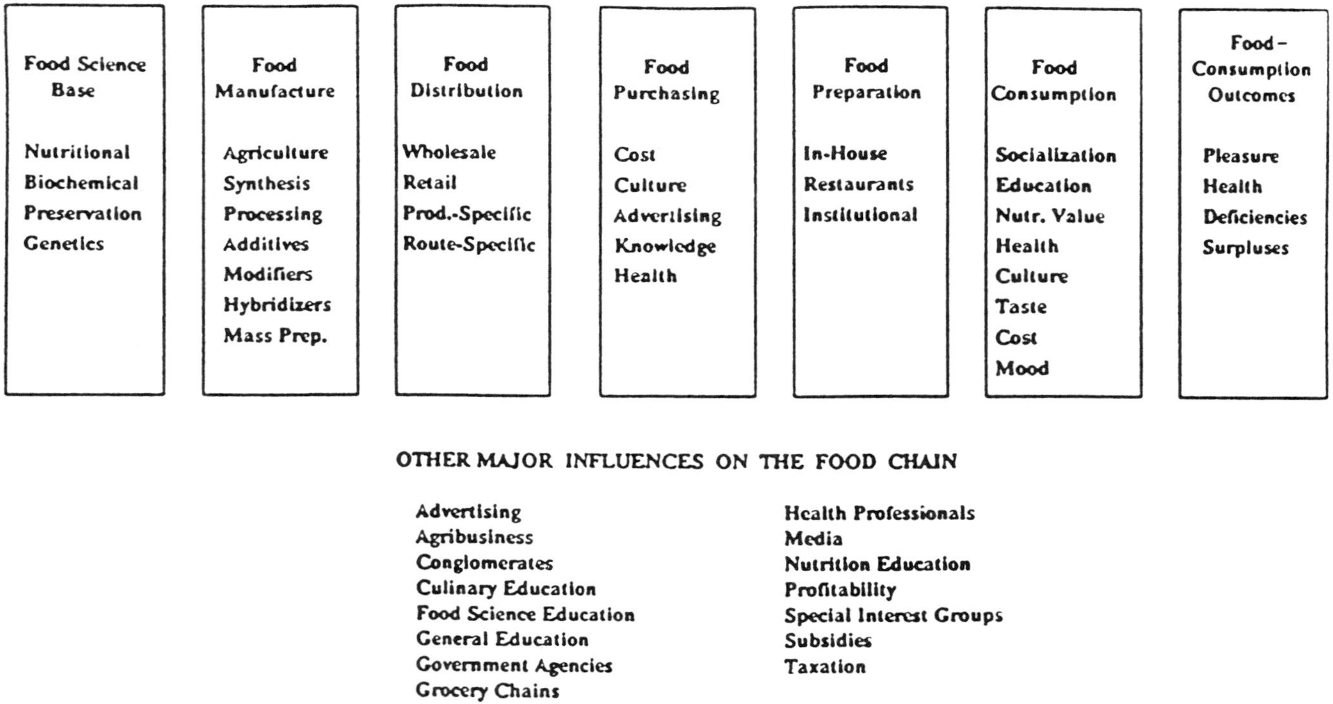

Food Science Base	Food Manufacture	Food Distribution	Food Purchasing	Food Preparation	Food Consumption	Food-Consumption Outcomes
Nutritional	Agriculture	Wholesale	Cost	In-House	Socialization	Pleasure
Biochemical	Synthesis	Retail	Culture	Restaurants	Education	Health
Preservation	Processing	Prod.-Specific	Advertising	Institutional	Nutr. Value	Deficiencies
Genetics	Additives	Route-Specific	Knowledge		Health	Surpluses
	Modifiers		Health		Culture	
	Hybridizers				Taste	
	Mass Prep.				Cost	
					Mood	

OTHER MAJOR INFLUENCES ON THE FOOD CHAIN

Advertising   Health Professionals
Agribusiness   Media
Conglomerates   Nutrition Education
Culinary Education   Profitability
Food Science Education   Special Interest Groups
General Education   Subsidies
Government Agencies   Taxation
Grocery Chains

Figure 5 Major elements in the food chain which determine eating patterns in a population. (From Ref. 13.)

Strategies to Affect Diet on a Populationwide Basis

Enhancing Awareness About the Health Effects of Elevated Blood Cholesterol

Mass Media. One strategy to effectively reduce dietary consumption of saturated fat and cholesterol deals with enhancing the health consciousness of the U.S. population in regard to blood cholesterol. A general public health education strategy deals with the explanation of the role of blood cholesterol in the causation of CHD and the acceptance of blood cholesterol as a major CHD risk factor. Mass media programs of the National Cholesterol Education Program, particularly through television and visual materials in public places (e.g., airports), have sought to improve and maintain awareness of the causes of and ways to reduce blood cholesterol levels, and awareness of the benefits of lowering high levels of blood cholesterol to lower the risk of heart disease (14). The rationale for such a national program was based on the results of community intervention programs, particularly the Stanford Three Community Study (15), which showed that eating behaviors could be improved by using mass media messages, especially those on television. The initial mass media messages emphasized that anyone can be affected by high cholesterol, which usually causes no symptoms, that something can be done to lower high blood cholesterol if it is detected, and that knowledge about one's blood cholesterol is important for someone interested in his or her health (14). Subsequent research suggests the need for follow-up messages dealing with specific dietary issues.

Screening for Blood Cholesterol. A more specific strategy deals with the recommendation that all adults know their serum cholesterol value, as promoted in the National Cholesterol Education Program's "Know Your Cholesterol Number" campaign. Screening for serum cholesterol, either as part of health care programs or by mass screenings at health fairs, is obviously part of the high-risk strategy to identify those with elevated blood cholesterol levels, allowing them to be targeted for intensive interventions. However, recommending screening to health care practitioners and the public for widespread implementation is also an effective means to enhance awareness of blood cholesterol in the general public (3). Since the average blood cholesterol level is in the 205 mg/dl (5.26 mmol/liter) range, over half of those screened may be given the specific message to consider a reduction in dietary saturated fat and cholesterol. In the Minnesota Heart Health Program, for example, a random sample of adults were offered a risk-factor screening and education program, which included a blood cholesterol level (16). At the end of a year, those who participated in the screening had significantly lower blood cholesterol levels than a randomly selected comparison group who were not screened. They also were more likely to select low-fat and low-cholesterol meals in local restaurants. This illustrates the potential role of mass screening as part of the population strategy as well as part of the high-risk approach.

There is a concern about screening for those people found to have desirable serum levels. Although the U.S. goal is to reduce the dietary consumption of fat and cholesterol populationwide, one study demonstrated a reduced inclination to change diets in those told their cholesterol levels were normal (17). Clearly, this demonstrates the need to provide appropriate counseling for all persons screened for blood cholesterol, regardless of cholesterol level.

Nutrition Education of the Public: Advertising and Food Labeling

Two major forces in nutrition education include advertising and food labeling, both of which are coming under increased regulation by the Food and Drug Administration (18). Advertising that is clear and factual seems an obvious means to transfer considerable nutritional information to the consumer as well as to influence purchasing behavior. In general, successful advertising campaigns are those that identify increased benefits (health, cost, etc.) for the consumer and require very brief effort to process the information (19). Since simply providing the information does not mean it will be used, any nutrition information must be highly visible and actively promoted (20). Advertising programs that emphasized the heart-healthy aspects of foods have been successful in boosting sales, but effects have been transient and limited to the period of the active campaign (21,22).

Similarly, new nutrition labeling seeks to show clearly a food's identity, ingredients, nutritional content, and standard portion size in easily understandable terms (18). Both this and better advertising should allow the consumer to avoid being misled in the purchase of foods high in saturated fat and cholesterol and will allow interested consumers to rather carefully regulate their intake of cholesterol-raising macronutrients. Approximately 42–45% of food shoppers report looking for food labels when shopping, with increased numbers of consumers who are concerned about nutrition and who are eating a special diet likely to look for labels (23).

Involvement of the Food Industry

The food industry can play a very positive role in populationwide dietary change by producing, distributing, and marketing products low in saturated fat and cholesterol. Key to the motivation to do this is the profitability of such products. The creation of a demand for new food products low in cholesterol and saturated fats has been effective in motivating the food industry to develop such products. A major challenge has been to develop foods that retain the taste and texture similar to those foods high in fat and cholesterol. Research and development in the creation of fat and sugar substitutes is an illustration of the industry's response to consumer demand.

Altering Practices of Institutions That Serve Food

Institutions with food service functions can play an important role in changing individual behaviors and attitudes with regard to eating. Breakfast and lunch programs in elementary and high schools, college food services, military mess halls, worksite cafeterias, airlines, etc. control some nutritional behaviors of large numbers of people through the choices that they offer. The providing of low-cholesterol, low fat choices and identifying them as such are first steps in a positive direction (24,25). The alteration of boarding school meals to emphasize foods low in cholesterol and saturated fat, for example, has been shown to lower blood cholesterol levels in boys (26).

Controlling the Cost of Healthful Foods

The cost of low-fat, low-cholesterol foods is subject to the laws of supply and demand like any other food. Adequate demand must be matched by adequate supply from the agricultural and food-manufacturing industries. Artificially raising or lowering costs can greatly affect consumption. Price supports of agricultural products may boost prices above those of market conditions. One particular strategy to increase the price of consumer goods is, of course, taxation. Taxation has been effective in reducing the use of tobacco products (27); price elasticities, in which an increase in price due to increased taxes leads to reduced consumption, are especially effective in persons with lower incomes and in the young. Taxation of high-fat, high-cholesterol foods has not been implemented and could be expected to meet considerable opposition.

BARRIERS TO THE SUCCESS OF THE POPULATION STRATEGY

Confusion About the Health Effects of Blood Cholesterol Lowering

Considerable discussion within the academic community has focused on the health effects of cholesterol lowering in general and the presence of low blood cholesterol levels in particular (12). This has led some authors to recommend against cholesterol screening in young adults (28). This could obviously diminish any effect of cholesterol screening on motivations to follow a cholesterol-lowering diet. Additional studies have suggested that reduced fat in the diet may not affect life expectancy, despite a projection of 42,000 fewer CHD deaths per year (29). This discussion, while helpful in the resolution of important scientific issues, is sure to confuse the public and the health professions alike until all issues are resolved and consensus is reached.

Misunderstanding of Food Labels and Health Classes Claims

Considerable confusion on the part of the public has resulted from "low-cholesterol" or "cholesterol-free" labels on high-fat products. Similarly, portion sizes have often been difficult to comprehend. New requirements that any health claim be substantiated, discontinuation of the use of meaningless words like "light," and the newly mandated food labeling should improve the public's ability to select foods beneficial to blood cholesterol levels and general health (18).

Costs of Food: Are Low-Fat Foods Expensive?

The market forces of supply and demand do not always favor access to healthful foods. One example of this can be seen in the economy of Hungary between 1960 and 1985 (Fig. 6). Market conditions were such that tomatoes, peppers, apples, and carrots increased in cost more rapidly than the average income increased, whereas the price of fatty products, cigarettes, spirits, and meat increased little (30). The purchase and consumption of fruits and vegetables can be impeded by imbalances in supply of these low-fat foods, price subsidies to reduce the cost of high-fat foods, or other factors.

Also, there is a perception in the United States that a diet low in saturated fat and cholesterol is an expensive diet, creating a barrier to dietary change. However, a recent examination of the costs of foods selected as part of a regimen to lower blood cholesterol has shown diets low in cholesterol and saturated fat to be *less* expensive (31). Patients who initiated a cholesterol-lowering diet spent $0.75 to $1.10 per day less on food than they did prior to the initiation of the diet.

Mass Media Messages for Foods High in Saturated Fat and Cholesterol

The free market system will continue to provide products high in saturated fat and cholesterol. Commercial messages to sell these products can influence eating behaviors, particularly those of children. Advertising on national television emphasizes food products for children; one study counted 9.6 such messages per hour, with more than half being for beverages and snack foods (32). Furthermore, those messages influence children's food preferences (33) and would therefore compete with messages promoting foods low in cholesterol and saturated fat.

Conflicts Between National Health Policy and National Agricultural Policy

The agriculture and food-manufacturing sectors have vested interests in the production of a wide variety of foods and commodities. National policies to support these industries do not always agree with policies to reduce dietary fat and cholesterol (34). These conflicts need to be resolved to allow employment and revenues in the agricultural sector while still encouraging a demand for production of foods low in cholesterol and saturated fat.

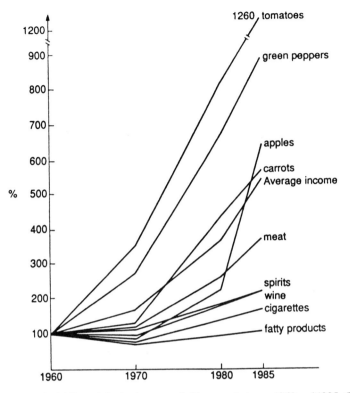

Figure 6   Differential price increases in Hungary between 1960 and 1985. (From Ref. 29.)

SUCCESS OF THE POPULATION STRATEGY TO LOWER BLOOD CHOLESTEROL LEVELS IN THE UNITED STATES

The strategies discussed have been implemented to varying degrees in the United States since the advent of the National Cholesterol Education Program in 1987, and probably well before that. Several lines of evidence suggest that these strategies have been successful.

First, the awareness that dietary saturated fat and dietary cholesterol are important determinants of blood cholesterol levels has increased in surveys of random samples of the American population (35). In fact, knowledge and attitudes supportive of the role of dietary factors in heart disease were more prevalent in the lay public than in physicians in the early 1980s.

Second, the proportion of American adults who have had their blood cholesterol measured has progressively increased since 1988, paralleled by increasing numbers who have been told that their blood cholesterol levels are high (36). Despite this, the number of these under treatment by a physician remains very low—only 11.7%. Thus, the effect of the screening and referral into treatment was rather small. The main effect of this screening may have been to increase awareness of blood cholesterol levels and to motivate dietary change.

Third, additional evidence in fact exists of a progressive change in the American diet toward lower levels of fat and saturated fat (Table 2) (37). Data over the past 65 years from a number of sources suggest a progressive fall in the percent of calories from saturated fat and a rise in the polyunsaturated:saturated fat ratio. American women have nearly reached the Year 2000 goal for dietary cholesterol (Table 1).

Fourth, evidence from several studies documents the fall in total cholesterol in population samples in the United States. Surveys of representative national samples between 1960 and 1991 demonstrate a reduction in the average serum cholesterol in American adults from 220 mg/dl in 1960–1962 to 205 mg/dl in 1988–1989 (38) (Fig. 7). This decline was observed over all age-sex-race groups. Similar findings were observed in the Minnesota Heart Study (39). The prevalence of high-risk cholesterol levels fell from 26 to 20% between 1976–1980 and 1988–1991 (40), consistent with the declining average blood cholesterol levels in the United States.

Finally, most of the decline in blood cholesterol levels in the 1960s, 1970s, and 1980s seems to be attributable to populationwide dietary changes rather than to interventions by physicians with high-risk individuals. While sizable increases in

Table 2 Estimated Average Intake of Fats for Adults in the United States, 1920–1985

Years	Estimated % of calories			P/S[a]
	Total fat	Saturated fat	Polyunsaturated fat	
1920–29	35.5	—	—	—
1930–39	41.2	—	—	—
1940–49	37.6	15.3	2.5	0.16
1950–59	40.5	16.6	4.3	0.26
1960–69	39.9	15.8	3.7	0.24
1970–79	37.8	13.8	5.1	0.37
1980–85	37.5	11.8	5.4	0.46

[a]Polyunsaturated fat–to–saturated fat ratio.
Source: Ref. 34.

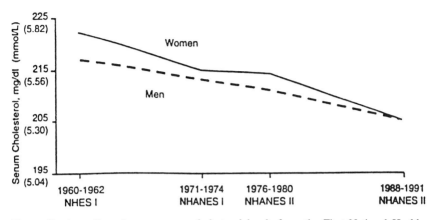

Figure 7 Age-adjusted mean serum cholesterol levels from the First National Health Examination Survey (NHES1), and the National Health and Nutrition Examination Surveys, I–III. (From Ref. 35.)

the use of lipid-lowering medications and other physician-mediated interventions were observed in the mid- to late 1980s, declines prior to that time occurred in lieu of much activity on the part of the medical care system (39,41). Dietary changes observed over these time periods are a more likely explanation for the observed declines in serum cholesterol levels.

Thus, the strong secular trends in increasing awareness of the importance of blood cholesterol, declining consumption of dietary cholesterol and saturated fat, and declining levels of serum cholesterol all provide evidence that the population approach to blood cholesterol lowering is not only feasible, but initially successful.

FUTURE CONSIDERATIONS

Any population strategy to lower blood cholesterol is characterized by slow and cumulative effects of numerous small efforts to provide information, increase awareness, offer healthy dietary alternatives, etc. While the trends in decreasing blood cholesterol levels in Americans have been impressive in magnitude and relative rapidity, there are no guarantees that these trends will continue. None of the current efforts to screen the U.S. population, to reduce the saturated fat and cholesterol in the diet, or to refer high-risk patients for intensive diet and drug therapy is entirely successful. Additional efforts to alter the American food chain, to provide better nutrition education, to limit misleading advertising, to reduce cost, and to increase healthy dietary alternatives are needed to facilitate further reductions in the average blood cholesterol level. The current average blood cholesterol of 205 mg/dl is nearing the interim goal of 200 mg/dl set for the Year 2000 (11). The population strategy, complemented by community interventions and the high-risk approach, should assist greatly in the attainment of this goal.

REFERENCES

1. Pearson TA, Jamison AT, Trejo-Gutierrez J. Cardiovascular disease. In: Jamison DT, Mosley WH, Measham AR, Bobadilla JC, eds. Disease Control Priorities in Developing Countries. Oxford: Oxford University Press, 1993:577–594.
2. Thom TJ. International mortality from heart disease. Rates and trends. Int J Epidemiol 1989; 18(suppl):S20–S28.
3. Carleton RA, Dwyer J, Finberg L, et al. Report of the Expert Panel on Population Strategies for Blood Cholesterol Reduction. Circulation 1991; 83:2154–2232.
4. Rose G. The Strategy of Preventive Medicine. Oxford: Oxford Medical Publications, 1992.
5. Malmros H. At Ratt Fett. Vasteras: ICA Bokforlag, 1979:58.
6. Malmros H. Diet, lipids, and atherosclerosis. Acta Med Scand 1980; 207:145–149.
7. Shaper AG, Marr J. Dietary recommendations for the country towards the postponement of coronary heart disease. Br Med J 1977; 1:867–871.
8. Kesteloot H, Joossens JV. Nutrition and international patterns of disease. In: Marmot M, Elliott P, eds. Coronary Heart Disease Epidemiology. From Aetiology to Public Health. Oxford: Oxford Medical Publications, 1992:152–165.
9. Byington R, Dyer AR, Garside D, et al. Recent trends of major coronary risk factors and CHD mortality in the United States and other industrialized countries. In: Havlik RJ, Feinlieb M, eds. Proceedings of the Conference on the Decline in Coronary Heart Disease Mortality. Washington, DC: U.S. Department of Health, Education, and Welfare, Public Health Service. National Institutes of Health. NIH Publication No. 79-1610, 1979:340–380.
10. Shekelle RB, Shryock AM, Paul O, et al. Diet, serum cholesterol, and death from coronary heart disease. N Engl J Med 1981; 304:65–70.
11. U.S. Department of Health and Human Services. Healthy People 2000—National Health Promotion and Disease Prevention Objectives. Public Health Services. DHHS Publication No. 91-50213, 1990.
12. Jacobs D, Blackburn H, Higgins M, et al. Report of the conference on low blood cholesterol: mortality associations. Circulation 1992; 86:1046–1060.
13. Carleton RA, Lasater TM. Population intervention to reduce coronary heart disease incidence. In: Pearson TA, Criqui M, Luepker R, Oberman A, eds. A Primer of Preventive Cardiology. Dallas: American Heart Association, 1994: (in press).
14. Bellicha T, McGrath J. Mass media approaches to reducing cardiovascular disease risk. Public Health Rep 1990; 105:245–252.
15. Farquhar JW, Maccoby N, Wood PD, et al. Community education for cardiovascular health. Lancet 1977; 1:1192–1195.
16. Murray DM, Luepker RV, Pirie PL, et al. Systematic risk factor screening and

education: A community-wide approach to prevention of coronary heart disease. Prev Med 1986; 15:661–672.

17. Kinlay S, Heller RF. Effectiveness and hazards of case finding for a high cholesterol concentration. Br Med J 1990; 300:1545–1547.

18. Kessler DA. The federal regulation of food labeling. Promoting foods to prevent disease. N Engl J Med 1989; 321:717–725.

19. Russo LE, Leclerc F. Characteristics of successful product information programs. J Social Issues 1991; 47:73–92.

20. Kendall A, Spicer D. Supermarket Nutrition Education in New York State. Ithaca, NY: Cornell University, 1993.

21. Levy AS, Matthews O, Stephenson M, Tenney JE, Schucker RE. The report of a nutrition information program on food purchases. J Public Policy Marketing 1985; 4:1–16.

22. Levy AS, Stokes RC. Efforts of a health promotion advertising campaign on sale of ready-to-eat cereals. Public Health Rep 1987; 102:398–403.

23. Schucker RE, Levy AS, Tenney JE, Matthews O. Nutrition shelf-labeling and consumer purchase behavior. J Nutr Educ 1992; 24:75–81.

24. Mayer JA, Heins JM, Vegel JM, et al. Promoting low-fat entree choices in a public cafeteria. J Appl Behav Analysis 1986; 19:387–402.

25. Zifferblatt SM, Wilbur CS, Pinsky JL. Changing cafeteria eating habits. J Am Diet Assoc 1980; 76:15–20.

26. Ellison RC, et al. The environmental component: changing school food service to promote cardiovascular health. Health Educ Q 1989; 16:285–297.

27. Warner KE. Tobacco taxation as health policy in the Third World. Am J Public Health 1990; 80:529–531.

28. Hulley SB, Newman TB, Grady D, et al. Should we be measuring blood cholesterol levels in young adults? JAMA 1993; 269:1416–1419.

29. Browner WS, Westenhouse J, Tice JA. What if Americans ate less fat? A quantitative estimate of the effect on mortality. JAMA 1991; 265:3285–3291.

30. Poulter N. The coronary heart disease epidemic: British and international trends. In: Poulter N, Sever P, Thom S, eds. Cardiovascular Disease Risk Factors and Intervention. Oxford: Radcliffe Medical Press, 1993:1–11.

31. Shaul J, Pearson TA, Jenkins P, et al. The cost of a cholesterol-lowering diet. Submitted for publication.

32. Story M, Faulkner P. The prime time diet. A content analysis of eating behavior and food messages in television program content and commercials. Am J Public Health 1990; 80:738–740.

33. Taras HC, et al. Television's influence on children's diet and physical activity. J Dev Behav Pediatr 1989; 10:176–180.

34. Jones WPT, Ralph A. National strategies for dietary change. In: Marmot M, Elliott P. Coronary Heart Disease Epidemiology. From Aetiology to Public Health. Oxford: Oxford Medical Publications, 1992:525–540.

35. Schucker B, Wittes JT, Santanello NC, et al. Change in cholesterol awareness and action: results of national physician and public surveys. Arch Intern Med 1991; 151:666–673.

36. Giles WH, Anda RF, Jones DH, et al. Recent trends in the identification and treatment of high blood cholesterol by physicians. Progress and missed opportunities. JAMA 1993; 269:1133–1138.

37. Stephen AM, Wald NJ. Trends in individual consumption of dietary fat in the United States, 1920–1984. Am J Clin Nutr 1990; 52:457–469.

38. Johnson CL, Rifkind BM, Sempos CT, et al. Declining serum total cholesterol levels among US adults. The National Health and Nutrition Examination Surveys. JAMA 1993; 269:3002–3008.

39. Burke GL, Sprafka JM, Folsom AR, et al. Trends in serum cholesterol levels from 1980 to 1987. The Minnesota Heart Survey. N Engl J Med 1991; 324:941–946.

40. Sempos CT, Cleeman JI, Carroll MD, et al. Prevalence of high blood cholesterol among US adults. An update based on guidelines from the Second Report of the National Cholesterol Education Program Adult Treatment Panel. JAMA 1993; 269:3009–3014.

41. Pearson TA. Influences on CHD incidence and case fatality: medical management of risk factors. Int J Epidemiol 1989; 18(suppl 1):S217–S222.

9

The Community Approach

Russell V. Luepker

University of Minnesota, Minneapolis, Minnesota

INTRODUCTION

The epidemic of ischemic coronary heart disease of this century is associated with the mass elevation in blood cholesterol levels. Underlying those increased levels are the widespread availability and consumption of high-fat, particularly animal fat, calorie-dense foods. A central strategy in combating this epidemic thus lies with healthier eating patterns in the population.

In the past decades, we have learned much about factors affecting eating behaviors. They are complex, involving cultural and environmental factors, learned behaviors, hunger and satiety, emotional needs, food availability, and others. These factors are in continuous flux as a national desire for slimness meets increasing weight, decreased exercise levels, and dramatic changes in food technology. For those concerned about the public health, the challenge of encouraging the consumption of a healthy balanced diet has been particularly difficult.

Nonetheless, knowledge is expanding and changes are occurring at several levels. At the macro or population level, national recommendations for healthy food consumption patterns have been made, widely discussed, and distributed. These and other factors have resulted in an increasing number of healthy products becoming available. Increasingly sophisticated labeling is now found on processed foods allowing informed consumers to make healthier choices. The population is more aware of the health effect of various foods, and while the relationship between this knowledge and eating behaviors is not always concordant, the average consumer is both more informed and interested.

At the clinical level, cholesterol screening has resulted in an identification of many individuals with increased blood levels who are actively treated with diet and drug approaches. The detection of hypercholesterolemia has dramatically increased since the release of the National Cholesterol Education Program.

This chapter discusses a third element in a national strategy to control elevated blood cholesterol levels: community strategies. While the broad population approach and the high-risk clinical approach are important, they will be ineffective without local or community-based strategies. The three strategies are complementary and, in our experience, necessary to confront the difficult problem of dietary change to lower population blood cholesterol levels.

This chapter will describe (1) the rationale for cholesterol reduction at the population level with an emphasis on community intervention strategies, (2) theories of eating behaviors and intervention, (3) the current status of community approaches to change, and (4) experience through examples from the community heart health programs.

RATIONALE

Blood cholesterol levels, particularly the low-density lipoprotein fraction, are causal for coronary atherosclerosis. Laboratory experiments, animal models, and epidemiological data consistently confirm this association. Clinical trials have shown that reduction in blood cholesterol levels by diet and/or drugs (1,2) leads to reduced nonfatal and fatal events from coronary heart disease (1,2). More recent studies utilizing quantitative angiography have demonstrated stabilization or event regression of atherosclerotic lesion with diet- or drug-related reduction in blood cholesterol (3,4).

The central role of eating patterns in the control of population levels of blood cholesterol is also well studied. The work of Hegsted et al. and Keys et al. demonstrated the relationship of the type of fat ingested to blood cholesterol levels (5,6). Epidemiological studies such as the Seven Countries Study demonstrate that these relationships described in the laboratory have populationwide effects (7). As shown in Figure 1, population levels of saturated fat intake are strongly associated with mortality from coronary heart disease.

Similarly, research has demonstrated the critical role of culture, food supplies, and environment in the habitual food intake of populations. While genetic factors clearly play an important role in an individual's level of blood cholesterol, populationwide factors, such as culture, are central to the expression of mass elevation in blood cholesterol. The role of culture and environment is best shown in migration studies where individuals move from low- to high-risk cultures where

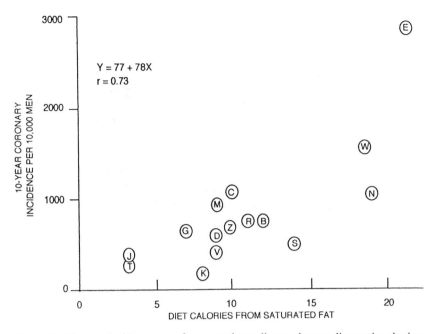

Figure 1 Ten-year incidence rate of coronary heart disease, by any diagnostic criterion, plotted against the percentage of dietary calories supplied by saturated fatty acids. (From Ref. 31.)

they assume the diet, blood cholesterols, and disease patterns of their new homes (8,9).

While national levels of disease and risk are high in most industrialized nations, it is also apparent that change does occur and is common. For example, countries with more affluent lifestyles and greater access to animal products have seen rising disease rates in recent years. This is particularly true of former Eastern Bloc countries but is also occurring in other areas of the world. At the same time, other countries, particularly the United States, have seen a dramatic decline in blood cholesterol levels associated with declining coronary heart disease incidents and mortality. These changes have been associated with a shift in national patterns of food consumption with decreasing use of high-fat animal products and rising vegetable oil use (10–12). Changes in dietary pattern are shown in Figure 2 and national blood cholesterols in Figure 3. Favorable and unfavorable changes can and do occur in disease patterns, blood cholesterol levels, and food intake patterns. These changes, nationally and internationally, suggest that the potential for continued advances is great.

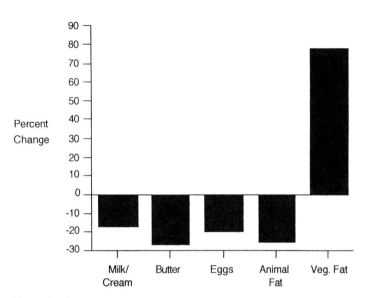

Figure 2 Food-consumption trends in selected items: 1963 and 1985. (From U.S. Department of Agriculture, 1987.)

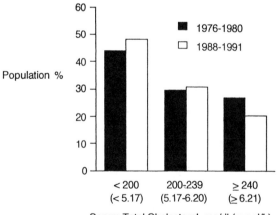

Figure 3 Age-adjusted serum total cholesterol levels of the U.S. population aged 20 through 74 years for 1976 through 1980. (From Ref. 11.)

CHANGING FOOD PATTERNS

While it is clear that food consumption habits have changed for the better in the United States over the last decades, it is also true that change continues, though not always in a favorable direction. Lifestyles in the United States associated with increased disposable income and two-career families have had an effect on eating patterns. For example, in 1985, 3.7 meals per week were eaten outside of the home (13). By 1987, 40% of the food expenditures were spent on foods at restaurants and fast food establishments (14). While these changes represent convenience to the consumer and demonstrate the flexibility of the food service industry in meeting family needs, they also change eating behaviors. Research has shown that people are less likely to be interested in healthy eating patterns when eating restaurant food. In a recent survey, 61% of those studied described improved eating patterns at home, while only 39% reported similar habits when eating out (15). Research in changing restaurant menus has demonstrated that diners are more resistant to healthy food when eating restaurant food.

In keeping with these trends, there has been a dramatic increase in prepared foods sold in supermarkets. The old TV dinner has become increasingly sophisticated and costly. The fat content of most of these prepared foods is driven by cost, availability, and palatability. The amount and type of fat available in these processed items is frequently unclear. At the same time, many items have been developed that are low in fat and cholesterol and so advertised. These healthful foods are widely recognized by the health-conscious consumer and have gained a market share. Unfortunately, they are commonly more costly than standard items and so are less accessible to those of lower income.

Finally, despite all the changes in food consumption, there is a steady increase in the prevalence of obesity among Americans. On the average, both men and women have gained significant weight in the last decade. While the reasons for this are not certain, there are data to suggest that increased use of calorie-dense foods in the setting of decreased physical activity underlies this growing problem (16).

LEVELS OF INTERVENTION

Population strategies to improve eating patterns can occur at at least three levels: the macro or national level, the clinical or individual level, and the community level. Given the complexity and difficulty of changing eating habits, all three are essential and complementary.

At the macro or national level, policy and regulatory strategies are central. The work of the Nutrition Committee of the National Academy of Sciences with its clear recommendations for healthy eating patterns is an excellent example of the setting of national goals and objectives (17). Similarly, in the area of cholesterol policy, the Population Panel of the National Cholesterol Education Program (NCEP) recommends eating pattern goals for all citizens and methods by which these strategies can be implemented (18). These policies provide the basis for national education campaigns and government regulation for the production and subsidization of food. For example, changes in recommendations for the nutrients in the school lunch program have followed these national policies.

At the clinical or individual level, recommendations for mass screening of blood cholesterol have led to identification of high-risk individuals requiring clinical interventions for hypercholesterolemia. The Adult Treatment Panel of the NCEP has defined guidelines for clinical practice in evaluation and treatment (19). At this level, professional education of physicians and other health professionals on the appropriate use of dietary and/or drug therapy is essential. In addition, methods for effective follow-up of individuals found to have hypercholesterolemia is an essential part of the high-risk strategy. An effective clinical approach to hypercholesterolemia is the second element in a populationwide program.

Community-level approaches recognize that national policies and regulations with effective clinical treatment strategies alone will not be enough to result in large populationwide changes. Local or community strategies are necessary and complementary because of the unique nature of eating behaviors and their complexity. Community approaches include the involvement of local activist groups, educational institutions, suppliers of food, and places where prepared food is sold. It must involve local leaders and groups that have an interest in health, food policy, and distribution.

THEORETICAL MODEL

Most of the research on eating patterns has concentrated on the most severe disorders. There has been considerable clinical investigation into anorexia and bulimia. At a broader level, rising levels of obesity in the population coupled with widespread interest in weight loss has led to both clinical and population studies of increased weight. These represent more extreme examples of dysfunctional eating patterns but also provide insight into population patterns.

Eating is complex behavior with many associated and etiological factors. These have been reviewed by Glanz and Mullis (20). First and foremost, eating fulfills a physiological need. This is well understood as guidelines for appropriate food intake for adequate growth, development, and health maintenance are widely distributed. The issue of adequate nutrition in many developing countries and among the poor in industrialized societies continues to be an issue. However, for most individuals living in societies where there is a surplus of calories, vitamins, and minerals, the availability of adequate nutrition to meet physiological needs is not an issue.

Beyond physiological needs, food serves emotional, social, and other needs. In our society, food meets a number of psychological needs. Children learn early in life that good behavior is rewarded with rich and tasty snacks. Adults reward themselves in similar ways. The type of food eaten and enjoyed represents learned behaviors. Early experiences in the setting of family, school, and culture result in food preferences. These frequently favor expensive and calorie-dense foods. People learn to prefer foods that are associated with happy times, affluence, and celebrations in their culture. Food intake is also a social phenomenon. People eating with groups tend to consume more food than they would if eating alone. Group dynamics also reinforce excessive eating and the use of rich, calorie-dense foods.

Finally, in our culture, food processor and commodity groups encourage the consumption of their products. This is done through the mass media, economic incentives, and other strategies. The products encouraged most vigorously tend to be costly processed foods and commodities such as dairy products and meat products. The recent decline in consumption of some of these foods has resulted in even more aggressive marketing of these products.

These and other observations have led us to a model based on social learning theory directing community-level interventions (21). The model considers environmental, personality, and behavioral attributes associated with eating behaviors.

At the environmental level, it is recognized that the physical and social environment is important in food selection and consumption. The current environment does not always support healthy eating patterns. A positive environment would facilitate and encourage healthy eating patterns in a number of ways. It would create opportunities for selecting and consuming healthy foods. It would reduce barriers such as cost and availability to the selection of those foods. It would provide role models to individuals who practice healthy eating behaviors and group or peer support that would encourage these changes. Building an environment for healthy food consumption is essential at the community level.

Personality factors at the individual level are also important. Individuals need adequate knowledge to make appropriate food selections. They need the perception of the importance of making those choices and to have the experience of control over their food intake. These cognitive aspects are obviously modified by individual factors. However, much can be done to better inform and educate the population regarding the importance of healthy eating patterns.

Finally, behavioral factors are important. It is necessary to increase the behavioral repertoire of the population. People need to have the skills to read food labels and appropriately select healthy items. They need the skills to cook foods in a healthy manner in the home. They need the skills to order healthy items in restaurants. They need to have self-monitoring and reinforcement strategies in order to maintain their commitment to healthy eating patterns.

This background and these understandings have led to a series of community-based interventions to analyze and change eating patterns towards healthier lifestyles. While the picture is incomplete, it does provide a framework for the experiences described below.

EXPERIENCES WITH COMMUNITY INTERVENTIONS

Overview

During the past two decades, much has been learned from the community cardiovascular disease-prevention programs in the United States and other industrialized countries. Those experiences have led to a number of generalizable insights into two community-based strategies.

In the Minnesota Heart Health Program, six goals were established to guide these efforts (20):

1. Increased availability of healthy food items in supermarkets and restaurants
2. Health information delivered to consumers at the point of purchase
3. Methods to promote selection of these healthy foods through mass and individual educational programs
4. Strategies to reinforce healthy choices at the site of purchase
5. Consistent messages to community leaders and health professionals
6. Establishment of healthy food choices as normative over and above the "health food" concept

To attain these goals, the community programs initially established focus groups, and later task forces, at the community level to generate ideas and implement programs.

Community leaders in the Pawtucket Heart Health Program and food vendor opinion leaders were engaged to determine the incentives and interventions for marketing more healthful foods (22). Food vendors were found to be sensitive to consumer demands and recognized increasing interest in fish, low-calorie products, and salads. This was particularly true among younger, elderly, and affluent customers. Food vendors looked for help in establishing a market for more healthful products. They hoped to gain publicity from making these products available. Cost was an important factor both in terms of profitability and maintaining low enough prices to be competitive. Finally, they stressed the importance of palatability or taste for healthier foods. This information helped the Pawtucket Program launch their efforts to convince food providers to become part of the program. In the Minnesota Heart Health Program, a similar strategy was developed. Leading restaurateurs, supermarket owners, and cafeteria managers were brought together in a community-based task force. They were given a modest budget for publicity and the services of a public health nutritionist. With these minimal resources, they were able to collaborate to develop healthy eating pattern campaigns, which were supported by food providers including the groups mentioned. This approach led other restaurants, supermarkets, and cafeterias to join in the program and its adoption. The involvement of local leaders from the food industry was an important element in the development of this program.

Chef Training

Early contacts with cafeterias and restaurants led to the observation that many lack sufficient expertise to change their food offerings. The problem resided in the kitchens where chefs were resistant to changes in the food preparation methods they had been taught and found to be successful. One approach to this problem, used in Minnesota, was an intervention that focused on the local training of chefs. This commonly occurred in local vocational technical schools, which were responsible for providing training in these skills. Collaboration with the faculty in these schools led to curricula that emphasized low-fat and low-salt cooking. It ranged from modification of standard recipes to the development of new recipes designed to emphasize healthy items. The issue of palatability and cost was also emphasized. These curricula were developed by study nutritionists who had previous experience in the commercial food preparation setting. The insertion of course materials into the training of new chefs resulted in a generation of capable food preparers who carried their skills to local cafeterias and restaurants.

Restaurants

Restaurants, as described in the background section, are the site of an increasing proportion of our meals. More and more U.S. food dollars are being spent in restaurants (20). People eating in restaurants are less likely to make healthful choices, although this is changing. Restaurant owners are very concerned about palatability and offering choices that consumers will purchase and enjoy. Finally, the restaurant business is one of tight profit margins, so cost is an important consideration. Both the Pawtucket Heart Health Program and the Minnesota Heart Health Program developed considerable experience with restaurant programs designed to provide and promote healthful choices (23,24). After negotiating with restaurant owners and managers, several specific steps were taken. Restaurant menu items and recipes were evaluated and scored based on fat content. Items currently on the menu that met study criteria were designated with a heart symbol and an explanation of the heart's meaning. A second step was the development with restaurant chefs of new recipes that could merit hearting. The wait staff were trained to explain the meaning of the hearts on the menu. Finally, table tents and other promotional information were provided to highlight the program for the consumers. Study nutritionists checked with the restaurants at regular intervals to ensure compliance and score new menu items.

This program, labeled "Dining à la Heart" in Minnesota, proved to be among the most recognized and successful of efforts. Substantial numbers of restaurants in each city were willing and anxious to participate. Low-fat milk, whole grain products, and margarine became readily available in most settings. Surveys of randomly selected adults found that 80% of citizens were aware of the program, and 77% said that the hearting of items led them to select those items or feel good about their choices. In addition, items that received the healthy markers increased in sales during a long-term follow-up (Fig. 4). It is clear that these restaurants had a high visibility and long-term effect in those communities that were part of the "Dining à la Heart."

However, there are also problems and challenges associated with restaurant programs. The high turnover of ownership and staff in the restaurant industry leads to increased training costs and the need for continuous recruitment as new owners must be convinced to join the program. In addition, regular review of the menu is costly and requires skilled personnel. In some areas, restaurants were willing to bear these costs, but in others they were not. Despite the success in increasing the market for healthful items, less healthful ones still sold better when there was direct competition (23). Finally, fast food restaurants and nationally owned chains, where menus are set centrally, were not willing to join the program.

In summary, restaurant programs have considerable potential for facilitating healthy choices among consumers. However, the time and effort needed to initiate and maintain these programs can be considerable and must be included in allocated resources.

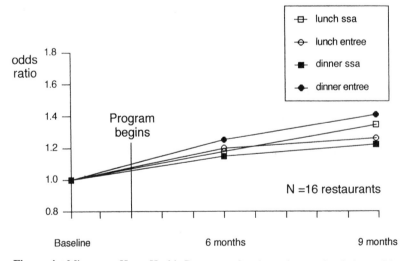

Figure 4 Minnesota Heart Health Program sales data: changes in choices of hearted versus nonhearted menu items for "Dining à la Heart." (Our data, unpublished.)

Cafeterias and Food Services

Development of programs with institutional food services and worksite cafeterias faced challenges similar to those of restaurants. However, many of these institutions were more amenable to change and had significant advantages over restaurants. Food service managers in these settings were more willing to test new items and provide healthful choices. Many agreed to feature one such item daily and added salad bars, low-fat dairy items, and more healthful desserts. Food service managers had the advantage of a regular clientele and an opportunity for systematic feedback on the served items. They were particularly amenable to new recipes and item-specific advice (20). Given the number of institutions that serve food to their workers, this can be a particularly effective way of involving a large population in more healthful eating patterns.

School Cafeteria

School cafeterias are a unique site for the development of healthful-eating programs. School breakfasts and lunches provide a major source of nutrients to the school-aged population and faculty in those institutions. School food programs are regulated by state and federal guidelines requiring specific nutrient content. The actual food served has been driven by the availability of commodity items, which are provided at low cost by the federal government. These commodity items are surplus food purchased by the government. They have emphasized high-fat animal products, which are increasingly shunned in the marketplace.

Working with schools can be particularly challenging because of regulations, the low-cost commodity foods, concern about adequate nutrition among growing children, and resistance by food service workers. Fortunately, many of these characteristics are changing, and work with school food services can be particularly rewarding. Expert panels of nutritionists and pediatricians led by the NCEP and health advocacy groups have presented evidence on the harmful effects of the usual American diet and the benefits of a balanced diet for children (25). The National Academy of Sciences report on foods adds scientific credibility to the balanced nutrition pattern for children (17). The regulatory environment has increasingly recognized recommendations for eating patterns that go beyond adequate calorie, vitamin, and mineral content. The widespread prevalence of obesity among school children and the effect of eating habits on chronic diseases of adulthood is now a matter of national concern. In addition, the food commodity program has become more flexible in offering alternatives to high-fat products, including low-fat dairy products, low-fat meats, fruits, and vegetables.

New school food service leaders are emerging who recognize the need for change. Activated parents in school districts have demanded more healthful choices for their children, putting pressure on school boards and school officials.

In some areas of the country, considerable progress has been made in changing the foods served to children in the schools. However, much remains to be accomplished and there are numerous opportunities within the school system.

Supermarkets

Most food that is consumed is still purchased in supermarkets and prepared at home. Thus, supermarkets are a natural target for community programs to improve eating patterns. There is and will continue to be a wide choice in supermarkets; the task is to highlight and encourage shoppers to buy the more healthful items.

The Minnesota Heart Health Program had several goals in its program, entitled "Shop Smart for Your Heart." They included:

1. Informing consumers about specific healthy foods
2. Promoting the selection of those healthy foods
3. Taste testing in the stores for low-fat, low-sodium foods
4. Encouraging better food-preparation skills
5. Providing a supportive environment in the supermarket for healthful food choices

To that end, several strategies were developed (26). A food shelf-labeling program was initiated which highlighted items that met criteria for low-fat and/or low-salt. These labels were placed on the product shelves and explained with billboards and posters throughout the store. Free samples and cooking demonstrations were performed in the store to promote the more healthful items. Short videotapes describing food-preparation ideas were placed by emphasized items. Finally, coupons distributed through local newspapers and other channels provided cost incentives for the purchase of these items. The store staff was involved in the development of these programs and underwent brief training. Sales tracking programs were developed to monitor the effects. Similar programs were undertaken in the Pawtucket Heart Health Program (27).

The supermarket program generated considerable awareness. In random surveys of adults, 47% were aware of the program and 87% of those stated that it influenced their choices of items. As shown in Figure 5, programs that emphasized certain items through educational messages increased sales of those items. The program was widely interpreted as a success and continues in supermarkets in different forms throughout the nation. The advent of national food labeling on products may institutionalize such programs.

Supermarket programs also had some of the same difficulties that restaurant programs experienced. A continuous monitoring program was needed. Products were changed on the shelves and the labels were not moved. Competing vendors would move their products into the area favorably labeled, although their product did not meet the criteria. New products are continually introduced in the market

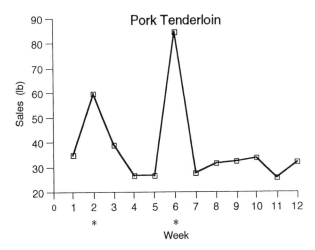

Figure 5 Minnesota Heart Health Program supermarket sales data for educational video presentations in store. *Video shown. (Our data, unpublished.)

and require shelf labeling. Products are reformulated and may gain or fall from approved status. Thus, an ongoing surveillance is necessary to operate these programs. Some supermarkets are willing to support these programs, but others feel financially unable to do so. Thus, supermarket-based programs have potential but are also a challenge to operate.

Education in School

In addition to the school cafeteria programs, direct education in the school of youth and their families can be an important communitywide force for nutritional change. Experience in Minnesota with the Hearty Heart Program was particularly informative. This program was aimed at third graders and involved a school curriculum and a home-based learning program (28). It included all schools in the community, thus engaging many students and parents.

Surveys of students involved in the Hearty Heart curriculum plus the home component demonstrated significantly reduced total fat and saturated fat intakes (28). The curriculum alone did not have this effect, although it did significantly increase knowledge. The combining of the third-grade program with school cafeteria measures such as those tested in the Childhood and Adolescent Trial of Cardiovascular Health may provide an even more potent combination for change (29).

In the later school years, programs have been tested for teaching teenagers nutrition patterns. In Pawtucket, a junior high school held a cookout at which nutrition-conscious purchasing and cooking methods were demonstrated. Association with a cholesterol measurement resulted in changed food habits and other positive outcomes (30). The addition of a contest to the program is believed to have enhanced its effectiveness.

Education Programs for Adults

As part of a communitywide program, education for adults on healthful diets, food selection, and cooking methods would seem to be both reasonable and important. Such programs have been attempted in a number of settings with limited success. They reached a small audience of those highly educated and already informed. Subsequent experience coupling healthful eating patterns with weight-loss programs have been much more successful.

SUMMARY

The populationwide increased blood cholesterol levels observed in the United States and other industrialized countries require populationwide approaches to effectively promote health. Broad programs of regulation and changes in the food supply coupled with clinical identification and treatment of high-risk individuals have important roles to play. However, programs at the community level are an essential complement to these approaches if we are to effectively lower blood cholesterol levels and stem the tide of cardiovascular disease.

ACKNOWLEDGMENTS

The author wishes to particularly acknowledge the work of Rebecca Mullis and Cheryl Perry, who, along with their colleagues and staff, developed the eating pattern programs for the Minnesota Heart Health Program. This work was supported in part by NIH R01 HL 25523, The Minnesota Heart Health Program.

REFERENCES

1. Lipid Research Clinics Program. The Lipid Research Clinics Coronary Primary Prevention Trial results. I. Reduction in incidence of coronary heart disease. JAMA 1984; 251:351–364.
2. Frick MH, Elo O, Haapa K, et al. Helsinki Heart Study: primary-prevention trial with gemfibrozil in middle-aged men with dyslipidemia: safety of treatment, changes in risk factors, and incidence of coronary heart disease. N Engl J Med 1987; 317:1237–1245.
3. Cashin-Hemphill L, Mack WJ, Pogoda JM, Sanmarco ME, Azen SP, Blankenhorn DH. Beneficial effects of colestipol-niacin on coronary atherosclerosis: a 4-year follow-up. JAMA 1990; 264:3013–3017.

4. Kane JP, Malloy MJ, Ports TA, Phillips NR, Diehl JC, Havel RJ. Regression of coronary atherosclerosis during treatment of familial hypercholesterolemia with combined drug regimens. JAMA 1990; 264:3007–3012.

5. Hegsted DM, McGandy RB, Myers ML, Stare FJ. Quantitative effects of dietary fat on serum cholesterol in man. Am J Clin Nutr 1965; 17:281–295.

6. Keys A, Anderson JT, Grande F. Serum cholesterol response to changes in the diet. IV. Particular saturated fatty acids in the diet. Metabolism 1965; 14:776–787.

7. Keys A, Menotti, A, Aravanis C, et al. The Seven Countries Study: 2,289 deaths in 15 years. Prev Med 1984; 13:141–154.

8. Kagan A, Harris BR, Winkelstein W, et al. Epidemiologic studies of coronary heart disease and stroke in Japanese men living in Japan, Hawaii and California: demographic, physical, dietary and biochemical characteristics. J Chron Dis 1974; 27: 345–364.

9. McGee DL, Reed DM, Yano K, Kagan A, Tillotson J. Ten-year incidence of coronary heart disease in the Honolulu Heart Program: relationship to nutrient intake. Am J Epidemiol 1984; 119:667–676.

10. Heart and Stroke Facts. Dallas, TX: American Heart Association, 1993.

11. Sempos CT, Cleeman JI, Carroll MD, et al. Prevalence of high blood cholesterol among US adults: an update based on Guidelines from the Second Report of the National Cholesterol Education Program Adult Treatment Panel. JAMA 1993; 269: 3009–3014.

12. CDC. Daily Dietary Fat and Total Food-Energy Intakes—Third National Health and Nutrition Examination Survey, Phase 1, 1988–91. MMWR 1994; 43:116–125.

13. National Restaurant Association: Meal Consumption Behavior. Washington, DC: NRA, 1985.

14. Sinclair V, ed. The Gallup Annual Report on Eating Out. Princeton, NJ: Gallup Organization, 1988.

15. National Restaurant Association and Gallup Organization. Changes in Consumer Eating Habits. Washington, DC: NRA, 1986.

16. Shah M, Jeffery RV. Is obesity due to overeating and inactivity, or to a defective metabolic rate? A review. Ann Behav Med 1981; 13:73–81.

17. National Research Council. Diet and Health: Implications for Reducing Chronic Disease Risk. Report of The Committee on Diet and Health; Food and Nutrition Board. Washington, DC: National Academy Press, 1989.

18. National Cholesterol Education Program. Report of the Expert Panel on Population Strategies for Blood Cholesterol Reduction. Washington, DC: National Heart, Lung, and Blood Institute, NIH Publication No. 90-3046, 1990.

19. National Cholesterol Education Program. Second Report of the National Cholesterol Education Program Expert Panel on Detection, Evaluation and Treatment of High Blood Cholesterol in Adults. Washington, DC: NIH Publication 93-3095, September 1993.

20. Glanz K, Mullis RM. Environmental interventions to promote healthy eating: a review of models, programs, and evidence. Health Educ Q 1988; 15:394–415.

21. Perry CL, Mullis RM, Maile MC. Modifying the eating behavior of young children. J Sch Health 1985; 55:399–402.

22.   Peterson G, Elder JP, Knisley PM, et al. Developing strategies for food vendor interventions: the first step. J Am Diet Assoc 1986; 86:659–661.
23.   Colby JJ, Elder JP, Peterson G, Knisley PM, Carleton RA. Promoting the selection of healthy food through menu item description in a family-style restaurant. Am J Prev Med 1987; 3:171–177.
24.   Mullis RM. Health promotion opportunities. Common Nutr 1983; 2:11–12.
25.   National Cholesterol Education Program. Report of the Expert Panel on Blood Cholesterol Levels in Children and Adolescents. Washington, DC: National Heart, Lung, and Blood Institute, NIH Publication No. 91-2732, 1991.
26.   Mullis RM, Hunt MK, Foster M, et al. The Shop Smart for Your Heart Grocery Program. J Nutr Educ 1987; 19:225–228.
27.   Hunt MK, Lefebvre RC, Hixson ML, Banspach SW, Assaf AR, Carleton RA. Pawtucket Heart Health Program Point-of-Purchase Nutrition Education Program in Supermarkets. AJPH 1990; 80:730–732.
28.   Perry CL, Luepker RV, Murray DM, et al. Parent involvement with children's health promotion: a one year follow-up of the Minnesota Home Team. Health Educ Q 1989; 16:171–180.
29.   Perry CL, Parcel GS, Stone E, et al. The Child and Adolescent Trial for Cardiovascular Health (CATCH): overview of the intervention program and evaluation methods. Cardiovasc Risk Factors 1992; 2:36–44.
30.   Gans KM, Levin S, Lasater TM, et al. Heart healthy cook-offs in home economics classes: an evaluation with junior high school students. J Sch Health 1990; 60: 99–102.
31.   Keys A. Seven Countries: A Multivariate Analysis of Death and Coronary Heart Disease. Cambridge, MA: Harvard University Press, 1980.

10

The Cholesterol-Lowering Diet

Margo A. Denke and Scott M. Grundy

University of Texas Southwestern Medical Center, Dallas, Texas

Many physicians are familiar with the recommendations for a cholesterol-lowering diet forwarded by the National Cholesterol Education Program (NCEP) (1) and the American Heart Association (2) (Table 1). These diets call for a reduction in total dietary fat intake and progressive reductions in saturated fat and dietary cholesterol as a means to reduce serum cholesterol levels and, therefore, coronary heart disease (CHD) rates. This chapter will review the scientific basis for these recommendations and will then address the 1990s issues of efficacy, cost benefits, and side effects of dietary therapy. Implementation of a cholesterol-lowering diet is detailed in another chapter.

SCIENTIFIC BASIS FOR A CHOLESTEROL-LOWERING DIET

A large body of epidemiological evidence has suggested that dietary intake of the population affects population rates of CHD. Several international epidemiological comparisons have shown that dietary composition correlates with CHD rates among countries (3,4). Moreover, studies of Japanese who moved to Hawaii or San Francisco and adopted the dietary habits of their new surroundings suggest that immigrants experience CHD rates characteristic of their adopted environment rather than their native environment (5). Case-control and longitudinal epidemiological studies have identified factors that are associated with, precede, and increase the probability of an individual developing the disease (6–9). Additional studies suggested that some factors, notably serum cholesterol levels and blood

Table 1 Dietary Therapy for CHD Prevention

Nutrient	Recommended daily intake	
	Step I diet	Step II diet
Total calories	To achieve and maintain desirable weight	
Total fatty acids	Less than 30% of total calories	
Saturated	Less than 10% of total calories	Less than 7% of total calories
Polyunsaturated	Up to 10% of total calories	
Monounsaturated	10–15% of total calories	
Carbohydrates	50–60% of total calories	
Protein	10–20% of total calories	
Cholesterol	Less than 300 mg	Less than 200 mg
Sodium	1650–2400 mg	
Alcohol[a]	Less than 30 g	

[a]13.6 g of alcohol are contained in one 12 oz beer, 5 oz of wine, or 1.5 oz of spirits.

pressure, are associated with differences in population intakes of saturated fat, cholesterol, and sodium (3,10). Several animal models have demonstrated that dietary composition affects blood cholesterol levels, blood pressure, and progression of atherosclerosis (11). Metabolic diet studies in humans have identified that it is the saturated fatty acid and dietary cholesterol content of the diet that increases serum cholesterol levels (12,13). Finally, intervention trials of subjects with hypercholesterolemia have demonstrated that lowering serum cholesterol levels with either diet or drug therapy will reduce rates of CHD (14,15). In sum, these parallel and complementary lines of investigation have clearly established a causal relationship between diet and CHD.

This book focuses on cholesterol lowering as a means to reduce CHD. In this regard, it will be important to review the evidence that a cholesterol-lowering diet per se decreases risk of CHD. Several large studies have been performed to evaluate the effects of a cholesterol-lowering diet on CHD rates. Initial studies employed institutionalized subjects where diet could be maximally controlled and CHD event rates carefully recorded. Later studies employed intensive dietary counseling in free-living, normal individuals. Since these studies were carried out in patients without CHD, they were called primary prevention trials.

Primary Prevention Diet Trials

In one of the first attempts at a diet-heart trial, 846 men in a Veterans Administration facility were randomized to dietary treatment of either a cholesterol-lowering diet or a standard, high-saturated-fat hospital diet (16). Adherence to the diet, defined as meals eaten, was 80%. Both diets resulted in some lowering in total cholesterol: the control diet reduced cholesterol from 234 to 218 mg/dl, whereas the treatment diet resulted in a fall from 234 to 192 mg/dl. No tachyphylaxis was observed in the cholesterol-lowering response to dietary therapy. Of the 422 men on the control diet, 71 had cardiovascular events during the 8-year study period, and of the 424 men consuming the treatment diet there were 54 events. When these differences were combined with the lower rates of stroke observed in the cholesterol-lowering diet group, the total cardiovascular event rate was significantly lower on the cholesterol-lowering diet. The VA Domiciliary Trial suggested that lipid lowering by diet was feasible, but a larger subject number was needed to definitely establish the beneficial effects of dietary therapy on CHD.

In the Finnish Mental Hospital Study (17), 4000 men maintained in two different hospitals were fed a cholesterol-lowering diet and a cholesterol-raising (control) diet for two 6-year periods in a crossover dietary design. The cholesterol-raising diet contained 34% calories from fat with 18% saturated fat and 480 mg of dietary cholesterol/day. The cholesterol-lowering diet contained 34% calories from fat with 9% saturated fat and 282 mg dietary cholesterol/day. In hospital N, where a cholesterol-lowering diet was first employed, men achieved an average serum cholesterol of 216 mg/dl. This rose to 267 mg/dl during the control diet. A less brisk dietary response was observed in hospital K, where the control diet achieved a mean cholesterol level of 268 mg/dl, which fell to 236 mg/dl on the treatment diet. Despite these differences in response, both hospitals achieved differences in age-adjusted death rates from CHD. Hospital N had a 5.1/1000 person-year rate during the treatment diet, which rose to 13.0 on the control diet ($p < 0.002$). Hospital K had a 15.2/1000 person-year rate on the control diet, which fell to 7.5 on the treatment diet ($p < 0.06$). Total mortality was slightly lower in the treatment diet (32 vs. 36/1000 person-years), but this difference did not achieve statistical significance.

The Finnish Mental Hospital Study showed that dietary therapy resulted in significant improvement in cardiovascular death rates from CHD. This improvement in CHD death rates occurred over a relatively short period of time and was reversible if dietary intake reverted to the control diet. However, concern was raised that since dietary therapy did not achieve similar lipid lowering at both hospitals, significant differences in death rates observed could be attributed to other factors besides dietary therapy.

In an even larger trial, the Minnesota Coronary Survey (18), 4393 men and 4664 women were recruited from residents of six mental hospitals and one nursing home. Subjects were randomized to either a control diet (39% fat, 18% saturated fat, 446 mg dietary cholesterol/day) or a treatment diet (38% fat, 9% saturated fat, and 166 mg dietary cholesterol/day). The control diet produced a change in cholesterol from 207 to 203 mg/dl, whereas the treatment diet lowered cholesterol from 207 to 175 mg/dl. This trial was confounded by social change: mental institution residents were encouraged to rejoin society, and the average duration of dietary therapy was only 292 days. The control diet group had 131 cardiac events (MI or sudden death) and the treatment group had 121 events, but this difference did not achieve statistical significance. More impressive differences in event rates were observed in younger subjects on prolonged dietary therapy. Men and women under age 60 on the treatment diet for more than 2 years had 6 events compared to 11 events in a similar-age control group. In summary, the Minnesota Coronary Survey was successful at achieving lipid lowering, but could not confirm the long-term benefits of dietary therapy as suggested by the Finnish Mental Hospital Study.

These three inpatient trials suggested that dietary change could reduce CHD events and supported the notion that a cholesterol-lowering diet should be tested in a free-living population. A one-year pilot project to test the feasibility of such a diet-CHD trial was authorized in the Diet Heart Feasibility Study (19); 1000 men from six centers across the United States were recruited for an outpatient partial metabolic study. Three diets were tested:

Diet B: 30% fat, 7% saturated fat, and 275 mg/day dietary cholesterol
Diet C: 34% fat, 7% saturated fat, but high in polyunsaturated fat and containing 275 mg/day dietary cholesterol
Diet D: a control diet meant to mimic the current intake of participants (37% fat, 12% saturated fat, and 316 mg/day dietary cholesterol)

Specially prepared meats, baked goods, oils, and dairy products were made available to participants. Dieticians worked extensively with the participants and their spouses to integrate these foods into meal planning. Participants, dietitians, and investigators were blinded to dietary assignment. Diet B achieved a cholesterol lowering from 230 to 205 mg/dl, Diet C from 230 to 202 mg/dl, and Diet D from 227 to 220 mg/dl. The same dietary protocol was conducted simultaneously in 192 men in Faribault State School and Hospital; in this setting, Diet B achieved a cholesterol lowering from 208 to 172 mg/dl, Diet C from 207 to 176 mg/dl, and Diet D from 205 to 201 mg/dl. The cholesterol lowering obtained varied significantly between centers and achieved 44–84% of that predicted from diet equations developed from metabolic studies (12,13). Although the Diet-Heart Feasibility Group concluded that a large-scale clinical dietary trial was possible, such a trial has never been undertaken because of considerations of population size and stability, duration of study, and costs. Instead, a more feasible cholesterol-lowering drug study was instituted (14).

The feasibility of a dietary trial among women was similarly examined by the Women's Health Trial (20,21). Three hundred and three women were randomized to a control or a very-low-fat diet. The latter reduced saturated fat intake from 13.7% of calories to 7% and decreased dietary cholesterol intake from 319 to 146 mg/day. After 24 months of dietary modification, the control women had total cholesterol levels averaging 223 mg/dl, whereas the women on the low-fat diet had a total cholesterol of 211 mg/dl. Based on 4-day diet records, these women achieved the cholesterol predicted by the equation of Keys et al. (22). However, because of similar considerations given the Diet-Heart Feasibility Study, a full-scale women's health trial was not conducted.

These relatively large trials of diet in primary prevention have clearly identified that dietary therapy is beneficial for cholesterol lowering and subsequent CHD rates. The effectiveness of dietary therapy on lowering CHD rates compares favorably with primary prevention drug trials: benefits of either type of intervention are directly related to the cholesterol lowering achieved (23). A complementary perspective on the role of diet in CHD can be made by reviewing the effects of dietary interventions in a population at greater risk for CHD events: patients with established CHD in need of secondary prevention.

Secondary Prevention Diet Trials

In the Oslo Diet-Heart Study (24), 412 men who had suffered a myocardial infarction were randomized to control and cholesterol-lowering diets. Patients receiving the latter were counseled to consume 39% of calories from fat with 9% saturated fat. The control group continued on its current diet, which was relatively high in saturated fat. Serum cholesterol levels fell from 296 to 285 mg/dl in the control group and from 296 to 244 mg/dl in the intervention group. After 5 years, there was a statistically significant reduction in reinfarction rates and new angina in the intervention group compared to the control group.

More recent secondary prevention trials have evaluated the changes in lesion size assessed by angiography. The first trial of this kind, the Leiden trial (25), tested a cholesterol-lowering vegetarian diet in 35 men and 4 women. The study design did not include a control group or a control diet. The reduction in calories from saturated fatty acid was from 11.6 to 6.6% and in dietary cholesterol was from 177 to 59 mg/day. Total cholesterol fell from 267 to 240 mg/dl and HDL remained essentially unchanged at 39 mg/dl. Twenty-one of 39 patients had angiographic progression of their disease, while 18 showed no lesion growth. The lack of lesion growth was observed in patients who had lower values for total/HDL cholesterol ratio and for those whose initially high total/HDL cholesterol ratio was significantly lowered by diet. The Leiden Trial suggested that dietary therapy, acting through changes in cholesterol and lipoprotein levels, could halt CHD lesion progression.

In the Lifestyle Heart Trial (26), the efficacy of significant lifestyle changes including a very-low-fat cholesterol-lowering vegetarian diet was compared to a usual care group in men and women. The control group had initial total cholesterol levels of 245 mg/dl, LDL cholesterol of 167 mg/dl, and HDL cholesterol of 52 mg/dl. These values fell to 232, 158, and 51 mg/dl, respectively, at 12 months. The treatment group had initial total cholesterol levels of 228 mg/dl, LDL cholesterol levels of 152 mg/dl, and HDL cholesterol levels of 39 mg/dl, which fell to 172, 95, and 38 mg/dl, respectively, during the intervention period. Subjects in the Lifestyle treatment group showed a reduction in lesion size during the trial, while control diet subjects showed progression in lesion size. The Lifestyle Heart Trial, however, was not a definitive endorsement for dietary therapy in secondary prevention of CHD because additional lifestyle changes besides diet were made (e.g., meditation, weight loss, stress reduction, smoking cessation) and patients were not randomized to either control or lifestyle group.

The Stars Trial (27) is a particularly noteworthy study on the role of diet in secondary prevention. Ninety men were randomized to usual care, dietary treatment alone, or diet plus cholestyramine. Dietary treatment consisted of 27% calories from fat, 8–10% saturated fat, and 100 mg cholesterol/1000 calories. Over the course of the trial the usual care group experienced a reduction in LDL cholesterol levels from 187 to 181 mg/dl, the cholesterol-lowering diet only group from 194 to 162 mg/dl, and the diet plus cholestyramine group from 204 to 130 mg/dl. Whereas 46% of men in the usual care group showed progression, progression was observed in only 15% of men in the diet group and 12% of men in the diet plus cholestyramine group. The Stars Trial clearly demonstrated that a cholesterol-lowering diet is beneficial for reducing the progression of CHD in patients with established CHD.

Summary

Several primary prevention trials in institutionalized men have shown that a cholesterol-lowering diet can reduce CHD event rates and perhaps CHD mortality over a 4- to 8-year period. This diet-heart connection has not been further tested in a large-scale trial of free-living subjects because of the expense, subject number, and duration of treatment needed to demonstrate significant differences from a control group. On the other hand, clear benefits from a cholesterol-lowering diet have been observed in trials of patients with established heart disease. It is likely that much of the beneficial effect of diet on CHD is mediated by diet's effect on lipid levels. The scientific evidence supporting the role of specific nutrient changes in a cholesterol-lowering diet will now be considered.

COMPONENTS IN A CHOLESTEROL-LOWERING DIET: RATIONALE

Total Fat

The Step I and Step II diet recommendations of the American Heart Association and the NCEP state that total fat should be 30% or less of total calories. The current American diet has 34–36% of total calories from fat (28–30). A reduction in total fat intake can be achieved in either of two ways. First, carbohydrates can be substituted isocalorically for fat, especially saturated fat; this exchange will promote lowering of LDL cholesterol levels (31–35). Alternatively, in overweight individuals, fats high in saturated fatty acids can be removed from the diet without caloric replacement. This change has two benefits: a reduction in saturated fat will lower LDL cholesterol levels and a decrease in total calories will promote weight reduction, which in the long term will increase HDL cholesterol levels.

Greater decreases in fat intake are favored by some authorities to maximize reduction in saturated fatty acids. When this approach is employed, care must be taken to maintain adequate intakes of iron and calcium. A very-low-fat diet will also lower HDL cholesterol levels if fat intake is reduced to less than 25% of calories. On the other hand, alternative diets containing 30–35% of total calories as fat also can be used in patients who are unable to adopt lower fat intakes. In these diets, saturated fats should be replaced principally by unsaturated fats (36); such diets have been traditional in certain Mediterranean countries where rates of CHD are low (37).

Saturated Fat

Three common saturated fatty acids having carbon chain lengths of 12 (lauric acid) (38), 14 (myristic acid) (13), and 16 (palmitic acid) (13,39–41) have cholesterol-raising properties. The predominant effect of saturated fat is to raise LDL cholesterol levels. Saturated fatty acids have little effect on HDL or triglycerides. In the American diet, animal fats provide about two thirds of the saturated fat (43).

Hydrogenation is a fat-processing technique that converts a vegetable oil to a more solid fat. This process increases the saturated fat content, but the saturated fat produced is stearic acid (carbon chain length 18), which does not raise LDL cholesterol levels (13,41,44).

Monounsaturated Fat

Cis-*Monounsaturated Fatty Acids*

In both types of therapeutic diet, monounsaturated fat can comprise up to 15% of total calories. Oleic acid is the major monounsaturated fatty acid in the diet and is present in most animal fats and vegetable oils. Oleic acid is the major fatty acid in olive oil, canola oil, and high-oleic forms of sunflower seed oil and safflower oil. For many years, oleic acid was considered to be "neutral" in its effect on serum cholesterol, neither raising nor lowering the cholesterol level (12,13). Recent evidence, however, suggests that both oleic acid and linoleic acid, substituted for saturated fat in the diet, result in LDL cholesterol reductions (40,45–50). The current American diet already contains 14–16% of calories as monounsaturated fatty acids. Much of it, however, comes from foods containing animal fats, which also are rich in saturated fat. When animal fats are decreased as part of the Step I and Step II diets, a larger portion of monounsaturated fat can come from vegetable oils. Although these oils are low in saturated fat, they still have a high caloric density and, if used in excess, can cause weight gain.

Trans-*Monounsaturated Fatty Acids*

Trans-monounsaturated fats are not specified in the current guidelines for a cholesterol-lowering diet, but there is recent evidence that *trans*-monounsaturated fatty acids contribute to the cholesterol-raising potential of the diet. *Trans* fatty acids occur naturally in some meat and dairy fats and are also a by-product of hydrogenation, where some polyunsaturated fatty acids can be converted to either a *cis* monounsaturated fat (e.g., oleic acid) or a *trans* fatty acid (e.g., elaidic acid, the optical isomer of oleic acid). Recent research indicates that *trans* fatty acids raise LDL cholesterol levels nearly as much as do cholesterol-raising fatty acids (51–55). *Trans* fatty acids account for about 3% of total calories in the American diet (56); this amount causes a definite increase in LDL cholesterol levels, but of course contributes less to raising cholesterol levels than the more abundant cholesterol-raising saturated fatty acids. Improvements in food technology in the future may reduce the *trans* fatty acid content of the American diet. In the meantime, patients with high cholesterol levels should limit their intake of foods high in *trans* fatty acids such as shortenings, margarines, and foods containing these fats. Parenthetically, tub or liquid margarine has a lower *trans* fatty acid content than the hard margarine, and even hard margarine is less cholesterol raising than butter. Soft margarine remains a much better choice as a spread than is butter (56), and liquid oil can be used in lieu of margarine for pan frying.

Polyunsaturated Fat

Omega-6 Polyunsaturated Fatty Acids

The major polyunsaturated fatty acids in the diet are of the omega-6 variety. The most common omega-6 polyunsaturated fatty acid is linoleic acid. Substitution of foods rich in linoleic acid for those high in saturated fats results in a decrease in LDL cholesterol levels (12,13,46). Several vegetable oils are rich in linoleic acid, including soybean oil, corn oil, and high-linoleic forms of safflower and sunflower seed oils. High intakes of linoleic acid were once advocated for cholesterol-lowering diets; however, concern about the possible adverse consequences of long-term ingestion of large amounts of linoleic acid has led to the recommendation that the average intake of polyunsaturated fats remain at the present 7% and that intake by an individual not exceed 10% of total calories (57).

Omega-3 Polyunsaturated Fatty Acids

Fish is the major source of omega-3 fatty acids in the U.S. diet. The predominant fatty acids in this class are eicosapentaenoic acid (EPA) and docosahexaenoic acid (DHA). They have minor effects on LDL cholesterol levels in patients with normal triglyceride levels. At high intakes, omega-3 fatty acids will reduce serum triglycerides substantially (58–61). In patients with high triglycerides, a fall in triglycerides may be accompanied by an increase in LDL cholesterol levels (62–64). Although some investigators suggest that omega-3 fatty acids will lower risk for CHD (65), this possibility has not been clearly established (66). Furthermore, whether long-term ingestion of large amounts of omega-3 fatty acids has undesirable effects is not known.

Consumption of omega-3 fatty acids in supplements should be distinguished from ingestion of fish. Whereas some ocean fish are rich in omega-3 acids, most other types of fish are low in fat and have little omega-3 fatty acids. Some epidemiological data (67) suggest that consumption of fish of any type, seemingly independent of omega-3 fatty acids, is associated with reduced CHD risk; whether this apparent connection is due to fish per se or to other factors is uncertain. In any case, since fish are low in saturated fat, they are a good source of dietary protein.

Cholesterol

The Step I diet calls for an average intake of dietary cholesterol of less than 300 mg/day. A further restriction to less than 200 mg/day is recommended with the Step II diet. Dietary cholesterol causes marked hypercholesterolemia and atherosclerosis in many laboratory animals, including nonhuman primates (11,68–70). Although high intakes of cholesterol in rabbits and some primates cause striking rises in serum cholesterol levels, controlled metabolic studies in humans indicate a more modest affect (71–76). The degree of rise, however, varies from person to person (77,78). There is some evidence that dietary cholesterol may augment the serum cholesterol–raising action of saturated fat (19,79–81). Epidemiological studies (82,83) further suggest that dietary cholesterol increases the risk for CHD beyond its serum cholesterol–raising effect. Mechanisms for this latter effect are unknown.

Protein

Protein intake in both Step I and Step II diets is approximately 15% of calories. In some laboratory animals (84), soy protein lowers cholesterol levels relative to animal protein; however, this effect has not been reproduced in humans at usual levels of protein intake (85–87). Thus, foods supplying any type of protein can be used in the therapeutic diets as long as the foods are not high in saturated fats. Some experts contend that unusually high intakes of protein consumed over many years may not be healthy and may contribute to the rate of decline in renal function observed in some conditions such as diabetes (88). For this reason, protein intake should be maintained as close to the recommended level as possible and not increased excessively when fat calories are reduced.

Carbohydrates

Carbohydrates should provide 55% or more of calories in both the Step I and Step II diets. These calories come from simple sugars (monosaccharides and disaccharides) and complex digestible carbohydrates (starches). In most people, when carbohydrates are substituted for saturated fat, the LDL cholesterol level will fall about as much as with monounsaturated or polyunsaturated fats (31,33,35). If fats are restricted to much below 30% of calories and are replaced by carbohydrates, there is a tendency for HDL cholesterol levels to fall and VLDL triglyceride levels to rise (89,90). Many authorities consider these latter changes to be harmless, although there is still some disagreement on this question.

Calorie Balance

Obesity often raises cholesterol levels in both VLDL and LDL fractions; it also raises triglycerides, lowers HDL cholesterol (91,92), and raises blood pressure in many people. Weight reduction reduces LDL cholesterol levels in many people, lowers VLDL cholesterol and triglycerides, and raises HDL cholesterol levels (93–99). Some patients are unusually sensitive to the cholesterol-raising effect of even mild obesity (100,101), and maintenance of desirable body weight will prevent this response. Weight reduction and maintenance are best achieved by combining caloric restriction and regular exercise (102). Restricting the intakes of calorie-dense fat and alcohol also may help to achieve weight loss (103,104).

Fiber

Indigestible carbohydrates and related polymers constitute dietary fiber. Fiber can be categorized by its water solubility. Water-insoluble fiber adds bulk to the stools and promotes normal colonic function. Insoluble dietary fiber such as wheat bran cellulose does not lower serum cholesterol levels, although it may have other health benefits. Soluble fiber, including pectins and certain gums, have some cholesterol-lowering potential (105–109). One of the gums, beta-glycan, is present in oat products and has been reported to produce a modest reduction in cholesterol levels (110).

Some authorities recommend a total dietary fiber intake of 20–30 g daily for adults (111). This recommendation is primarily to achieve normal gastrointestinal function and possibly to provide other health benefits. About 25% (6 g) of this probably should be soluble fiber. A total daily intake of 20–30 g of dietary fiber and 6 g of soluble fiber can be achieved with the recommended five or more servings of fruits and vegetables and six or more servings of grain products daily. This level of intake should produce a 3–5% reduction of LDL cholesterol levels beyond the Step I or Step II diets. Inclusion of fiber-rich foods in the diet may facilitate adherence to a low-saturated fat, low-cholesterol diet (112).

Vitamins

Recent investigations suggest that oxidation of LDL in the artery wall increases its atherogenicity (113). Several vitamins including vitamin C, vitamin E, and beta-carotene have antioxidant properties (114–118), and if the LDL-oxidation theory is proven, these vitamins could help to prevent atherosclerosis (119,120). Antioxidant vitamin supplements are under investigation for a variety of health benefits, but insufficient data are available to recommend their use at this time. The recommended daily allowance of all the major vitamins will be consumed if the patient is consuming a well-balanced diet. Dark-green vegetables, deep-yellow vegetables, fruits, and their juices are good sources of antioxidant vitamins.

Alcohol

Alcohol intake among Americans averages 5% of total calories (121), but this value varies widely among individuals. Alcohol apparently has few adverse effects when consumed in moderation. In fact, moderate alcohol consumption in populations is associated with lower CHD rates than abstinence (122,123); this association, however, may not be causal, and the consumption of alcohol is not specifically recommended. A high intake of alcohol, moreover, is well known to have many harmful effects, including liver damage and cirrhosis, cardiomyopathy, adverse psychosocial consequences, and motor vehicle accidents; consuming more than two drinks per day also will raise blood pressure in many people. The Dietary Guidelines for Americans issued by the U.S. Department of Agriculture and the U.S. Department of Health and Human Services recommend no more than two drinks per day for men and no more than one drink per day for women. This report reaffirms this recommendation. A drink is defined as 5 ounces of wine, 12 ounces of beer, or 1½ ounces of 80 proof liquor.

Alcohol can affect lipoprotein metabolism in several ways. It increases serum triglyceride concentrations in many but not all people (124) and also raises HDL cholesterol levels (125,126). In most people, it does not affect LDL cholesterol concentrations. Whether the alcohol-induced rise in HDL affords any protection against CHD is not known; thus, because of uncertainty about the benefit of alcohol on HDL levels and well-known adverse effects, alcohol intake should not be prescribed as part of dietary therapy to prevent CHD.

Sodium

Blood pressure levels are correlated with dietary sodium intakes (127). An analysis of recent studies suggests that moderate degrees of salt (sodium chloride) restriction will lower the average blood pressure and reduce CHD risk (128–130). In addition, a certain percentage of the populations appears to be "salt-sensitive," particularly individuals who are older, overweight, and hypertensive (127,131–

133). These individuals will have greater than average falls in blood pressure when salt intake is reduced. Since no practical and reliable tests are available to identify salt-sensitive people, moderate salt restriction has been recommended for the general public by the National High Blood Pressure Education Program (134), the Diet and Health Committee of the National Academy of Sciences (57), the Dietary Guidelines for Americans from the American Heart Association (2), and other organizations. The National High Blood Pressure Education Program also recommends salt restriction for hypertensives. The average daily consumption of salt in the United States is 8–12 g/day, although intakes vary widely. This intake is far greater than the approximately 500 mg daily requirement for sodium. Achieving the recommended 2400 mg/day restriction becomes difficult at higher calorie levels, especially when trying to select certain low-fat foods. As a result, patients with high blood cholesterol who do not have concomitant hypertension should make lipid lowering the first priority in dietary modification.

CURRENT CONTROVERSIES

Although the value of a cholesterol-lowering diet is well established, disputes concerning diet's efficacy, side effects, and cost continue to exist. Some of these controversies arise from the assumption that disease prevention should be without cost. However, this can never be the case. In the broad picture, disease prevention requires financial commitment to educational programs that will stimulate awareness about the disease and to programs aimed at reducing the chances of disease. In response to these educational programs, individuals must in turn commit time, effort, and the additional expense necessary to follow through with therapy.

Whether or not disease prevention should be undertaken depends on the prevalence of the disease, the effectiveness of available therapies, and the therapy's side effects and cost. Since CHD is by far and away the number one cause of morbidity and mortality in the United States, efforts at disease prevention are easily justified (135). We conclude this chapter with details on the effectiveness of diet, its risks, and costs.

Effectiveness of Dietary Therapy

Population Response

Although the multiple dietary trials summarized in the section "Scientific Basis for a Cholesterol-Lowering Diet" illustrate that dietary therapy effectively lowers cholesterol levels, generalization of these findings to the population at large requires further consideration. Specifically, a population strategy for lowering cholesterol levels relies on multiple methods to disseminate educational information concerning the risks of high blood cholesterol and specific information concerning a cholesterol-lowering diet. The NCEP population Panel Report (136) summarized the strategies employed so far to promote cholesterol awareness and change among the U.S. population.

Recent evidence from the NHANES III survey of cholesterol levels in the nation suggests that population messages have been effective (137). Specifically, the mean total cholesterol level for American adults has fallen from 220 mg/dl in 1960 to 213 mg/dl 1978 to 205 mg/dl in 1990. Although the dietary intake data from NHANES III are not yet available, it is highly likely that the cholesterol lowering observed is in part due to changes in dietary intake. Whereas the 1972–1976 LRC (29) tabulated the mean saturated fat intake of 2771 men and 2674 women as 15% of calories from fat, the 1977–1978 USDA (30) survey of 10,522 men and 15,153 women showed dietary saturated fat intake at 14% of calories. A recent published survey of women (28) calculated that only 13% of calories came from saturated fat.

These survey results of reduced intake of nutrients that raise serum cholesterol levels mirror changes in the food marketplace. Consumer demand for lower-fat and lower-saturated-fat products has fueled wide availability of fat-modified products. Whereas the NHANES II (43) dietary intake suggested that red meats and dairy products contributed substantially to the saturated fat content of the diet, the CSFII (28) suggests that the major source of saturated fat in the diet is vegetable oil products. Grocery stores now allot ample shelf space to lower-fat milk and dairy products. Meat markets carry select beef (previously termed USDA "good") in lieu of fattier grades. The trim cut on meats has been reduced from the typical ½″ seen in the 1970s to ⅛″ or less (138). In addition, food companies have developed no-fat or low-fat products to substitute for higher-fat foods such as salad dressings, mayonnaise, crackers, chips, and baked goods.

These observations suggest that it is overly pessimistic to assume that Americans will not change their diet. Although it is true that some individuals are unwilling to change dietary habits, the marketplace indicates that the majority of consumers are motivated to follow a cholesterol-lowering diet. In turn, these changes in food intake have likely already led to the lower cholesterol levels observed in NHANES III.

The population approach to a cholesterol-lowering diet has been successful at achieving small reductions in the nation's mean cholesterol level. These changes rely on an entirely different philosophy than the high-risk strategy approach (1). In the latter, individual patients at high risk for CHD are identified for an intensive treatment plan, which involves individualized counseling. The cholesterol-lowering response is individually assessed for each patient and has little relevance to mean changes within the population.

Individual Responsiveness

Dietary response is, by definition, the change in blood cholesterol that occurs as a result of a change in the composition of dietary intake. Evaluation of dietary responsiveness for an individual patient relies upon baseline cholesterol measurements, dietary assessment, and endpoint measurements of cholesterol levels and dietary adherence. There must be a significant change in dietary composition in order to affect a significant change in dietary responsiveness.

Individual responsiveness to a defined dietary change is highly variable. Whereas initial work suggested that response to a cholesterol-lowering diet could be separated into three categories of hypo-, normo-, and hyperresponders (139), more recent evidence suggests that dietary responsiveness is normally distributed (140,141). This means that the average serum cholesterol lowering achieved by diet is only an average: some individuals will have greater cholesterol lowering response to diet while others have a lesser one.

The incidence of "hyporesponders" to a cholesterol-lowering diet can be estimated from trials such as the Diet Heart Feasibility Study (19), where 10% of institutionalized patients and 25% of free-living subjects achieved less than a 5% reduction in total cholesterol with dietary therapy. The reasons for this insensitivity to diet can only be speculated. Certainly, adherence can be a factor in outpatient studies. However, adherence cannot explain the 10% incidence of nonresponse observed in institutionalized subjects, so clearly genetic and environmental factors also play a role in altering diet responsiveness.

The genetic and environmental factors that determine responsiveness in individuals who are complying to a cholesterol-lowering diet have not been fully elucidated. A few factors appear to be good candidates since these factors are known to influence lipoprotein metabolism.

For example, it has been known for some time that saturated fatty acids contribute 70–80% of the cholesterol-lowering response to diet (12,13). In an elegant animal model where the effects of chronic diet on LDL clearance rates was examined, chronic ingestion of saturated fatty acids reduced LDL clearance rates (142). It seems plausible, then, that individuals with reduced LDL receptor activity would have an enhanced responsiveness to a cholesterol-lowering diet because removing saturated fatty acids from the diet would result in a measurable increase in LDL receptor activity. In a study in men with moderate hypercholesterolemia from our laboratory, men with low fractional catabolic rates of LDL cholesterol on a high-saturated-fat diet achieved a 30 mg/dl LDL lowering from a Step I diet, while subjects with a high fractional catabolic rate on a high-saturated-fat diet showed a diminished response of only 7 mg/dl (140).

A few other factors that affect lipoprotein metabolism have been correlated with dietary responsiveness. For example, obesity induces hepatic overproduction of triglycerides and apoproteins (143). Lean premenopausal women have been shown to be more diet sensitive than obese premenopausal women (144). Another example of a candidate factor is apo E phenotype. The four common apo E phenotypes have different binding affinities to the LDL receptor and, in this manner, could also alter dietary response (145). However, whereas some investigations have shown a definite association between apo E phenotype and response (146,147), others have failed to confirm an association (148,149).

Side Effects of Dietary Therapy

There are side effects to any intervention. The purpose of dietary therapy is to reduce LDL cholesterol levels, but dietary therapy could have other effects that are undesirable. For example, a cholesterol-lowering diet may not provide a nutritionally adequate diet. There is no evidence that this is the case. In fact, NHANES II men and women who reported adherence to a cholesterol-lowering diet had increased nutrient density compared to men and women not reporting a cholesterol-lowering diet (150). Therefore, concerns about the nutrient adequacy of a cholesterol-lowering diet cannot be substantiated.

Another potential side effect of the cholesterol-lowering diet is that the diet, while lowering LDL cholesterol levels, will also lower HDL cholesterol levels (151). Claims have been made that a cholesterol-lowering diet is ineffective at reducing CHD because the beneficial effects of lowering LDL has been offset by the deleterious effects of lowering HDL (152). Since epidemiological evidence suggests an even more potent role for HDL in predicting CHD (153), it is difficult to ignore this issue.

Whereas diets high in saturated fat raises LDL cholesterol levels, diets low in fat lower HDL cholesterol levels. Carbohydrates tend to raise triglycerides and lower HDL, but this effect is only significant with very-low-fat diets (i.e., less than 25% of calories from fat or more than 60% of calories from carbohydrates) (154). In normolipidemic subjects, the magnitude of triglyceride raising from very-low-fat diets is typically 30–100 mg/dl, with a concomitant reduction of HDL of 3–8 mg/dl (155). While some have argued that these adverse changes with high-carbohydrate, very-low-fat diets are transient, most studies reveal that these effects persist. The chronic nature of this effect is supported by a population study of young boys in whom the effects of percent calories from fat vs. percent calories from carbohydrates were graded and significant across populations with diverse dietary intakes (156). Therefore, if individuals follow a very-low-fat diet, LDL levels will fall because of the reduced saturated fat intake and the HDL cholesterol levels will fall because of the reduced total fat intake.

Despite the "adverse" effects of high-carbohydrate, very-low-fat diets on triglycerides and HDL cholesterol levels, population studies do not support an increase in CHD risk on these diets (157). Reasons for this discrepancy are multiple:

1. Such populations may have low prevalence rates of other CHD risk factors such as hypertension, obesity, or physical inactivity (158).
2. Such populations have low LDL cholesterol levels from their low-saturated-fat, low-cholesterol diet (159).

3. Studies of lipoprotein composition suggest that high-carbohydrate, very-low-fat diets produce triglyceride-enriched VLDL unlike the dense, cholesterol-enriched VLDL observed in heritable states of abnormal triglyceride metabolism.

Clearly a better understanding of the role of HDL and VLDL in atherogenesis will be needed before the long-term consequences of a high-carbohydrate, very-low-fat diet can be predicted (161).

Parenthetically, the current dietary recommendations do not specifically recommend a very-low-fat diet. The total fat recommendation is 30% or less of total calories. Since the HDL-lowering effect of diet is seen with very-low-fat diets, the HDL lowering of a cholesterol-lowering diet could be avoided by maintaining fat intake at 25–30% of calories. A simple dietary maneuver to achieve a higher fat intake in patients already following a very-low-fat diet is to add low-saturated fatty-acid sources of fat to the diet, e.g., nuts and unhydrogenated vegetable oils.

Cost-Effectiveness of Diet Therapy

In the 1990s age of cost-effectiveness, interventions must withstand a cost analysis. Different costs should be estimated for the population approach as opposed to the high-risk approach, since the latter includes the additional costs of dietary counseling and physician monitoring. From a population standpoint, there is evidence that a cholesterol-lowering diet may cost an extra 30 cents to one dollar per person per day for the extra fruits and vegetables in the diet (162). Since many individuals are overweight, these costs may be lower if caloric restriction is used as a means to comply with a cholesterol-lowering diet. Therefore, from a population standpoint, a cholesterol-lowering diet adds some costs beyond the expense of the educational programs.

Concerning the benefits, it has been calculated that diet therapy could reduce CHD mortality by 5–20% (163). This benefit has been further translated into a 3–4 months of life saved per person. Whereas these estimates may diminish enthusiasm for a cholesterol-lowering diet, the benefit of diet therapy will not be equally distributed among the population (164). Evidence presented in the first section of this chapter suggests that individuals with higher initial serum cholesterol levels and individuals with established coronary disease have the most to gain with diet therapy. Therefore, whereas the mean gain in lifespan may be only 3–4 months, high-risk individuals are likely to achieve even greater benefits from diet modification.

Standard cost-benefit analyses on diet therapy for the high-risk approach have been evaluated only on a limited basis. In an interesting analysis of the cost-effectiveness of oat bran for the high-risk approach (165), the cost per year of life saved was $17,800 as compared to $4,500 for smoking cessation and $29,000 for drug therapy with moderate hypertension. Still, this analysis compares favorably to the cost of annual pap smear screening at $33,693 per year of life saved (166).

SUMMARY

There is substantial evidence that a cholesterol-lowering diet is effective not only for achieving cholesterol lowering but at reducing CHD. The cholesterol-lowering diet is soundly based in science. It is effective and cost effective for both a population and a high-risk approach.

REFERENCES

1. The Expert Panel. Summary of the second report of the National Cholesterol Education Program (NCEP) expert panel on detection, evaluation and treatment of high blood cholesterol in adults (Adult Treatment Panel II). JAMA 1993; 269(23): 3015–3023.
2. The Nutrition Committee of the American Heart Association. Dietary guidelines for healthy american adults. A statement for physicians and health professionals. Circulation 1988; 77(3):721A–724A.
3. Keys A. Seven Countries: A Multivariate Analysis of Death and Coronary Heart Disease. Cambridge, MA: Harvard University Press, 1980.
4. Liu K, Stamler J, Trevisan M, Moss D. Dietary lipids, sugar, fiber and mortality from coronary heart disease: bivariate analysis of international data. Arteriosclerosis 1982; 2:221–227.
5. Kagan A, Harris BR, Winkelstein, Jr W, et al. Epidemiological studies of coronary heart disease and stroke in Japanese men living in Japan, Hawaii and California: demographic, physical, dietary and biochemical characteristics. J Chron Dis 1974; 27:345–364.
6. Posner BM, Cobb JL, Belanger AJ, Cupples A, D'Agostino RB, Stokes III J. Dietary lipid predictors of coronary heart disease in men: The Framingham Study. Arch Intern Med 1991; 151:1181–1187.
7. Kushi LH, Lew RA, Stare FJ, et al. Diet and 20-year mortality from coronary heart disease: The Ireland-Boston Diet-Heart Study. N Engl J Med 1985; 312:811–818.
8. Rosenberg L. Case-control studies of risk factors for myocardial infarction among women. In: Eaker ED, Packard B, Wenger NK, Clarkson TB, Tyroler HA, eds. Coronary Heart Disease in Women. New York: Haymarket Doyma, 1987:70–77.
9. Stokes III J, Kannel WB, Wolf PA, Cupples LA, D'Agostino RB. The relative importance of selected risk factors for various manifestations of cardiovascular disease among men and women from 35 to 64 years old: 30 years of follow-up in the Framingham Study. Circulation 1987; 75(suppl V):V65–V73.
10. Shekelle RB, Shyrock AM, Paul O, et al. Diet, serum cholesterol, and death from coronary heart disease: The Western Electric Study. N Engl J Med 1981; 304:65–70.
11. Clarkson TB, Shively CA, Weingand KW. Animal models of diet-induced atherosclerosis. In: Beynen AC, West CE, eds. Comparative Animal Nutrition. Vol. 6. Use of Animal Models for Research in Human Nutrition. Basel: S. Karger AG, 1988: 56–82.
12. Keys A, Anderson JT, Grande F. Prediction of serum-cholesterol responses of man to changes in fats in the diet. Lancet 1957; (Nov. 16):955–966.

13. Hegsted DM, McGandy RB, Myers ML, Stare FJ. Quantitative effects of dietary fat on serum cholesterol in man. Am J Clin Nutr 1965; 17:281–295.

14. Lipid Research Clinics Program. The Lipid Research Clinics Coronary Primary Prevention Trial Results. II. The relationship of reduction in incidence of coronary heart disease to cholesterol lowering. JAMA 1984; 251(3):365–374.

15. Frick MH, Elo O, Haapa K, et al. Helsinki Heart Study: primary-prevention trial with gemfibrozil in middle-aged men with dyslipidemia. Safety of treatment, changes in risk factors, and incidence of coronary heart disease. N Engl J Med 1987; 317(20):1237–1245.

16. Dayton S, Pearce ML, Hashimoto S, Dixon WJ, Uwamie T. A controlled clinical trial of a diet high in unsaturated fat in preventing complications of atherosclerosis. Circulation 1969; XXXIX(suppl II):II-1–II-60.

17. Turpeinen O. Effect of cholesterol-lowering diet on mortality from coronary heart disease and other causes. Circulation 1979; 59(1):1–7.

18. Frantz ID, Jr, Dawson EA, Ashman PL, et al. Test of effect of lipid lowering by diet on cardiovascular risk. The Minnesota Coronary Survey. Arteriosclerosis 1989; 9(1):129–135.

19. The National Diet-Heart Study Final Report. Circulation 1968; XXXVII & XXXVIII(suppl I):I-1–I-428.

20. Insull W, Jr, Henderson MM, Prentice RL, et al. Results of a randomized feasibility study of a low-fat diet. Arch Intern Med 1990; 150:421–427.

21. Henderson MM, Kushi LH, Thompson DJ, et al. Feasibility of a randomized trial of a low-fat diet for the prevention of breast cancer: dietary compliance in the Women's Health Trial Vanguard Study. Prev Med 1990; 19:115–133.

22. Keys A, Parlin RW. Serum cholesterol response to changes in dietary lipids. Am J Clin Nutr 1966; 19:175–181.

23. Muldoon MF, Manuck SB, Matthews KA. Lowering cholesterol concentrations and mortality: a quantitative review of primary prevention trials. BMJ 1990; 301: 309–314.

24. Leren P. The Oslo Diet-Heart Study: eleven-year report. Circulation 1970; XLII: 935–942.

25. Arntzenius AC, Kromhout K, Barth JD, et al. Diet, lipoproteins, and the progression of coronary atherosclerosis. N Engl J Med 1985; 312(13):805–811.

26. Ornish D, Brown SE, Scherwitz LW. Can lifestyle changes reverse coronary heart disease? The Lifestyle Heart Trial. Lancet 1990; 336:129–133.

27. Watts GF, Lewis B, Brunt JNH, et al. Effects of coronary artery disease of lipid-lowering diet, or diet plus cholestyramine, in the St Thomas' Atherosclerosis Regression Study (STARS). Lancet 1992; 339:563–569.

28. Human Nutrition Information Service. National Food Consumption Survey: Continuing Survey of Food Intakes by Individuals, Women 19–50 Years and Their Children 1–5 Years, 4 Days. Hyattsville, MD. U.S. Department of Agriculture, CSFII Report No. 85-4, 1985.

29. U.S. Department of Health and Human Services, National Institutes of Health. The Lipid Research Clinics Population Studies Data Book, Volume II, The Prevalence Study—Nutrient Intake. Bethesda, MD: NIH Publication No. 82-2014, 1982.

30. Fischer DR, Morgan KJ, Zabik ME. Cholesterol, saturated fatty acids, polyunsatu-

rated fatty acids, sodium, and potassium intakes of the United States population. J Am Coll Nutr 1985; 4(2):207–224.

31. Grundy SM. Comparison of monounsaturated fatty acids and carbohydrates for lowering plasma cholesterol. N Engl J Med 1986; 314:745–748.

32. Mensink RP, Katan MB. Effect of monounsaturated fatty acids versus complex carbohydrates on high-density lipoproteins in healthy men and women. Lancet 1987; 1:122–125.

33. Grundy SM, Florentine L, Nix D, Whelan MF. Comparison of monounsaturated fatty acids and carbohydrates for reducing raised levels of plasma cholesterol in man. Am J Clin Nutr 1988; 47:965–969.

34. Mensink RP, deGroot MJM, van den Broeke LT, Severijnen-Nobels JP, Demacker PNM, Katan MB. Effects of monounsaturated fatty acids v complex carbohydrates on serum lipoproteins and apoproteins in healthy men and women. Metabolism 1989; 38:172–178.

35. Ginsberg HN, Barr SL, Gilbert BA, et al. Reduction of plasma cholesterol levels in normal men on an American Heart Association Step-1 Diet or a Step-1 Diet with added monounsaturated fat. N Engl J Med 1990; 322:574–579.

36. Grundy SM, Vega GL. Plasma cholesterol responsiveness to saturated fatty acids. Am J Clin Nutr 1988; 47:822–824.

37. Keys A, Menotti A, Karvonen MJ, et al. The diet and 15-year death rate in the Seven Countries Study. Am J Epidemiol 1986; 124:903–915.

38. Denke MA, Grundy SM. Comparison of effects of lauric acid and palmitic acid on plasma lipids and lipoproteins. Am J Clin Nutr 1992; 56:895–898.

39. Keys A, Anderson JT, Grande F. Serum cholesterol response to changes in the diet. IV. Particular saturated fatty acids in the diet. Metabolism 1965; 14:776–787.

40. Mattson FH, Grundy SM. Comparison of effects of dietary saturated, monounsaturated, and polyunsaturated fatty acids on plasma lipids and lipoproteins in man. J Lipid Res 1985; 26:194–202.

41. Bonanome A, Grundy SM. Effect of dietary stearic acid on plasma cholesterol and lipoprotein levels. N Engl J Med 1988; 318;1244–1248.

42. Grundy SM, Denke MA. Dietary influences on serum lipids and lipoproteins. J Lipid Res 1990; 31:1149–1172.

43. Block G, Dresser CM, Hartman AM, Carroll MD. Nutrient sources in the American diet: quantitative data from the NHANES II Survey. II. Macronutrients and fats. Am J Epidemiol 1985; 122(1):27–40.

44. Denke MA, Grundy SM. Effects of fats high in stearic acid on lipid and lipoprotein concentrations in man. Am J Clin Nutr 1991; 54:1036–1040.

45. Mensink RP, Katan MB. Effect of a diet enriched with monounsaturated or polyunsaturated fatty acids on levels of low-density and high-density lipoprotein cholesterol in healthy women and men. N Engl J Med 1989; 321:436–441.

46. Mensink RP, Katan MB. Effects of dietary fatty acids on serum lipids and lipoproteins: a meta-analysis of 27 trials. Arterio Thromb 1992; 12:911–919.

47. Dreon DM, Vranizan KM, Krauss RM, Austin MA, Wood PD. The effects of polyunsaturated fat vs. monounsaturated fat on plasma lipoproteins. JAMA 1990; 263: 2462–2466.

48. Berry EM, Eisenberg S, Haratz D, et al. Effects of diets rich in monounsaturated

fatty acids on plasma lipoproteins—the Jerusalem Nutrition Study: high MUFAs vs high PUFAs. Am J Clin Nutr 1991; 53:899–907.

49. McDonald BE, Gerrard JM, Bruce VN, Corner EJ. Comparison of the effect of canola oil and sunflower oil on plasma lipids and lipoproteins and on in vivo thromboxane A2 and prostacyclin production in healthy young men. Am J Clin Nutr 1989; 50:1382–1388.

50. Valsta LM, Jauhiainen M, Aro A, Katan MB, Mutanen M. Effects of a monounsaturated rapeseed oil and a polyunsaturated sunflower oil diet on lipoprotein levels in humans. Arterio Thromb 1992; 12:50–57.

51. Mensink RP, Katan MB. Effect of dietary trans fatty acids on high-density and low-density lipoprotein cholesterol levels in healthy subjects. N Engl J Med 1990; 323: 439–445.

52. Mensink RP, Zock PL, Katan MB, Hornstra G. Effects of dietary cis and trans fatty acids on serum lipoprotein(a) levels in humans. J Lipid Res 1992; 33:1493–1501

53. Nestel P, Noakes M, Belling B, et al. Changes in very low density lipoproteins with cholesterol loading in man. Metabolism 1982; 31:398–405.

54. Zock PL, Katan MB. Hydrogenation alternatives: effects of trans fatty acid and stearic acid versus linoleic acid on serum lipids and lipoproteins in humans. J Lipid Res 1992; 33:399–410.

55. Lichenstein AH, Ausman LM, Carrasco W, Jenner JL, Ordovas JM, Schaefer EJ. Hydrogenation impairs the hypolipidemic effect of corn oil in humans: hydrogenation, trans fatty acids, and plasma lipids. Arterio Thromb 1993; 13:154–161.

56. Hunter JAW, Applewhite TH. Reassessment of trans fatty acid availability in the US diet. Am J Clin Nutr 1991; 54:363–369.

57. National Research Council, Committee on Diet and Health, Food and Nutrition Board, and Commission on Life Sciences. Diet and Health: Implications for Reducing Chronic Disease Risk. Washington, DC: National Academy Press, 1989.

58. Harris WS. Fish oils and plasma lipid and lipoprotein metabolism in humans: a critical review. J Lipid Res 1989; 30:785–807.

59. Nestel PJ, Connor WE, Reardon MF, Connor S, Wong S, Boston R. Suppression by diets rich in fish oil of very low density lipoprotein production in man. J Clin Invest 1984; 74:82–89.

60. Phillipson BE, Rothrock DW, Connor WE, Harris WS, Illingworth DR. Reduction of plasma lipids, lipoproteins, and apoproteins by dietary fish oils in patients with hypertriglyceridemia. N Engl J Med 1985; 312:1210–1216.

61. Connor WE. Hypolipidemic effects of dietary omega-3 fatty acids in normal and hyperlipidemic humans: effectiveness and mechanism. In: Simopoulos AP, Kifer RR, Martin RE, eds. Health Effects of Polyunsaturated Fatty Acids in Seafoods. Orlando, FL: Academic Press, 1986:173–210.

62. Failor RA, Childs MT, Bierman EL. The effects of w3 and w6 fatty acid-enriched diets on plasma lipoproteins and apoproteins in familia combined hyperlipidemia. Metabolism 1988; 37:1021–1028.

63. Schectman G, Kaul S, Kissebah AH. Effect of fish oil concentrate on lipoprotein composition in NIDDM. Diabetes 1988; 37:1567–1573.

64. Nozaki S, Garg A, Vega GL, Grundy SM. Postheparin lipolytic activity and plasma lipoprotein response to w-3 polyunsaturated fatty acids in patients with primary hypertriglyceridemia 1-3. Am J Clin Nutr 1991; 53:638–642.
65. Connor WE. Omega-3 fatty acids and heart disease. In: Evaluation of Publicly Available Scientific Evidence Regarding Certain Nutrient-Disease Relationships. Bethesda, MD: Life Science Research Office, FASEB, 1991.
66. Leaf A, Weber PC. Medical progress: cardiovascular effects of n-3 fatty acids. N Engl J Med 1988; 318:549–556.
67. Kromhout D, Bosschieter EB, De Lezzene Coulander C. The inverse relation between fish consumption and 20-year mortality from coronary heart disease. N Engl J Med 1985; 312:1205–1209.
68. McGill HC Jr, McMahan CA, Kruski AW, Mott, GE. Relationship of lipoprotein cholesterol concentrations to experimental atherosclerosis in baboons. Arteriosclerosis, 1981; 1:3–12.
69. McGill HC Jr, McMahan CA, Kruski AW, Kelley JL, Mott GE. Responses of serum lipoproteins to dietary cholesterol and type of fat in the baboon. Arterio 1981; 5:337–344.
70. Rudel LL, Parks JS, Bond MG. LDL heterogeneity and atherosclerosis in nonhuman primates. Ann NY Acad Sci 1985; 454:248–253.
71. Beveridge JMR, Connell WF, Mayer GA, Haust HL. The response of man to dietary cholesterol. J Nutr 1960; 71:61–65.
72. Connor WE, Hodges RE, Bleiler RE. The serum lipids in men receiving high cholesterol and cholesterol-free diets. J Clin Invest 1961; 40:894–901.
73. Mattson FH, Erickson BA, Kligman AM. Effect of dietary cholesterol on serum cholesterol in man. Am J Clin Nutr 1972; 25:589–594.
74. Nestel P, Tada N, Billington T, Huff M, Fidge N. Changes in very low density lipoproteins with cholesterol loading in man. Metabolism 1982; 31:398–405.
75. Beynen AC, Katan MB. Reproducibility of the variations between humans in the response of serum cholesterol to cessation of egg consumption. Atherosclerosis 1985; 57:19–31.
76. Hegsted DM. Serum-cholesterol response to dietary cholesterol: A reevaluation. Am J Clin Nutr 1986; 44:299–305.
77. Slater G, Mead J, Dhopeshwarkar G, Robinson S, Alfin-Slater RB. Plasma cholesterol and triglycerides in men with added eggs in the diet. Nutr Rep Int 1976; 14:249–260
78. Porter MW, Yamanaka W, Carlson SD, Flynn MA. Effect of dietary egg on serum cholesterol and triglyceride of human males. Am J Clin Nutr 1977; 30:490–495.
79. Clifton PM, Kestin M, Abbey M, Drysdale M, Nestel PJ. Relationship between sensitivity to dietary fat and dietary cholesterol. Arteriosclerosis 1990; 10:394–401.
80. Katan MB, Burns MAM, Glatz JFC, Knuiman JT, Nobels A, deVries JHM. Congruence of individual responsiveness to dietary cholesterol and to saturated fat in man. J Lipid Res 1988; 29:883–892.
81. Schonfeld G, Patsch W, Rudel LL, Nelson C, Epstein M, Olsen RE. Effects of dietary cholesterol and fatty acids on plasma lipoproteins. J Clin Invest 1982; 69:1072–1080.

82. Stamler J, Shekelle R. Dietary cholesterol and human coronary heart disease: the epidemiological evidence. Arch Pathol Lab Med 1988; 112:1032–1040.
83. Shekelle RB, Stamler J. Dietary cholesterol and ischemic heart disease. Lancet 1989; 1:1177–1179.
84. Huff MW, Hamilton RMG, Carroll KK. Plasma cholesterol levels in rabbits fed low fat, cholesterol-free, semi-purified diets: effects of dietary proteins, protein hydrolysates and amino acid mixtures. Atherosclerosis 1977; 28:187–195.
85. Sirtori CR, Agradi E, Conti F, Mantero O, Gatti E. Soybean-protein diet in the treatment of type-II hyperlipoproteinemia. Lancet 1977; 1:275–277.
86. Carroll KK, Giovannetti PM, Huff MW, Moase O, Roberts DC, Wolfe BM. Hypercholesterolemic effect of substituting soybean protein for animal protein in the diet of healthy young women. Am J Clin Nutr 1978; 31(8):1213–1321.
87. Grundy SM, Abrams JJ. Comparison of actions of soy protein and casein on metabolism of plasma lipoproteins and cholesterol in humans. Am J Clin Nutr 1983; 38:245–252.
88. Zeller K, Whittaker E, Sullivan L, Raskin P, Jacobson HR. Effect of restricting dietary protein on the progression of renal failure in patients with insulin-dependent diabetes mellitus. N Engl J Med 1991; 324:78–84.
89. Knuiman JT, West CE, Katan MB, Hautvast JGA. Total cholesterol and high density lipoprotein cholesterol levels in populations differing in fat and carbohydrate intake. Arteriosclerosis 1987; 7:612–619.
90. Brinton EA, Eisenberg S, Breslow JL. A low-fat diet decreases high density lipoprotein (HDL) cholesterol levels by decreasing HDL apolipoprotein transport rates. J Clin Invest 1990; 85:144–151.
91. Denke MA, Sempos CT, Grundy SM. Excess body weight: an under-recognized contributor to high blood cholesterol in Caucasian American men. Arch Int Med 1993; 153(9):1093–1103.
92. Denke MA, Sempos CT, Grundy SM. Excess body weight: an under-recognized contributor to high blood cholesterol in American women. Arch Int Med. 1994; 154: 401–410.
93. Tran ZV, Weltman A. Differential effects of exercise on serum lipid and lipoprotein levels seen with changes in body weight: a meta analysis. JAMA 1985; 254:919–924.
94. Stamler J, Farinaro E, Mojonnier LM, Hall Y, Moss D, Stamler R. Prevention and control of hypertension by nutrition-hygienic means: long-term experience of the Chicago Coronary Prevention Evaluation Program. JAMA 1980; 243:1819–1823.
95. Caggiula AW, Christakis G, Farrand M, et al. The multiple risk factor intervention trial (MRFIT) IV. Intervention on blood lipids. Prev Med 1981; 10:443–475.
96. Rössner S, Björvell H. Early and late effects of weight loss on lipoprotein metabolism in severe obesity. Atherosclerosis 1987; 64:125–130.
97. Wolf RN, Grundy SM. Influence of weight reduction on plasma lipoproteins in obese patients. Arteriosclerosis 1983; 3:160–169.
98. Wood PD, Stefanick ML, Dreon DM, et al. Changes in plasma lipids and lipoproteins in overweight men during weight loss through dieting as compared with exercise. N Engl J Med 1988; 319:1173–1179.

99. Wood PD, Stefanick ML, Williams PT, Haskell WL. The effects of plasma lipoproteins of a prudent weight-reducing diet, with or without exercise, in overweight men and women. N Engl J Med 1991; 325:461–466.

100. Anderson JW, Lawler A, Keys A. Weight gain from simple overeating. II. Serum lipids and blood volume. J Clin Invest 1957; 36:81–88.

101. Sims EA, Goldman RF, Gluck CM, et al. Experimental obesity in man. Trans Assoc Am Phys 1968; 81:153–170.

102. NIH Technology Assessment Conference Panel. Methods for voluntary weight loss and control. Ann Int Med 1992; 116(11):942–949.

103. Tremblay A, Lavallee N, Almeras N, Allard L, Despres JP, Bouchard C. Nutritional determinants of the increase in energy intake associated with a high-fat diet. Am J Clin Nutr 1991; 53:1134–1137.

104. Haskell WL, Camargo C, Williams PT, et al. The effect of cessation and resumption of moderate alcohol intake on serum high-density-lipoprotein subfractions. N Engl J Med 1984; 310(13):805–810.

105. Bell LP, Hectorn KJ, Reynolds H, Hunninghake DB. Cholesterol lowering effects of soluble-fiber cereals as part of a prudent diet for patients with mild to moderate hypercholesterolemia. Am J Clin Nutr 1990; 52:1020–1026.

106. Anderson JW, Garrity TF, Wood CL, Whitis SE, Smith BM, Oeltgen PR. Prospective, randomized controlled comparison of the effects of low-fat and low-fat plus high-fiber diets on serum lipid concentrations 1-3. Am J Clin Nutr 1992; 56:887–894.

107. Everson GT, Daggy BP, McKinley C, Story JA. Effects of psyllium hydrophilic mucilloid on LDL-cholesterol and bile acid synthesis in hypercholesterolemic men. J Lipid Res 1992; 33:1183–1192.

108. Lepre F, Crane S. Effect of oatbran on mild hyperlipidemia. Med J Aust 1992; 157: 305–308.

109. Whyte JL, McArthur R, Topping D, Nestel P. Oat bran lowers plasma cholesterol levels in mildly hypercholesterolemic men. J Am Diet Assoc 1992; 92:446–449.

110. Ripsin CM, Keenan KM, Jacobs DR, Jr., et al. Oat products and lipid lowering. A meta-analysis. JAMA 1992; 268(21):3074.

111. Life Sciences Research Office (LSRO). Physiological Effects and Health Consequences of Dietary Fiber. Bethesda, MD: Federation of American Societies for Experimental Biology, 1987:236.

112. Swain JF, Rouse IL, Curley CB, Sacks FM. Comparison of the effects of oat bran and low-fiber wheat on serum lipoprotein levels and blood pressure. N Engl J Med 1990; 322:147–152.

113. Steinberg D, Parthasarathy S, Carew TE, Khoo JC, Witztum JL. Beyond cholesterol. Modifications of low-density lipoprotein that increase its atherogenecity. N Engl J Med 1989; 320(14):915–924.

114. Jialal I, Vega GL, Grundy SM. Physiological levels of ascorbate inhibit the oxidative modification of low density lipoprotein. Atherosclerosis 1990; 82:185–191.

115. Jialal I, Norkus EP, Cristol L, Grundy SM. Beta-carotene inhibits the oxidative modification of low-density lipoprotein. Biochim Biophys Acta 1991; 1086: 134–138.

116. Jialal I, Grundy SM. Preservation of the endogenous antioxidants in low density lipoprotein by ascorbate but not probucol during oxidative modification. J Clin Invest 1991; 87:597–601.

117. Jialal I, Grundy SM. Effect of dietary supplementation with alpha-tocopherol on oxidative modification of low density lipoproteins. J Lipid Res 1992; 33:899–906.

118. Dieber-Rotheneder M, Puhl H, Waeg G, Striegl G, Esterbauer H. Effect of oral supplementation with D-alpha-tocopherol on the vitamin E content of human low density lipoproteins and resistance to oxidation. J Lipid Res 1991; 32(8):1325–1332.

119. Steinberg D, Berliner JA, Burton GW, et al. Antioxidants in the prevention of human atherosclerosis. Summary of the proceedings of an NHLBI workshop: September 5–6, 1991, Bethesda, MD. Circulation 1992; 85(6):2338–2345.

120. Gaziano JM, Manson JE, Buring JE, Hennekens CH. Dietary antioxidants and cardiovascular disease. Ann NY Acad Sci 1992; 669:249–259.

121. U.S. Surgeon General. Report on Nutrition and Health. DHHS(PHS) Publication No. 88-50210. Public Health Service U.S. Department of Health and Human Services. Washington DC: U.S. Government Printing Office, 1988.

122. Klatsky AL, Friedman GD, Siegelaub AB. Alcohol and mortality, a ten-year Kaiser-Permanente Experience. Ann Intern Med 1981; 95:139–145.

123. Steinberg D, Pearson TA, Kuller LH. The David Conference, alcohol and atherosclerosis. Ann Intern Med 1991; 114:967–976.

124. Ginsberg H, Olefsky J, Farquhar JW, Reaven GM. Moderate ethanol ingestion and plasma triglyceride levels—a study in normal and hypertriglyceridemic persons. Ann Intern Med 1974; 80:143–149.

125. Ostrander LD, Jr, Lamphiear DE, Block WD. Relationship of serum lipid concentrations to alcohol consumption. Arch Intern Med 1974; 134(3):451–456.

126. Crouse JR, Grundy SM. Effects of alcohol on plasma lipoproteins and cholesterol and triglyceride metabolism in man. J Lipid Res 1984; 25:486–496.

127. Stamler J, Rose G, Elliott P, et al. Findings of the International Cooperative INTERSALT Study. Hypertension 1991; 17(suppl I):I-1–I-7.

128. Law MR, Frost CD, Wald NJ. By how much does dietary salt reduction lower blood pressure? I—Analysis of observational data among populations. BMJ 1991; 302:811–815.

129. Frost CD, Law MR, Wald NJ. II—Analysis of observational data within populations. BMJ 1991; 302:815–818.

130. Law MR, Frost CD, Wald NJ. III-Analysis of data from trials of salt reduction. BMJ 1991; 302:819–824.

131. Van Brummelen P, Koolen MI. Differences in sodium sensitivity in human hypertensives. Clin Invest Med 1987; 10:581–585.

132. Weinberger MH, Cohen SJ, Miller JZ, Luft FC, Grim CE, Fineberg NS. Dietary sodium restriction as adjunctive treatment of hypertension. JAMA 1988; 259(17):2561–2565.

133. Rocchini AP, Key J, Bondie D, et al. The effect of weight loss on the sensitivity of blood pressure to sodium on obese adolescents. N Engl J Med 1989; 321(9):580–585.

134. National High Blood Pressure Education Program. Working Group Report on Primary Prevention of Hypertension. Arch Intern Med 1993; 153:186–208.

135. N.I.H. Consensus Development Conference. Lowering blood cholesterol to prevent heart disease. JAMA 1985; 253(14):2080–2086.

136. The Expert Panel. Report of the Expert Panel of the National Cholesterol Education Program on Population Strategies for Blood Cholesterol Reduction. U.S. Department of Health and Human Services, National Institutes of Health, NIH Publication No. 90-3046, Bethesda, MD, November 1990.

137. Johnson CL, Rifkind BM, Sempos CT, et al. Declining serum total cholesterol levels among US adults: The National Health and Nutrition Examination Surveys. JAMA 1993; 269(23):3002–3008.

138. Savell JW, Harris JJ, Cross HR, Hale DS, Beasley LC. National beef market basket survey. J Anim Sci 1991; 69:2883–2893.

139. Beynen AC, Katan MB, VanZutphen LFM. Hypo- and hyperresponders: Individual differences in the response of serum cholesterol concentration to changes in diet. Adv Lipid Res 1987; 22:115–171.

140. Denke MA, Grundy SM. Individual responses to a cholesterol lowering diet in fifty men with moderate hypercholesterolemia. Arch Int Med 1994; 154:317–325.

141. Denke MA. Individual responsiveness to a cholesterol lowering diet in postmenopausal women with moderate hypercholesterolemia. Arch Int Med 1994; 154:1977–1982.

142. Spady DK, Dietschy JM. Dietary saturated triglycerides suppress hepatic low density lipoprotein receptors in the hamster. Proc Natl Acad Sci USA 1985; 82:4526–4530.

143. Kesaniemi YA, Grundy SM. Increased low density lipoprotein production associated with obesity. Arteriosclerosis 1983; 3:170–177.

144. Cole TG, Bowen PE, Schmeisser D. Differential reduction of plasma cholesterol by the American Heart Association Phase 3 Diet in moderately hypercholesterolemic, premenopausal women with different body mass indexes. Am J Clin Nutr 1992; 55:385–394.

145. Rubinstein A, Gibson JC, Paterniti Jr, JR, et al. Effect of heparin-induced lipolysis on the distribution of apolipoprotein E among lipoprotein subclasses. J Clin Invest 1985; 75(2):710–721.

146. Tikkanen MJ, Huttunen JK, Ehnholm C, Pietinen P. Apolipoprotein E_4 homozygosity predisposes to serum cholesterol elevation during high fat diet. Arteriosclerosis 1990; 10(2):285–288.

147. Gylling H, Miettinen TA. High cholesterol and fat intake have similar respective effects on serum cholesterol, LDL kinetics and cholesterol metabolism. Circulation 1991; 84(suppl II):II-68.

148. Uusitupa MI, Mäkinen ES, Pietinen P, Aro A, Kesäniemi. Hypercholesterolemic response to fat modified diets—effects of apo E phenotypes. Circulation 1991; 84 (suppl II):II-680.

149. McPherson R, Martin L, Connelly PW, Tall A, Milne R, Marcel Y. Plasma cholesteryl ester transfer protein (CETP) response to cholesterol feeding varies according to apo E phenotype. Circulation 1991; 84(suppl II):II-680.

150. Schectman G, McKinney WP, Pleuss J, Hoffman RG. Dietary intake of Americans reported adherence to a low cholesterol diet (NHANES II). AJPH 1990; 80(6): 698–703.

151. Denke MA. Dietary determinants of high-density-lipoprotein-cholesterol levels. Cardiovasc Risk Factors 1993; 3(5):4–8.

152. Hunninghake DB, Stein EA, Dujovne CA, et al. The efficacy of intensive dietary therapy alone or combined with lovastatin in outpatients with hypercholesterolemia. N Engl J Med 1993; 328(17):1213–1219.

153. Gordon DJ, Probstfield JL, Garrison RJ, et al. High-density lipoprotein cholesterol and cardiovascular disease. Four prospective American studies. Circulation 1989; 79:8–15.

154. Nestel PJ, Carroll KF, Havenstein N. Plasma triglyceride response to carbohydrates, fats and caloric intake. Metabolism 1970; 19(1):1–18.

155. Harris WS, Connor WE, Inkeles SB, Illingworth DR. Dietary omega-3 fatty acids prevent carbohydrate-induced hypertriglyceridemia. Metabolism 1984; 33(11): 1016–1019.

156. West CE, Sullivan DR, Katan MB, Halferkamps IL, VanDer Torre HW. Boys from populations with high-carbohydrate intake have higher fasting triglyceride levels than boys from populations with high-fat intake. Am J Epidemiol 1990; 131(2): 271–282.

157. Gordon DJ, Rifkind BM. High-density lipoprotein—the clinical implications of recent studies. N Engl J Med 1989; 321(19):1311–1316.

158. Knuiman JT, West CE, Katan MB, Hautvast JGAJ. Total cholesterol and high density lipoprotein cholesterol levels in populations differing in fat and carbohydrate intake. Arteriosclerosis 1987; 7(6):612–619.

159. Miller GJ, Miller NE. Plasma-high-density-lipoprotein concentration and development of ischemic heart-disease. Lancet, 1975; (Jan 4):16–19.

160. Ginsberg HN. Lipoprotein physiology and its relationship to atherogenesis. Endocrinol Metab Clin North Am 1990; 19(2):211–228.

161. Sacks FM, Willett WW. More on chewing the fat. The good fat and the good cholesterol. N Engl J Med 1991; 325(24):1740–1742.

162. Scott LW, Foreyt JP, Young J, Reeves RS, O'Malley MP. Are low-cholesterol diets expensive? J Am Diet Assoc 1979; 74(5):558–561.

163. Browner WS, Westenhouse J, Tice JA. What if Americans ate less fat? A quantitative estimate of the effect on mortality. JAMA 1991; 265(24):3285–3291.

164. Denke MA. Drug treatment of hyperlipidemia in elderly patients. Curr Opin Lipidol 1993; 4(1):56–62.

165. Kinosian BP, Eisenberg JM. Cutting into cholesterol: cost-effective alternatives for treating hypercholesterolemia. JAMA 1988; 259(15):2249–2254.

166. Fahs MC, Mandelblatt J, Schechter C, Muller C. Cost effectiveness of cervical cancer screening for the elderly. Ann Int Med 1992; 117(6):520–527.

11

Implementing Dietary Change

**Penny M. Kris-Etherton, Sharon L. Peterson,
Madeleine Sigman-Grant, and Lori Beth Dixon**

Pennsylvania State University, University Park, Pennsylvania

Suzanne M. Jaax and Lynne W. Scott

Baylor College of Medicine and The Methodist Hospital, Houston, Texas

INTRODUCTION

Dietary therapy is the primary treatment for patients with an elevated low-density lipoprotein (LDL) cholesterol level (1). The recommended diets for the treatment of hypercholesterolemia are the Step I and Step II Diets (1). These diets are designed to progressively reduce consumption of saturated fatty acids (saturated fat) and cholesterol, decrease total fat, and facilitate achieving and maintaining a healthy body weight. The Step I diet is also recommended for the population at large to shift the distribution of blood cholesterol levels to a lower range. Together these approaches represent a major coordinated effort of the National Cholesterol Education Program (NCEP) to reduce the risk of coronary heart disease (CHD) in the United States.

Approximately 52 million Americans over the age of 20 are candidates for dietary therapy (2). This estimate, based on NHANES III data (collected from 1988 through 1991), indicates that presently 29% of the U.S. adult population (32% of men and 27% of women) requires dietary therapy for elevated blood cholesterol levels (Fig. 1). Although many Americans still are candidates for dietary therapy, considerable progress has been made in recent years to lower the blood cholesterol levels of the population at large. For example, data from NHANES II (1976–1980) indicated that 36% of adults previously required dietary therapy for elevated blood cholesterol levels (2). Moreover, average blood cholesterol levels in the United States decreased 8 mg/dl (from 213 to 205 mg/dl) from 1976–1980 to 1988–1991,

Percent

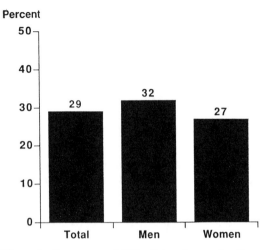

Figure 1   Percentage of U.S. adults who require dietary therapy. (From Ref. 5.)

and since 1960–1962 they have dropped by 12–17 mg/dl, or 6–8% (3). A major reason for the impressive population decrease in blood cholesterol levels in recent years is the activities of the NCEP and its participating organizations (3), which have resulted in changes in dietary intake patterns. Briefly, total fat intake was 40–42% of calories in the late 1950s to the mid-1960s, which decreased to 36% of calories in 1984 (4). In the early 1950s, saturated fat intake was 18–20% of calories, whereas in 1984 it represented 13–14% of energy intake (4). Over this same time period, polyunsaturated fat intake increased from 2–4% to 7.5% of calorie intake (4). More recent dietary data from the NHANES III study indicate that as of 1990, total fat and saturated fat have decreased to 34% and 12% of calories, respectively (5). Despite the progress that has been made in favorably changing food-consumption practices since the 1950s, saturated fat and total fat intake still are above recommended levels (6). In fact, less than 15% of the population meets the saturated fat and total fat recommendations of the Step I diet (7). Thus, major efforts are needed to develop effective strategies for implementing the Step I and Step II diets.

This chapter describes different approaches that can be used to help patients modify their dietary behaviors in a manner that facilitates achieving the recommendations of a Step I or Step II diet. Three different strategies are described, all of which are relatively easy to implement. To facilitate dietary change, the strategy selected should be guided largely by the personal preference of the patient. Various nutritional-assessment activities and key food-selection guidance also are discussed. Because of the need to educate the population about healthy dietary practices to lower blood cholesterol levels, this chapter briefly describes approaches that have been recommended to implement dietary changes nationwide. Since it is widely recognized that the atherosclerotic process begins in childhood, this chapter also will discuss the importance of selective screening of children and adolescents who have a family history of premature cardiovascular disease or at least one parent with high blood cholesterol.

STEP I AND STEP II DIETS

The nutrient specifications of Step I and Step II diets are defined in Table 1. The major objectives of these diets are to reduce saturated fat, cholesterol, and total fat and to achieve and maintain a healthy body weight. To achieve the recommendations of the Step I diet, saturated fat and cholesterol must be reduced by approximately 25% compared to a typical American diet. A Step II diet requires greater decreases in saturated fat and cholesterol; both must be decreased by almost 50%. For both diets, total fat should provide ≤30% of calories, representing a reduction of approximately 15%.

Sample one-day menus for a Step I diet, a Step II diet, and the typical American diet appear in Table 2. These menus illustrate how patients can progress from a typical American diet to a Step I and a Step II diet. Differences in food choices among the diets are shown in bold print. Both the Step I and Step II diets provide foods from all food groups as well as familiar and popular foods. In addition, lean red meat can be included on a Step I diet (as much as 6 oz/day) and a Step II diet (as much as 5 oz/day). Other favorite foods such as skim milk dairy products, breads and cereals, modified baked products, and desserts can be included. Neither diet requires completely changing a patient's typical cuisine. The nutrient specifications of these diets represent average values; on some days a patient's intake may exceed the specific recommendations, and on other days the intake will be less than the maximal amount recommended.

A Step II diet is a moderately aggressive therapeutic diet. Reducing saturated fat to <7% of calories and cholesterol to <200 mg/day represents a significant decrease in habitual intake for many patients and will require a number of dietary changes. It is important to note that a Step II diet has no lower limit for saturated fat, cholesterol, and total fat, so for some patients these nutrients/dietary constituents can be progressively decreased far below the maximally recommended amount. A Step II diet that provides minimal amounts of saturated fat (<3% of calories), cholesterol (<100 mg/day) and total fat (<10% of calories) should be an option for highly motivated patients (e.g., those with cardiovascular disease or as a possible alternative to pharmacological therapy). Together with other major lifestyle changes, this very aggressive low-fat diet has been shown to markedly lower total and LDL cholesterol levels and actually result in regression of atherosclerosis (8). While this diet has been shown to be efficacious, it nonetheless requires major dietary and lifestyle changes. The emphasis of a very low-fat, very low-saturated-fat and cholesterol diet is low-fat breads, cereals, and other grain products,

Table 1 Dietary Therapy for High Blood Cholesterol

Nutrient[a]	Recommended intake	
	Step I diet	Step II diet
Total fat	30% or less of total calories	
SFA	8–10% of total calories	Less than 7% of total calories
PUFA	Up to 10% of total calories	
MUFA	Up to 15% of total calories	
Carbohydrates	55% or more of total calories	
Protein	Approximately 15% of total calories	
Cholesterol	Less than 300 mg/day	Less than 200 mg/day
Total calories	To achieve and maintain desirable weight	

[a]Calories from alcohol not included.
SFA = saturated fatty acids; MUFA = monounsaturated fatty acids; PUFA = polyunsaturated fatty acids.
Source: Ref. 1.

fruits, vegetables, legumes, and skim milk dairy products. Meat, poultry, and fish are used sparingly (as a condiment, if at all), and consumption of fats and oils is markedly curtailed. Social activities and eating out are challenging for many patients. Nonetheless, for a patient who has been successful in achieving the recommendations of a Step I diet and then a Step II diet, further reductions in saturated fat, cholesterol, and total fat can be considered. All patients, even those who are motivated and committed, will benefit from the expertise of a registered dietitian in implementing a Step I and a Step II diet.

When total fat is reduced, calories usually are reduced, and this can be an effective strategy for achieving weight loss. Achieving and maintaining a desirable body weight is important in the treatment of an elevated LDL cholesterol level because of the adverse effects of overweight and obesity on many plasma lipids and lipoproteins. A slow, gradual weight loss of ½–1 lb/week is recommended to achieve and maintain a healthy weight. Reducing fat from 37 to 30% of calories (on a 2000 kcal diet) decreases calories by 140 per day. This change alone would lead to an almost ½-lb weight loss per week. If a patient does not need to lose weight, additional calories should be added with foods high in carbohydrate. (Note that the one-day menus in Table 2 are isocaloric.) By increasing the carbohydrate content of the diet, there is a further reduction in the percentage of calories from fat.

There is no specific ATP II (NCEP's Second Adult Treatment Panel) recommendation for dietary fiber, however, the combination of whole grain breads, cereals, fruits, and vegetables will provide at least 20 g/day of dietary fiber/2000 kcal. Many nutritionists recommend 20–30 g of dietary fiber/day, of which 25% (6 g) should be provided by soluble fiber. A number of studies (9–11) have shown that soluble fiber (in oat products, psyllium seed husks, pectin, and locust bean gums) can lower plasma total and LDL cholesterol levels by as much as 5% beyond the level achieved by a Step I diet. The total fiber contents of the one-day menus of the Step I and Step II diets are 32 and 31 g, respectively. The total, insoluble, and soluble fiber contents of some common foods appears in Table 3.

Patients frequently are surprised to find how easily lean red meat can be included in a Step I or Step II diet. Table 4 lists the total fat, saturated fat, and cholesterol content of lean and higher fat cuts of red meat, poultry, and fish. It is apparent that not all red meats are high in fat and chicken and fish are not always low in fat. In fact, depending on the chicken part and method of preparation, it may actually be higher in fat and saturated fat than red meats. For example, top round (lean red meat) has less fat and saturated fat than dark meat roasted chicken (without skin), dark meat roasted turkey (with skin), light meat roasted chicken (with skin), and lean ground turkey. Baked mackerel is considerably higher in fat and twice as high in saturated fat as top round. Tuna packed in oil is appreciably higher in fat than any lean cut of beef or pork listed and provides as much saturated fat as does top round and lean ham. Another misconception is that extra-lean

Table 2 Sample One-Day Menus for Three Diets for Males 25–49 Years

Average American diet	Step I diet	Step II diet
Breakfast	Breakfast	Breakfast
Bagel, plain (1 medium)	Bagel, plain (1 medium)	Bagel, plain (1 medium)
Cream Cheese, regular (1 T)	**Cream Cheese, low-fat (2 tsp)**	**Margarine (2 tsp)**
Cereal, shredded wheat (1 cup)	Cereal, shredded wheat (1½ cup)	Cereal, shredded wheat (1½ cup)
Banana, (1 small)	Banana, (1 small)	Banana, (1 small)
Milk, 2% (1 cup)	**Milk, 1% (1 cup)**	**Milk, skim (1 cup)**
Orange juice (¾ cup)	Orange juice (¾ cup)	Orange juice (¾ cup)
Coffee (1 cup)	Coffee (1 cup)	Coffee (1 cup)
Cream, half and half (1 T)	**Milk, 1% (1 oz)**	**Milk, skim (1 oz)**
Lunch	Lunch	Lunch
Minestrone Soup (1 cup)	Minestrone Soup (1 cup)	Minestrone Soup (1½ cup)
Roast Beef Sandwich	Roast Beef Sandwich	Roast Beef Sandwich
Whole Wheat Bread (2 slices)	Whole Wheat Bread (2 slices)	Whole Wheat Bread (2 slices)
Lean Roast Beef (3 oz)	Lean Roast Beef (3 oz)	Lean Roast Beef (3 oz)
American Cheese, regular (¾ oz)	**American Cheese, low-fat (¾ oz)**	**American Cheese, low-fat (¾ oz)**
Lettuce (1 leaf)	Lettuce (1 leaf)	Lettuce (1 leaf)
Tomato (3 slices)	Tomato (3 slices)	Tomato (3 slices)
Mayonnaise, regular (1 T)	**Mayonnaise, low-fat (2 tsp)**	**Margarine (2 tsp)**
Fruit and Cottage Cheese Salad	Fruit and Cottage Cheese Salad	Fruit and Cottage Cheese Salad
Cottage Cheese, regular (4% fat) (½ cup)	**Cottage Cheese, 2% (½ cup)**	**Cottage Cheese, 1% (½ cup)**
Peaches, canned in juice (½ cup)	Peaches, canned in juice (½ cup)	Peaches, canned in juice (½ cup)
Apple juice, unsweetened (6 oz)	Apple juice, unsweetened (1 cup)	Apple juice, unsweetened (1 cup)

Dinner

Salmon (3 oz)
 Vegetable oil (1 tsp)
Baked Potato (1 medium)
 Margarine (2 tsp)
Green Beans (½ cup), seasoned with
 margarine (½ tsp)
Carrots (½ cup), seasoned with
 margarine (½ tsp)
White Dinner Roll (1 medium)
 Margarine (1 tsp)
Ice Cream, regular (½ cup)
Iced Tea, unsweetened (1 cup)

Snack

Popcorn (3 cups)
 Margarine (1 T)

Dinner

Salmon (3 oz)
 Vegetable oil (1 tsp)
Baked Potato (1 medium)
 Margarine (2 tsp)
Green Beans (½ cup), seasoned with
 margarine (½ tsp)
Carrots (½ cup), seasoned with
 margarine (½ tsp)
White Dinner Roll (1 medium)
 Margarine (1 tsp)
Ice Milk (1 cup)
Iced Tea, unsweetened (1 cup)

Snack

Popcorn (3 cups)
 Margarine (1 T)

Dinner

Flounder (3 oz)
 Vegetable oil (1 tsp)
Baked Potato (1 medium)
 Margarine (2 tsp)
Green Beans (½ cup), seasoned with
 margarine (½ tsp)
Carrots (½ cup), seasoned with
 margarine (½ tsp)
White Dinner Roll (1 medium)
 Margarine (1 tsp)
Frozen Yogurt (1 cup)
Iced Tea, unsweetened (1 cup)

Snack

Popcorn (3 cups)
 Margarine (1 T)

Nutritional Analysis

Calories	2546	2518	2533
Total fat, % kcal	37	29	28
SFA, % kcal	13	8.6	6.6
Cholesterol, mg	241	150	150
Dietary fiber, g	28	32	31

Bold type indicates foods that are changed.
SFA = saturated fatty acids.

Table 3 Fiber Content of Some Common Foods
(g/serving)

Food	TDF	IF	SF
Vegetables (½ cup)			
Asparagus	2.4	2.0	0.4
Broccoli	2.8	1.5	1.3
Carrots	3.2	1.7	1.5
Cauliflower	2.6	1.7	0.9
Corn	1.9	1.7	0.2
Lettuce	0.9	0.7	0.2
Potato	2.0	1.0	1.0
Spinach	2.2	1.7	0.5
Legumes (½ cup)			
Black-eyed peas	3.9	3.5	0.4
Garbanzo beans	3.5	3.1	0.4
Kidney beans	6.2	4.7	1.5
Lentils	15.8	14.0	1.8
Pinto beans	6.9	4.7	2.2
Fruits			
Apple, small	2.8	1.8	1.0
Banana, ½	1.9	1.4	0.5
Orange, medium	1.9	0.8	1.1
Peaches, canned (½ cup)	1.9	1.1	0.8
Pears, canned (½ cup)	3.7	2.9	0.8
Pineapple, canned (½ cup)	2.1	1.8	0.3
Cereals (1 oz)			
Cornflakes	0.5	0.4	0.1
Oat bran	4.0	2.0	2.0
Oatmeal	2.5	1.3	1.2
Puffed rice	0.4	0.3	0.1
Wheat bran	10.0	9.0	1.0

TDF = Total dietary fiber; IF = insoluble fiber; SF = soluble
fiber.
Source: Ref. 72.

ground beef is relatively low in fat (as is ground turkey and even lean ground
turkey). A 3 oz serving of extra-lean ground beef provides 13.5 g of total fat and
5.3 g of saturated fat (20% of the daily total fat allowance and 25% of the saturated
fat allowance on a 2000-kcal Step I Diet) (93% fat free ground beef is available in
some supermarkets). A similar portion of ground turkey provides almost the same
amount of total fat and less saturated fat (i.e., 14% of the daily allowance). Table 4
lists four cuts of red meat and three types of ham that contain less total fat than
skinless dark meat chicken. Poultry skin contributes appreciably to the amount of

total fat in chicken and turkey. In general, removing the skin decreases the amount of total fat in a serving of chicken or turkey by 50%.

FOOD SOURCES OF TOTAL FAT AND SATURATED FAT

A fundamental precept in developing strategies to reduce consumption of total and saturated fat is to identify the major sources of these nutrients in the diet. Food disappearance data provide a useful estimate of food consumption trends. Based on food disappearance data (12), the three primary sources of fat in the U.S. diet in 1968 were fats/oils (40% of total fat), meat/poultry/fish (37%), and dairy products (12%). In 1988, these foods still were the primary sources of fat, however, the contribution of fats and oils had increased to 47%, whereas meats/poultry/fish had declined to 32%; dairy products remained unchanged. The decline in contribution of the meat group to total fat intake reflects an increased consumption of poultry (by 28 lb per capita) and fish (by 2 lb per capita) and a decrease in red meat (18 lb per capita) between 1967 and 1989 (13). With respect to dairy product consumption, the most significant change has been the steady substitution of low-fat and skim milk for whole milk. In fact, low-fat and skim milk comprised over half of all milk consumed in 1988 compared to 84% whole milk in 1967 (13). Despite the increase in low-fat and skim milk consumption, more cream products were consumed in 1988 than in 1979 (7.2 lb vs. 5 lb per person-year) (13). During the past 21 years, animal sources of fat have declined from 65 to 53%, while plant sources have increased from 35 to 47%. In fact, the major change in fat in the food supply has been in the type of fat used. Polyunsaturated and monounsaturated fat consumption has increased appreciably. Specifically, there has been almost a 2-fold increase in production of polyunsaturated fat from vegetable sources (6). It is important to note that disappearance data do not necessarily represent actual food consumption because they do not account for losses that occur after the wholesale/retail level. For example, it has been estimated that as much as one fourth of the fat in the food supply is not consumed (14) (a result of frying fat that is discarded, external fat that is trimmed from meat, food/fat that is not eaten, etc.). Nonetheless, they do provide interesting information about the foods Americans are purchasing.

In recent years, there has been a remarkable increase in the availability of fat- and calorie-modified foods and beverages on the market. Since 1978, there has been a greater than twofold increase in the number of adult Americans who consume such products (Fig. 2). The Healthy People 2000 goal of having 5000 products on the market that are reduced in fat and saturated fat by the year 2000 has already been met. Presently, there are in excess of 5600 fat-modified products available. Moreover, as noted by Light (15), this number can be expected to increase as new ingredients are introduced to reduce or replace fat. With the growing usage of fat-modified products by the public, it is likely that a further reduction in total fat and saturated fat consumption will occur. It will be important

Table 4 Total Fat and Saturated Fat Content of Lean and Higher-Fat Beef, Pork, Poultry, and Fish

Type of meat, poultry, fish (3 oz, cooked)	Total fat (g)	Saturated fat (g)
Lean beef (choice grade, 0″ fat trim, broiled or braised)[a]		
Top round	5.0	1.7
Top sirloin	6.7	2.6
Bottom round	7.5	2.5
Chuck arm pot roast	7.5	2.7
Higher-fat beef		
Chuck blade roast (choice, 0″ fat trim)	12.6	4.9
Extra-lean ground beef	13.5	5.3
Regular ground beef	16.7	6.6
Prime rib (1/4″ fat trim)	17.9	7.6
Lean pork (roasted)		
Lean ham, cured	4.7	1.5
Arm picnic, cured	6.0	2.1
Regular ham, cured	7.7	2.7
Higher-fat pork		
Sirloin	11.3	3.9
Center rib	11.8	4.1
Loin	11.9	4.1
Boston blade	14.4	5.0
Lean poultry		
Turkey w/out skin, light meat roasted	2.7	1.0
Chicken w/out skin, light meat roasted	3.9	1.1
Turkey w/out skin, dark meat roasted	6.2	2.1
Turkey with skin, light meat roasted	7.1	2.0
Chicken w/out skin, dark meat roasted	8.3	2.3
Lean ground turkey	8.4	2.5
Chicken with skin, light meat roasted	9.3	2.7
Higher-fat poultry		
Chicken w/out skin, dark meat fried	9.9	2.7
Turkey with skin, dark meat roasted	9.9	3.0
Chicken with skin, light meat fried	10.4	2.8
Ground Turkey	11.4	3.1
Chicken with skin, dark meat roasted	13.5	3.8
Lean fish		
Lobster, boiled	0.5	0.1
Baked cod	0.7	0.1
Baked haddock	0.8	0.1
Shrimp, boiled	0.9	0.2
Crab, boiled	1.3	0.1
Tuna, packed in water	1.5	0.3

Table 4 (Continued)

Type of meat, poultry, fish (3 oz, cooked)	Total fat (g)	Saturated fat (g)
Lean fish (continued)		
Swordfish, baked	4.4	1.2
Canned salmon	5.1	1.3
Higher-fat fish		
Shrimp, fried	10.4	1.8
Mackerel, baked	15.0	3.3
Tuna, packed in oil	19.4	1.5

[a]The value listed on the Nutrition Label is separable lean, 1/8″ fat trim.
Source: Ref. 73.

to establish the extent to which the availability and usage of these new food products affects the fat and saturated fat intake of our population.

Beyond identifying food-consumption trends, it is important to determine the major food sources of fat and saturated fat in the diets of different population groups. This latter information will help identify what foods should be modified by different groups. Several studies have been conducted (see Tables 5–7) to determine the major sources of total fat and saturated fat in the diets of men (16), women (17–19), and children (20–23). In addition, comparisons of ethnic groups have been made (Table 6). Despite the differences in the populations studied, the

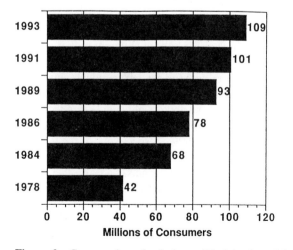

Figure 2 Consumption of calorie-modified foods and beverages by adult Americans, 1978–1991. (From Ref. 76.)

Table 5 Major Sources of Total Fat and Saturated Fat in Adult Diets[a]

Subjects	Sources of total fat	Sources of saturated fat	Ref.
12,842 men in MRFIT ages 35 to 57	High fat meats (24%)[b] Low- and medium-fat meats (12%) Fats (10%) Baked goods/desserts (9%) Saturated fats/oils (9%)	High-fat meats (25%) High-fat dairy products (13%) Low- and medium-fat meats (12%) Baked goods/desserts (10%) Saturated fats/oils (9%)	16
Low-income women ages 19 to 50	Meat, poultry, fish (36%) Grain products (21%) Milk and milk products (14%) Fats and oils (10%)		17
173 women ages 45 to 69	Fats and oils (19%) Red meats (16%) Milk (13%) Grains (7%)		18

[a]Foods listed are ranked on the basis of their contribution to total fat or saturated fat, respectively. It should be noted that foods were classified differently among the studies.
[b]Values in parentheses represent the percent contribution to the total or saturated fat intake.

dietary-assessment methodologies employed, and the year the studies were conducted, the major sources of total fat and saturated fat in the diets of the different population groups (including children) are remarkably similar. The major sources of total fat in the diets of men, women, and children and Caucasian, African American, and Hispanic men and women are: meat, milk and dairy products, and fats and oils. Because of the relatively high milk intake of children, milk is a more prominent contributor of total fat to the diet. With respect to saturated fat, again, there are striking similarities among groups, with full-fat dairy products appearing to contribute proportionally more to saturated fat than total fat. Thus, it is apparent that meats and dairy products are the major contributors of fat and saturated fat to the diet of Americans. Similarly, Ernst (24), in a recent review, reported that meat, poultry, fish, and eggs contributed 50–55 and 60%, respectively, to the total fat and saturated fat intake. Fats and oils contributed 33% of the total fat and 10% of the saturated fat intake. Grains, vegetables, fruits, sweets, and all other foods contributed 12–17% to the total fat and 30% to the saturated fat content of the diet. Collectively, this information gives us a good idea of the major sources of total fat and saturated fat in the American diet. Subtle differences in the contributions of different foods reported in recent studies (Tables 5–7) reflect the specific food-classification systems of different investigators. For example, Gorbach et al. (18) reported red meat and milk contributed only 29% of the total fat intake of the 173

Table 6 Major Sources of Total Fat and Saturated Fat in Adult Diets Based on Ethnicity[a]

Subjects	Sources of total fat			Sources of saturated fat		
	Caucasian	African-American	Hispanic	Caucasian	African-American	Hispanic
2,134 women ages 19 to 50	Grain mixtures (8%)[b] Meat mixtures (8%) Beef (8%) Sweet grains (8%) Salad dressing (8%)	Franks/Bacon (9%) Grain mixtures (8%) Beef (8%) Poultry (7%) Eggs (6%)	Meat mixtures (9%) Grain mixtures (8%) Beef (8%) Sweet grains (8%) Eggs (6%)	Cheeses/Cream (11%) Beef (9%) Grain mixtures (8%) Meat mixtures (7%) Butter/Margarine (6%)	Beef (10%) Franks/Bacon (9%) Grain mixtures (9%) Cheeses/Cream (7%) Meat mixtures (6%)	Whole milk (11%) Cheeses/Cream (10%) Beef (9%) Meat mixtures (9%) Sweet grains (7%)

[a]Foods listed are ranked on the basis of their contribution to total fat or saturated fat, respectively. It should be noted that foods were classified differently among the studies.

[b]Values in parentheses represent the percent contribution to the total or saturated fat intake.

Source: Ref. 19.

Table 7 Major Sources of Total Fat and Saturated Fat in Children's Diets[a]

Subjects	Sources of total fat	Sources of saturated fat	Ref.
185 (5–14 years old) in the Bogalusa Heart Study	Whole milk (18%)[b] Desserts (13%) Fats (11%) Beef (10%) Pork (10%)	Milk (26%) Desserts (13%) Beef (9%) Pork (9%)	20
476 females, 534 males (10–19 years old)	*Females* Fats (30%) Meat, fish, poultry (25%) Milk, cheese, yogurt (16%) *Males* Meat, fish, poultry (29%) Fats (28%) Milk, cheese, yogurt (18%)	*Females* Milk, cheese, yogurt (25%) Meat, fish, poultry (24%) Fats (23%) *Males* Meat, fish, poultry (27%) Milk, cheese, yogurt (27%) Fat (22%)	21
138 males & females (5th to 12th graders) in Texas	*>38% Fat Diet* Beefsteak/Roast/Ribs (9%)[b] Whole milk, milkshakes (8%) Pizza (6%) Ice cream (6%)	*<30% Fat Diet* Pizza (10%) Whole milk/milkshakes (8%) Fried potatoes (6%) Hamburgers/cheeseburgers (5%)	22
205 low-income Latinos (4–7 years old) in New York City		Whole milk (45%) Cheese (8%) Beef/Veal (6%) Pork/Lamb/Sausage (3%)	23

[a]Foods listed are ranked on the basis of their contribution to total fat or saturated fat, respectively. It should be noted that foods were classified differently among the studies.

[b]Values in parentheses represent the percent contribution to the total fat or saturated fat intake.

women in their study versus ~50–55% reported by some studies. Other investigators included poultry and fish in the meat grouping and other dairy products in the milk group, which explains the different results. Another important consideration when reviewing the data that report the food contributors to total and saturated fat in the diet is to appreciate that while a good estimate of food group sources can be obtained, less is known about how they are consumed. For example, cheese can be eaten in many different ways (e.g., as a snack, on a sandwich, in a mixed dish). Current information would be very valuable in understanding better how certain population groups are specifically consuming their major fat sources. Nonetheless, the information at hand is sufficient to give us a good estimate of the major food group sources of total fat and saturated fat in the diet.

Food-consumption patterns have changed over the years. Decreased consumption of total and saturated fat is due to a reduced intake of red meat and whole milk dairy products. With the growing availability of fat-modified products, it will be important to continually monitor the effects of changing food-consumption patterns on the nutrient profile of the diet and total fat and saturated fat intake. Additional studies are needed to characterize the specific food changes of different population groups and how these changes affect their nutrient intake.

ASSESSMENT: ENERGY NEEDS ASSESSMENT AND DIETARY ASSESSMENT

Assessing patients' energy needs, usual dietary intake, and adherence to a Step I or Step II diet are important activities in the implementation of dietary therapy. An energy needs assessment is required to plan a diet that either meets energy needs or leads to a slow weight loss. Dietary assessment initially provides essential information about a patient's usual diet, which is used to prescribe either a Step I or a Step II diet. Ongoing dietary assessment thereafter also is important to assess adherence to the prescribed diet. This section describes relatively simple methods that can be used to assess patients' energy needs and their habitual diets as well as their adherence to Step I and Step II diets.

Assessment of Energy Needs

To assess calorie requirements for weight maintenance, information about a patient's frame size and activity level is needed. Assessment of a patient's weight status also is required when prescribing a diet for either weight maintenance or weight loss. Simple techniques for assessing energy needs, frame size, activity level, and weight status follow.

Identification of Calories Needed to Maintain Weight
Tables 8 and 9 can be used to estimate the number of calories needed to maintain present body weight. Calories can be increased or decreased to achieve a weight

Table 8   Calorie Needs of Adult Males[a]

Height without shoes[b]	Frame size	Desirable weight (lb)	Calorie level based on physical activity			
			Very low	Low	Moderate	High
5'5"	Small	129 (124–133)	1700	1950	2200	2600
	Medium	137 (130–143)	1800	2050	2350	2750
	Large	147 (138–156)	1900	2200	2500	2950
5'6"	Small	133 (128–137)	1750	2000	2250	2650
	Medium	141 (134–147)	1850	2100	2400	2800
	Large	152 (142–161)	2000	2300	2600	3050
5'7"	Small	137 (132–141)	1800	2050	2350	2750
	Medium	145 (138–152)	1900	2200	2450	2900
	Large	157 (147–166)	2050	2350	2650	3150
5'8"	Small	141 (136–145)	1850	2100	2400	2850
	Medium	149 (142–156)	1950	2250	2550	3000
	Large	161 (151–170)	2100	2400	2750	3200
5'9"	Small	145 (140–150)	1900	2200	2450	2900
	Medium	153 (146–160)	2000	2300	2600	3050
	Large	165 (155–174)	2150	2500	2800	3300
5'10"	Small	149 (144–154)	1950	2250	2550	3000
	Medium	158 (150–165)	2050	2350	2700	3150
	Large	169 (159–179)	2200	2550	2850	3400
5'11"	Small	153 (148–158)	2000	2300	2600	3050
	Medium	162 (154–170)	2100	2450	2750	3250
	Large	174 (164–184)	2250	2600	2950	3500
6'0"	Small	157 (152–162)	2050	2350	2650	3150
	Medium	167 (158–175)	2150	2500	2850	3350
	Large	179 (168–189)	2350	2700	3050	3600
6'1"	Small	162 (156–167)	2100	2450	2750	3250
	Medium	171 (162–180)	2200	2550	2900	3400
	Large	184 (173–194)	2400	2750	3150	3700
6'2"	Small	166 (160–171)	2150	2500	2800	3300
	Medium	176 (167–185)	2300	2650	3000	3500
	Large	189 (178–199)	2450	2850	3200	3800
6'3"	Small	170 (164–175)	2200	2550	2900	3400
	Medium	181 (172–190)	2350	2700	3100	3600
	Large	193 (182–204)	2500	2900	3300	3850

[a]Adapted from 1959 Metropolitan Life Insurance Company, New York City. These tables are based on 1959 rather than 1983 Metropolitan Life Insurance Company height-weight tables because the earlier tables specify lower weights, which are more appropriate to health-related concerns.
[b]Table adjusted for measurement of height without shoes.

Table 9 Calorie Needs of Adult Females[a]

Height without shoes[b]	Frame size	Desirable weight (lb)	Calorie level based on physical activity			
			Very low	Low	Moderate	High
5'0"	Small	106 (102–110)	1400	1600	1800	2100
	Medium	113 (107–119)	1450	1700	1900	2250
	Large	123 (115–131)	1600	1850	2100	2450
5'1"	Small	109 (105–113)	1400	1650	1850	2200
	Medium	116 (110–122)	1500	1750	1950	2300
	Large	126 (118–134)	1650	1900	2150	2500
5'2"	Small	112 (108–116)	1450	1700	1900	2250
	Medium	119 (113–126)	1550	1800	2000	2400
	Large	129 (121–138)	1700	1950	2200	2600
5'3"	Small	115 (111–119)	1500	1750	1950	2300
	Medium	123 (116–130)	1600	1850	2100	2450
	Large	133 (125–142)	1750	2000	2250	2650
5'4"	Small	118 (114–123)	1550	1750	2000	2350
	Medium	127 (120–135)	1650	1900	2150	2550
	Large	137 (129–146)	1800	2050	2350	2750
5'5"	Small	122 (118–127)	1600	1850	2050	2450
	Medium	131 (124–139)	1700	1950	2250	2600
	Large	141 (133–150)	1850	2100	2400	2800
5'6"	Small	126 (122–131)	1650	1900	2150	2500
	Medium	135 (128–143)	1750	2050	2300	2700
	Large	145 (137–154)	1900	2200	2450	2900
5'7"	Small	130 (126–135)	1700	1950	2200	2600
	Medium	139 (132–147)	1800	2100	2350	2800
	Large	149 (141–158)	1950	2250	2550	3000
5'8"	Small	135 (130–140)	1750	2050	2300	2700
	Medium	143 (136–151)	1850	2150	2450	2850
	Large	154 (145–163)	2000	2300	2600	3100
5'9"	Small	139 (134–144)	1800	2100	2350	2800
	Medium	147 (140–155)	1900	2200	2500	2950
	Large	158 (149–168)	2050	2350	2700	3150
5'10"	Small	143 (138–148)	1850	2150	2450	2850
	Medium	151 (144–159)	1950	2250	2550	3000
	Large	163 (153–173)	2100	2450	2750	3250

[a]Adapted from 1959 Metropolitan Life Insurance Company, New York City. These tables are based on 1959 rather than 1983 Metropolitan Life Insurance Company height-weight tables because the earlier tables specify lower weights, which are more appropriate to health-related concerns.
[b]Table adjusted for measurement of height without shoes.

gain or loss. Using these tables requires the following information: height, frame size, and activity level. Height can be measured easily. Frame size and activity level can be estimated using techniques described below.

Frame Size.   Frame size can be estimated from the ratio of height (cm) to wrist circumference (cm). The wrist is measured distal to the styloid process (where it bends) on the right arm. Table 10 shows the values for the classification of frame size (i.e., small, medium, or large) for men and women.

Activity Level.   Activity level can be estimated very simply using the Lipid Research Clinics physical activity status questionnaire and scoring method (Table 11) (25). Patients are asked four questions about their level of physical activity. The first two questions rate patients' level of physical activity (both at work and in leisure activities) relative to their peers. The third and fourth questions address the issue of frequency of exercise or hard physical labor. If the answer to question 3 is yes, then patients are asked whether they participate in these activities at least three times a week. The four-point scoring method in Table 11 classifies individuals into the following physical activity categories: high activity, moderate activity, low activity, and very low activity.

Assessment of Weight Status

An assessment of a patient's weight status is required to establish the calorie level of the Step I or Step II diet. Whereas some patients require a diet to maintain their weight, many others need to lose weight.

Different techniques are used in practice to assess a patient's weight status. Body mass index (BMI) has become the recommended technique because of its simplicity and accuracy. The Metropolitan Height-Weight tables also are used in practice and provide information about whether a patient's weight falls within a general, recommended range. However, it is important to note that the weight ranges presented in the tables are quite large and thus provide only a "rough" estimate of weight status.

Table 10   Determination of Frame Size from Height and Wrist Circumference Measurements

Males	Females
$r > 10.4$ small	$r > 11.0$ small
$r = 9.6$–10.4 medium	$r = 10.1$–11.0 medium
$r < 9.6$ large	$r < 10.1$ large

r = height (cm)/wrist circumference (cm)
Source: Ref. 74.

Table 11 Assessment of Physical Activity (The LRC Questionnaire and Scoring Method)

INSTRUCTIONS

Answer the following four questions. Use the scoring system described below to classify patients' activity levels as: high activity, moderate activity, low activity, or very low activity.

QUESTIONS

1. Thinking about the things you do at work, how would you rate yourself as to the amount of physical activity you get compared with others of your age and sex?
   a. Much more active   c. About the same   e. Much less active
   b. Somewhat more active d. Somewhat less active f. Not applicable
2. Now, thinking about the things you do outside of work, how would you rate yourself as to the amount of physical activity you get compared with others your age and sex?
   a. Much more active   c. About the same   e. Much less active
   b. Somewhat more active d. Somewhat less active
3. Do you regularly engage in strenuous exercise or hard physical labor?
   a. Yes (answer question #4)
   b. No (stop)
4. Do you exercise or labor at least three times a week?
   a. Yes
   b. No

SCORING

Use answer to question 2 to assess self-rating of physical activity during leisure and answers to questions 3 and 4 to assess participation in strenuous exercise or labor. The answer to question 1 can provide additional information that may be needed to rate physical activity level.

| | | Self-Rating of Physical Activity During Leisure (relative to others of own age and gender) | | |
		Much more	Somewhat more/ about the same	Somewhat less/ much less
Strenuous Exercise or Labor	Yes	High activity	Moderate activity	
	No	Low activity		Very low activity

Source: Ref. 25.

BMI is determined from height and weight measurements using the following equation:

$$BMI = \frac{Wt\ (kg)}{Ht^2\ (cm)}$$

If weight is in pounds and height is in inches, the following equation can be used to calculate BMI (26):

$$BMI = \frac{Wt\ (lb)}{Ht^2\ (in)} \times 705$$

The nomogram in Figure 3 also can be used to determine BMI. Table 12 presents the BMI classifications for males and females. In general, an acceptable BMI is 19–25 kg/m² for men and women ages 19–34 years and 21–27 kg/m² for persons over 35 years of age. A BMI of >25 kg/m² for younger men and women and >27–31 kg/m² for men and women 35 years old is classified as overweight.

Dietary Assessment

Dietary assessment is the first step in developing an eating pattern for a patient who requires dietary therapy to lower LDL cholesterol levels. It provides information about a patient's usual food-consumption patterns and nutrient intake. This information is needed to prescribe the appropriate therapeutic diet. Dietary assessment performed over time provides important information about *changes* in food consumption practices and the intake of key nutrients. The information is used to assess patients' adherence to the Step I or Step II diet and, therefore, is essential for making subsequent treatment decisions.

Assessing a patient's consumption of key nutrients (e.g., saturated fat, cholesterol, and total fat) and dietary-intake patterns (e.g., usual food selections, number of meals and snacks eaten/daily, where they are eaten and with whom, the nutrient composition of the meals) typically is done using one of the following assessment tools: food-frequency instruments, 24-hour recalls, or food records. It is important to recognize the relative merits and limitations of each dietary-assessment instrument and the information provided when collecting dietary intake data.

A food-frequency questionnaire captures usual food-intake patterns and is quick and easy to administer. However, it is important to recognize that this method may not reflect actual intake because of difficulties in estimating frequency. Thus, it may appear that a patient is following a Step I or Step II diet, but in fact is not. To collect more accurate dietary information, we recommend food-frequency questionnaires be administered (orally) by someone who is knowledgeable and skilled in dietary-assessment techniques (e.g., someone who can ask appropriate follow-up questions and provide clarification to questions as needed). When patients provide accurate information, a food-frequency instrument also can

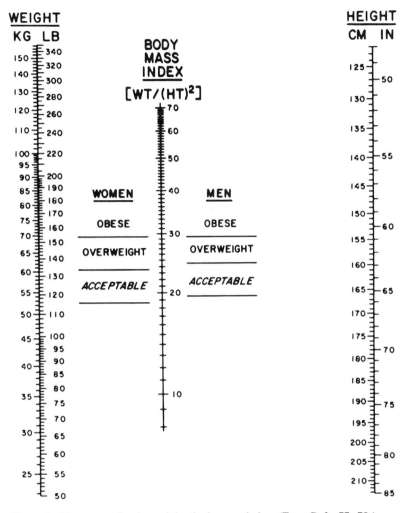

Figure 3 Nomogram for determining body mass index. (From Refs. 77, 78.)

be used by patients as a self-assessment tool. MEDFICTS (*M*eat, *E*ggs, *D*airy, *F*ish, *I*n Baked Goods, *C*onvenience Foods, *T*able Fats, and *S*nacks) is a food-frequency instrument that was developed specifically to assess adherence to a Step I or Step II diet. A copy appears in the ATP II report (1). Other dietary assessment approaches include Rate-Your-Plate (developed and validated by the Pawtucket Heart Health Program) (27), the Northwest Lipid Research Clinic's Fat Intake Scale (28), and a food-scoring tool to assess dietary adherence to a cholesterol-

Table 12 Body Mass Index Classification

	Women <35 years old	Men <35 years old	Men and women >35 years old
Underweight	<15	<16	<16
Desirable range	19–24	20–25	21–27
Overweight	24–29	25–30	27–31
Obese	>29	>30	>32

Source: Ref. 75.

lowering diet developed by nutritionists at Rush-Presbyterian St. Lukes Medical Center. (See the Resources section of this chapter for information about where to obtain a copy of the latter two instruments.)

A 24-hour recall can provide more quantitative information but requires that the day analyzed be representative of the patient's usual food intake. It may be necessary for a registered dietitian or other health professional to prompt patients to remember all foods and beverages consumed, provide guidance in estimating portion sizes, identify types of foods eaten (e.g., type of meat and margarine eaten), and explain their food-preparation techniques. This will increase the accuracy of information collected as will repeated 24-hour recalls.

Food records (collected for 3, 4, or 7 days including 1 or 2 weekend days) can provide a good estimate of a patient's usual intake. They are time-consuming for both the patient and the person analyzing the records but can provide more comprehensive information if all foods eaten are recorded accurately. Frequently, additional information (points of clarification) must be collected from the patient prior to analyzing the records.

Dietary assessment is a critical activity in implementing the ATP II Guidelines. Various methods can be used to collect information about a patient's habitual nutrient intake and food-consumption patterns as well as changes over time. This information provides the basis for all treatment decisions targeted to lowering LDL cholesterol levels (see below).

STRATEGIES FOR IMPLEMENTING THE STEP I AND STEP II DIETS

Individualized dietary therapy is the key to achieving life-long management of a blood cholesterol–lowering diet. In addition, both health professionals and patients must be willing to make a long-term commitment to this endeavor. Permanent dietary changes require both time and effort. Whereas some patients are able to modify their food-consumption practices relatively quickly and easily, others may need 6–12 months or even longer. Many patients will experience some degree of relapse back to established dietary habits requiring follow-up counseling sessions over time. Registered dietitians and other qualified nutrition professionals have the expertise to work closely and effectively with patients to implement the Step I and Step II diets. These nutrition experts recognize the importance of individualized diet therapy and, furthermore, have the requisite knowledge and skills needed to identify and help patients develop and implement an appropriate dietary-modification strategy.

In general, three approaches are used most often to implement Step I and Step II diets: (1) counting grams of fat and saturated fat, (2) using the Exchange Lists from the American Heart Association (29), and (3) applying relatively simple fat-reduction strategies to food choices. These approaches can be applied singly or in combination—two or even three can be used by one patient. Inherent in each of these strategies is initial and ongoing dietary assessment, goal setting, monitoring, evaluation, and, if indicated, modifying the treatment plan (e.g., changing the specific implementation strategy). Frequent follow-up and monitoring visits are of great assistance to patients (especially in the early stages of their treatment programs) in achieving the goals of their prescribed diets. The initial phase of the dietary-therapy program is critical to the patients' overall success in making long-term dietary changes. It is essential that practitioners be available to answer questions and be prepared to consider a different dietary strategy (or a combination of strategies), if indicated, for achieving adherence to either diet.

The application of approaches that can be used in practice to achieve the recommendations often requires first estimating a patient's energy needs to maintain current weight (as described above). Since achieving and maintaining desirable weight is a recommendation of both Step I and II diets, weight reduction is important for patients who are overweight or obese. A reasonable strategy for most patients who need to lose weight is to reduce calorie intake by 500 calories per day to achieve a gradual weight loss of ½–1 pound per week while increasing daily physical activity. A greater restriction may be necessary for very obese patients.

The strategies described below for implementing the Step I and Step II diets are for patients who are at their desirable weight as well as for those who must lose weight. Strategies 1 and 2 require identifying the appropriate calorie level for weight maintenance or weight loss. The application of the third strategy will result in reducing calories (e.g., mainly fat calories) in the diet. For patients who do not need to lose weight, weight maintenance is the goal; calories can be added back to the diet with high-carbohydrate foods.

Strategy 1: Counting Grams of Fat and Saturated Fat (and Achieving the Recommended Cholesterol Allowance)

Just as some individuals like to count calories, some patients like to count grams of fat. While fewer actually count grams of saturated fat in foods they select, they could learn how to do this. Table 13 lists the maximum daily intake of fat and saturated fat allowable in diets that provide 1600–3000 kcal. Since most foods sold

Table 13 Maximum Daily Intake of Fat and Saturated Fat to Achieve the Step I and Step II Diets[a]

	Total calorie level							
	1600	1800	2000	2200	2400	2600	2800	3000
Total fat, g[b]	53	60	67	73	80	87	93	100
Saturated fat—Step I, g[c]	18	20	22	24	27	29	31	33
Saturated fat—Step II, g[c]	12	14	16	17	19	20	22	23

[a]Average daily energy intake for women is 1800 kcal, for men, 2500 kcal.
[b]Total fat of both diets = 30% of calories (estimated by multiplying calorie level of the diet by 0.3 and dividing the product by 9 kcal/g).
[c]Recommended intake of saturated fat on the Step I diet should be 8–10% of total calories, and less than 7% for the Step II diet.
Source: Ref. 1.

in the supermarket have grams of fat and saturated fat listed on the label, it is easy for patients to calculate their fat and saturated fat intake. Patients who choose to count grams of fat and saturated fat will benefit from having resource books/booklets that provide this information on all types of foods including those eaten in restaurants and other places away from home (see the Resources Section).

To achieve the recommended cholesterol allowance, patients should limit major food sources of dietary cholesterol. Table 14 lists the cholesterol content of selected foods. The predominant source of cholesterol in the diet of most Americans is egg yolk, which provides approximately 211 mg of cholesterol per yolk.

Strategy 2: Use the Exchange Lists from the American Heart Association

Many individuals have used exchange lists to plan diets for weight loss or for managing conditions such as diabetes. They are familiar with this approach and find it easy for meal planning. Table 15 lists the number of servings per day from food groups for Step I and Step II diets for different calorie levels. (Guidelines for estimating energy needs are presented above.) When using this approach, patients still must be taught how to make appropriate food choices within each food group. In addition, many individuals are unaware of standard portion sizes and need guidance on estimating serving sizes.

Strategy 3: Apply One or More Fat-Reduction Techniques to Meal Planning

There are various fat-reduction techniques that can be used to reduce total fat, saturated fat, and cholesterol in the diet (30): (a) substituting low-fat foods for

Table 14 Cholesterol Content of Selected Foods

Source	Cholesterol content (mg)
Red meat, lean, 3 oz cooked	
Beef	77
Lamb	78
Pork	79
Veal	128
Organ meats, 3 oz cooked	
Liver	270
Brains	1746
Poultry, without skin, 3 oz cooked	
Chicken	
Light	72
Dark	79
Turkey	
Light	59
Dark	72
Fish, 3 oz cooked	
Cod	
Haddock	
Salmon	74
Shellfish, 3 oz cooked	
Clams	57
Crabmeat	
Alaskan king	45
Blue crab	85
Lobster	61
Oysters	93
Scallops	35
Shrimp	166
Egg yolk, 1 large	211
Dairy products	
Cheese, Cheddar, 1 oz	30
Mozzarella, part skim, 1 oz	15
Milk, 8 oz	
Whole	33
2%	20
1%	10
Skim	4

Source: Adapted from Ref. 1.

Table 15 Number of Servings per Day from Food Groups for Different Calorie
Levels

Food group	Step I				Step II			
	2500	2000	1600	1200	2500	2000	1600	1200
Meat, poultry, and fish	6 oz	6 oz	6 oz	6 oz	5 oz	5 oz	5 oz	5 oz
Eggs (per week)	3	3	3	3	1	1	1	1
Dairy products	4	3	3	2	3	2	2	2
Fat	8	6	4	3	8	7	5	3
Bread, cereal, pasta, and starchy vegetables	10	7	4	3	10	8	5	4
Vegetables	4	4	4	4	5	4	4	4
Fruit	5	3	3	3	7	4	3	3
Optional foods[a]	2	2	2	0	2	2	2	0

[a]Optional foods include fat-modified desserts, fat-free or low-fat sweets, and alcoholic beverages. If foods from the Optional Foods group are not used, add two portions from the Bread, Cereal, Pasta, and Starchy Vegetables group plus one portion from the Fat group.
Source: Ref. 29.

higher-fat counterparts (e.g., skim milk in place of whole milk), (b) decreasing the quantity of high-fat foods, (c) replacing high-fat foods with foods lower in fat (e.g., legumes for red meat), and (d) changing preparation techniques (e.g., broiling instead of frying). The impact of the application of these strategies on total fat and saturated fat in the diets of men, women, and children was examined using computer-modeling techniques (31–34). Smith-Schneider et al. (34) applied the following fat-reduction strategies to a representative 7-day menu for 25- to 50-year-old men:

Lean meats and lower-fat cheeses were substituted for higher-fat meat and cheeses.
Skim milk was substituted for higher-fat milks.
Fat-modified products were substituted for higher-fat foods.

The impact of these single and multiple strategies on total fat and saturated fat in the diet is shown in Figure 4. One strategy (substituting lean meats and lower-fat cheeses for higher-fat counterparts) nearly achieved the total fat and saturated fat goal of the Step I diet. A combination of two techniques reduced total fat to 28–29% of calories and saturated fat to 9.5% of calories. Application of all three resulted in a diet that provided 24% of calories from fat and 7.5% of calories from saturated fat. As expected, with the application of each fat-reduction strategy, there was a further decrease in total fat, saturated fat, and calories (by 200–400 kcal). It is important to note that use of just two strategies achieved the recommended

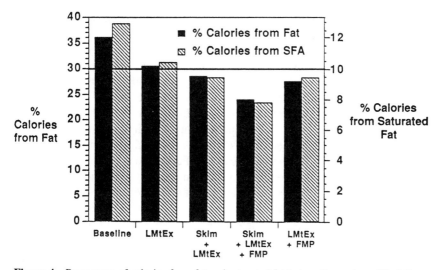

Figure 4 Percentage of calories from fat and saturated fat in baseline and modified diets for men. LMtEx = Lean meat exchanges and lower-fat cheeses; Skim = skim milk; FMP = fat-modified products. (From Ref. 34.)

nutrient goals. This provides flexibility in diet planning for individuals, especially with respect to a Step I diet, as not all fat-reduction strategies need to be applied. Certain foods need not be eliminated as long as other dietary changes have been made. Since a Step II diet requires further reductions in saturated fat, all three fat-reduction strategies need to be applied (see Fig. 4). No single strategy met the nutrient specifications of the Step II diet mainly because of the magnitude of the dietary changes required. However, with the increasing availability of fat-modified products, incorporation of a greater number of these products into the diet would give slightly greater flexibility in food choices from the meat and milk groups.

Computer-modeling exercises have identified several challenges for health professionals when providing information about how to lower fat and saturated fat. Choice of a single versus multiple strategies results in varying levels of dietary fat depending on gender and age (32,34). Combining two modifications such as substitution of lean meats and cheeses for higher-fat ones *plus* use of skim milk for whole milk resulted in lowering total fat from 36 to ~28% of calories for men, from 37 to ~30% of calories for women, and from 34 to ~23% of calories for preschool-aged children. Use of additional fat-reduction techniques (such as substituting fat-modified products for higher-fat ones) could decrease total fat to <20% of calories in children's diets, and calories could be reduced by as much as ~350. Thus, extensive use of multiple fat-reduction techniques can present a unique challenge in meeting the energy and nutrient needs of children. Hence, when providing nutrition guidance for one family member, general recommendations for the entire family must be given.

Diet planning for patients on Step I and Step II diets that must also be reduced in sodium can be challenging, since many fat-modified products are relatively high in sodium. Achieving recommended levels of sodium (2400 mg/day) in addition to meeting fat, saturated fat, and cholesterol recommendations may limit the use of convenience foods, commercially processed foods, and out-of-home meals. To achieve consumption of no more than 2400 mg of sodium, women ages 25–50 years may need to eliminate processed meats and high-sodium foods such as vegetable juices, soups, and crackers (33). Furthermore, maintaining sodium levels below 2400 mg requires proportionately more dairy products, fruit, and lean red meat and proportionately fewer servings of grain products, fish, convenience foods, and soups (30). The latter foods are frequently recommended for use in cholesterol-lowering diets. Thus, integrating cholesterol-lowering diet recommendations with other contemporary recommendations, while theoretically achievable, will require personalized nutrition guidance.

In summary, different techniques can be used to implement the Step I and Step II diets. These approaches enable practitioners to individualize dietary therapy, thereby simplifying implementation of the Step I and Step II diets for patients. All three approaches are relatively easy to apply in practice. Moreover, patients (and nutritionists) can select the approach most appropriate for them. Choices and simple implementation strategies will benefit both patients and practitioners in successfully implementing dietary therapy for elevated blood cholesterol levels.

Perception of Fat-Reduction Strategies

Several investigators have examined perceptions about the ease of implementing different fat-reduction strategies. Adults in the United Kingdom ($n = 390$) completed surveys that targeted attitudes and beliefs about nine behaviors associated with reducing fat intake (35). Respondents were stratified into a low-fat (31 ± 4% energy from fat), a medium-fat (39 ± 2% energy from fat), or a high-fat group (46 ± 3% energy from fat). For all three groups, attitude scores for making various changes were positive except for reducing red meat, which was neutral. Intention scores for reducing red meat were the only ones that significantly differed from the midpoint in a negative direction, indicating the respondents were least likely to make this change.

Participants in the Multiple Risk Factor Intervention Trial (MRFIT) (16) reported that "easy" dietary changes were increasing fish and poultry, skim milk, and low-fat milk, polyunsaturated margarines and oils, fruits, low-fat breads and cereals, and reducing egg yolks. "Difficult" changes were reducing high-fat beef and pork, high-fat cheeses, high-fat crackers, snacks, and desserts and increasing use of vegetarian meals. Fat-reduction behaviors perceived as most difficult for 658 adults in New York (36) included decreasing consumption of ice cream, potato chips, chocolate, high-fat desserts, and cheese as snacks.

Participants in the Women's Health Trial, an intervention to reduce fat intake from 39 to 20% of calories, identified strategies that were easiest and hardest to implement and maintain for 2 years (37). Decreasing fats and oils was one of the easiest changes to make initially. Fat from fats and oils decreased from ~20 to ~7 g/day after 3 months of intervention; however, the average consumption increased to 9 g/day after 24 months, indicating some recidivism in this population with this particular food group. Fat consumption from red meat decreased from 16 to 5 g/day in "high-level" performers and was essentially maintained for 24 months. In "low-level" performers, fat from red meats decreased ~3 g. Likewise, decrease in the fat-intake pattern from grains/baked goods was similar. These results suggest changes in red meat and grains/baked goods are easier for some women to make than others. In contrast, the decrease in fat consumption from dairy products appeared to be the easiest change to make and sustain because even subjects who were classified as "low-level" performers reduced and sustained fat consumption from dairy products almost as much as did "high-level" performers. With the exception of low-fat cheese, substitution of low-fat foods was easily adopted and maintained (30). Women had greater difficulty increasing their consumption of complex carbohydrates. While there was a modest (~10–12 g) increase in consumption of carbohydrates from fruit, there were only negligible increases in consumption of carbohydrates from grains and baked goods and no change in carbohydrate intake from vegetables.

Based on the results of these studies, strategies that involve reducing consumption of high-fat red meat may be difficult. In addition, it appears that reducing cheese consumption and using fat-modified cheese are challenging. Decreasing fat consumption from fats and oils appears to be a change that is somewhat difficult to maintain. Whereas one of the goals of a Step I or Step II diet is to increase carbohydrate intake from grain products, fruits, and vegetables, intervention studies suggest that this also is a difficult change to implement. However, a recent epidemiological study has shown that individuals classified in the lowest quartile for fat consumption substitute certain carbohydrate-rich foods such as fruits and vegetables for fat (38). Thus, this study (together with food disappearance data showing decreases in red meat and whole milk consumption) has shown that recommended food intake behaviors can be and are being implemented. Nonetheless, changes in certain food groups are more difficult to make.

It is critical for practitioners to appreciate the ease or difficulty people have in implementing dietary changes. Practitioners will be better prepared to assist patients in achieving dietary goals if they ask patients to consider how easy each suggested change will be. More intensive counseling and greater assistance in identifying realistic changes should be a part of each patient's treatment plan.

GUIDELINES FOR SELECTION OF FOOD

Patients need guidelines for selecting food in supermarkets and restaurants to help them plan diets that meet total fat, saturated fat, and cholesterol recommendations.

Selection of Food in Supermarkets

Consumers both want and need nutrition information at the point where they make most food-purchasing decisions. Thus, supermarkets offer a potentially important opportunity to positively influence food choices. A variety of point-of-purchase nutrition education programs are available in many supermarkets, including shelf-labeling programs, print materials, videotapes, taste-testing, games and activities, and shopping tours.

Shelf-labeling programs identify specific products as advantageous because they are lower in calories, sodium, fat, saturated fat, cholesterol, or sugar. Color codes often highlight the recommended foods. Shelf labels have been identified as a key aspect of successful supermarket nutrition programs (39). Reasons for their effectiveness include ease of identification of foods included in the program, being located at the point where consumers make most of their food purchasing decisions, and targeting information that consumers deem important.

Use of print materials such as brochures, posters, and recipes is another common educational strategy used in supermarkets. For example, the National Cancer Institute's 5-a-Day program now distributes brochures and recipes to promote consumption of fruits and vegetables. Print materials are less expensive to produce and distribute than television or radio messages. Unfortunately, print materials have not proven very effective in eliciting desired changes in food-purchasing behavior. Five studies that utilized this approach and measured changes in product sales or product usage reported either minimal changes (41), limited changes (42), or no changes at all (43–45).

Because supermarket nutrition programs must compete with commercial advertising, use of supplemental media in combination with print materials may prove more successful. Examples of such activities include videotapes, taste tests, and games. Taste tests provide consumers the opportunity to sample a healthful food product or recipe with the goal of encouraging its incorporation into the consumer's diet. No studies have evaluated the impact of taste tests on food-purchasing behavior.

Games and activities that encourage active participation may be more effective in promoting nutrition programs when compared to passive approaches (46). The Minnesota Heart Health Program developed a "bingo-like" game called The Shop Smart Game, which used illustrations of healthful foods instead of numbers on the game cards (47). With the purchase of a targeted food item, customers received a game piece. This strategy resulted in an increased purchase of 13 of the 20 foods; however, for only two foods was the increase significant.

Videos provide visual reinforcement of information and require little maintenance, however, they can be expensive to create and install. Two studies designed to examine the impact of videotapes have reported conflicting results. Dougherty et al. (48) were unable to demonstrate significant changes in product sales with the use of videotaped nutrition messages. Mullis et al. (49), on the other hand, reported significant increases in product sales. In comparing these studies, the latter used shorter messages that ran continuously, and the videotape was located next to the specific food item being targeted.

Supermarket tours are a relatively simple and inexpensive way to assist consumers in making healthful food choices. They are interactive and provide immediate feedback on specific products as well as "hands-on" practice with interpreting food labels (50). Research on the impact of supermarket tours is limited. Scott and Pollard (51) reported increased use of reduced-fat or low-fat dairy products by participants.

In general, point-of-purchase nutrition programs in supermarkets have been more successful in improving nutrition knowledge and attitudes than in changing food-purchasing behaviors (52). Nonetheless, some programs have led to food behavior changes. The characteristics of these successful programs include utilizing a variety of in-store activities and the strong support of supermarket personnel. In addition, relevant messages are developed from credible sources and there is a long-term commitment to the program and to educating consumers about nutrition (39). Supermarkets offer a win-win opportunity for consumers and nutrition educators that remains virtually untapped.

Selection of Food When Eating Out

The following suggestions can help individuals choose foods lower in fat, saturated fat, and cholesterol when eating out.

Select roasted, grilled, or baked meats instead of fried meats or casseroles. The latter are usually high in fat because they contain ground meat, butter, sour cream, oil, and/or cheese.

Select lean red meats, such as sirloin or tenderloin of beef, filet mignon (without bacon), loin pork chops, ham steak, or leg of lamb with the fat cut off, rather than higher-fat cuts, such as prime rib, prime grade steaks (i.e., T-bone steaks, rib-eye steaks, New York strip steaks) or ribs.

Ask that meat fat and poultry skin be removed before cooking.

Request that meat, poultry, and fish be broiled or grilled without added fat.

Ask to have fish "baked with a splash of wine," "poached," or "shallow-poached" (partially cooked on top of the stove and finished in the oven). Fish that is "broiled dry" may not be tasty.

Choose foods that are grilled, poached, roasted, steamed, or broiled dry (or with an oil low in saturated fat).

Avoid/Limit deep-fried foods; ask that a very small amount of oil be used in stir-frying foods.

Avoid/Limit foods made with rich sauces or cheese that cannot be served on the side, such as beef stroganoff, veal parmigiana, and macaroni and cheese.

Select hard rolls, hard breadsticks, saltine crackers, or sliced bread. These are lower in fat than cornbread, muffins, and dinner rolls (brushed with margarine or butter).

Eat bread, rolls, and crackers without butter or margarine.

Either skip dessert, split a serving with your dinner companion(s), or select fruit, angel food cake, sherbet, frozen yogurt, or gelatin (without sour cream, cream cheese, or whipped topping) instead of pie, cake, cookies, mousse, cheesecake, or ice cream. Most fat in fruit pie and cobbler is in the crust—reduce the fat by eating the filling and only a small amount of the crust. Most whipped toppings, even those labeled "nondairy," are high in fat and saturated fat.

Eat a large salad or a low-calorie soup before a meal to decrease appetite.

Order a vegetarian rather than a meat pizza.

Request less cheese or no cheese be used on pizza.

Ask what type of oil is used in food preparation; preferred oils include canola, corn, olive, safflower, soybean, and sunflower.

Share an entree (appetizer, salad, or dessert) with dining companion.

THE NUTRITION LABEL: AN IMPORTANT NUTRITION EDUCATION TOOL FOR IMPLEMENTATION OF BLOOD CHOLESTEROL–LOWERING DIETS

As a result of the passage of the Nutrition Labeling and Education Act of 1990, nutrition labels now appear on most foods, except fresh produce. Nutrition labels are optional on fresh meat, fish, and poultry products. Nutrition labels facilitate planning diets to meet specified nutrient criteria. For patients following Step I and Step II diets, the nutrition label provides grams of fat and saturated fat and milligrams of cholesterol per serving. The following information also appears: grams of total carbohydrate, dietary fiber, sugars, and protein, and milligrams of sodium. The percent Daily Value is the label reference value that illustrates how a serving of a food fits into a 2000-kcal diet. For macronutrients, cholesterol, and sodium, it is based on contemporary dietary recommendations. For vitamins and minerals, it is based on the U.S. Recommended Dietary Allowance (RDA) value [which is now the Reference Daily Intake (RDI)]. The amount of polyunsaturated and monounsaturated fat also can be listed, should a manufacturer decide to do so. Patients need to know that the amount of saturated fat, polyunsaturated fat, and monounsaturated fat (if the latter two are listed) will not equal the total fat on the label. This is because the *trans*-unsaturated fats will not be included with the other unsaturated fats on the nutrition label. The standard label format is shown in Figure 5.

Nutrient content claims and health claims can be included on the food labels. A nutrient content claim is a label word or phrase used on a food package that describes the amount of a nutrient in a serving of food (53). The nutrient content claims for fat, saturated fat, and cholesterol can be made only if a serving of food meets specified criteria. These criteria are presented in detail in Tables 16, 17, and 18.

Nutrition Facts

Serving Size 1 cup (228 g)

Servings Per Container 2

Amount Per Serving

Calories 260   Calories from Fat 120

	% Daily Value*
Total Fat 13 g	**20%**
Saturated Fat 5 g	**25%**
Cholesterol 30 mg	**10%**
Sodium 660 mg	**28%**
Total Carbohydrate 31 g	**10%**
Dietary Fiber 0 g	**0%**
Sugars 5 g	
Protein 5 g	

Vitamin A   4%   •   Vitamin C   2%

Calcium   15%   •   Iron   4%

*Percent Daily Values are based on a 2,000 Calorie diet. Your daily values may be higher or lower depending on your calorie needs:

		Calories	2,000	2,500
Total Fat	Less than		65 g	80 g
Sat Fat	Less than		20 g	25 g
Cholesterol	Less than		300 mg	300 mg
Sodium	Less than		2,400 mg	2,400 mg
Total Carbohydrate			300 g	375 g
Dietary Fiber			25 g	30 g

Calories per gram:

Fat 9   •   Carbohydrate 4   •   Protein 4

Figure 5   Sample nutrition label format. (From Ref. 53.)

Table 16 Nutrient Content Claim for Fat

Free	Low	Reduced	Other
Less than 0.5 g fat per Reference Amount and no added ingredient that is a fat or generally understood by consumers to contain fat unless marked by an asterisk referring to the statement "adds a negligible amount of fat," or "adds a dietarily insignificant amount of fat." Main dish products and meal products containing less than 0.5 g fat per labeled serving. Also labeled free of fat, no fat, zero fat, without fat, nonfat, trivial source of fat, negligible source of fat, dietarily insignificant source of fat.	3 g or less fat per Reference Amount when Reference Amount is more than 30 g or more than 2 tbsp. 3 g or less fat per serving and per 50 g when Reference Amount is 30 g or less or 2 tbsp or less. Main dish products and meal products containing 3 g or less total fat per 100 g and not more than 30 percent calories from fat. Also labeled low in fat, contains a small amount of fat, low source of fat, little fat.	At least 25 percent less fat per Reference Amount. Main dish products and meal products containing at least 25 percent less fat per 100 g. Also labeled reduced in fat, fat reduced, less fat, lower fat, lower in fat. Claim Example: "Reduced fat—50 percent less fat than our regular brownies. Fat content has been reduced from 8 g to 4 g per serving."	— percent fat free permitted if product meets requirements for "low fat." — percent lean permitted for meat and poultry products and main dish and meal products if products meet requirements for "low fat." 100 percent fat free meets requirements for "fat free" if product contains less than 0.5 g fat per 100 g and contains no added fat. Light or lite product has ⅓ fewer calories or 50 percent less fat per Reference Amount. Claim Example: "Light mayonnaise, 50 percent less fat than regular mayonnaise."

Source: Ref. 53.

242

Table 17 Nutrient Content Claim for Saturated Fat

Free	Low	Reduced
Less than 0.5 g saturated fat per Reference Amount and the level of trans fatty acids does not exceed 1 percent of the total fat; containing no ingredient that is saturated fat or is generally understood by consumers to contain saturated fat unless marked by an asterisk referring to the statement "adds a trivial amount of saturated fat," "adds a negligible amount of saturated fat," or "adds a dietarily insignificant amount of saturated fat." Main dish products and meal products containing less than 0.5 g saturated fat per labeled serving and the level of trans fatty acids does not exceed 1 percent of the total fat. Also labeled free of saturated fat, no saturated fat, zero saturated fat, without saturated fat, trivial source of saturated fat, negligible source of saturated fat, dietarily insignificant source of saturated fat.	1 gram or less saturated fat per Reference Amount and no more than 15 percent of calories from saturated fat. Main dish products and meal products containing 1 g or less saturated fat per 100 g and less than 10 percent of calories from saturated fat. Also labeled low in saturated fat, contains a small amount of saturated fat, low source of saturated fat, a little saturated fat.	At least 25 percent less saturated fat per Reference Amount. Main dish products and meal products containing at least 25 percent less saturated fat per 100 g. Also labeled reduced in saturated fat, saturated fat reduced, less saturated fat, lower saturated fat, lower in saturated fat. Claims Example: "Reduced saturated fat. Contains 50 percent less saturated fat than the national average for nondairy creamers. Saturated fat reduced from 3 to 1.5 g per serving."

Source: Ref. 53.

Table 18 Nutrient Content Claim for Cholesterol

Free	Low	Reduced	Other
Less than 2 mg cholesterol per Reference Amount and 2 g or less saturated fat per Reference Amount; containing no ingredient generally understood by consumers to contain cholesterol unless marked by an asterisk referring to the statement "adds a negligible amount of cholesterol," "adds a trivial amount of cholesterol," or "adds a dietarily insignificant amount of cholesterol." Main dish product containing 19.5 g or less of total fat per labeled serving, less than 2 mg cholesterol per labeled serving, and less than 2 g saturated fat per labeled serving.	20 mg or less cholesterol per Reference Amount and 2 g or less saturated fat per Reference Amount when Reference Amount is more than 30 g or more than 2 tbsp; contains 13 g or less total fat per Reference Amount. 20 mg or less cholesterol per Reference Amount and per 50 g; and 2 g or less saturated fat per Reference Amount when Reference Amount is 30 g or less or 2 tbsp or less; contains 13 g or less total fat per Reference Amount. When fat exceeds 13 g per Reference Amount (or per 50 g if Reference Amount is 30 g or less or 2 tbsp or less), the product must declare the total amount of fat in a serving next to claim.	At least 25 percent less cholesterol and 2 g or less saturated fat per Reference Amount per 50 g if Reference Amount is 30 g or less or 2 tbs or less. When fat exceeds 13 g per Reference Amount, per labeled serving, or per 50 g if Reference Amount is 30 g or less or 2 tbsp or less, the product must declare the total amount of fat in a serving next to claim. For main dish products, at least 25 percent less cholesterol per 100 g, and 2 g or less saturated fat per 100 g (product must declare the total amount of fat in a serving if product contains more than 19.5 g fat).	Cholesterol claims only allowed when food contains 2 g or less saturated fat per Reference Amount.

Meal product containing 26 g or less total fat per serving, less than 2 mg cholesterol per labeled serving, and less than 2 g saturated fat per labeled serving. When fat exceeds 13 g per Reference Amount, per labeled serving, or per 50 g if Reference Amount is 30 g or less or 2 tbsp or less, the product must declare the amount of total fat next to claim; also, for main dish products with more than 19.5 g fat and meal products with more than 26 g fat per labeled serving. Also labeled free of cholesterol, zero cholesterol, without cholesterol, no cholesterol, trivial source of cholesterol, negligible source of cholesterol, dietarily insignificant source of cholesterol.	Main dish products containing 19.5 g or less total fat per labeled serving, 20 mg or less cholesterol per 100 g, and 2 g or less saturated fat per 100 g. Meal products containing 26 g or less total fat per labeled serving, 20 mg or less cholesterol per 100 g, and 2 g or less saturated fat per 100 g. Also labeled low in cholesterol, contains a small amount of cholesterol, low source of cholesterol, little cholesterol.	For meal products at least 25 percent less cholesterol per 100 g and 2 g or less saturated fat per 100 g (product must declare the total amount of fat in a serving if the food contains more than 26 g fat). Also labeled reduced in cholesterol, cholesterol reduced, less cholesterol, lower cholesterol, lower in cholesterol. "Reduced Cholesterol" Claim Example (product has more than 13 g fat): "This pound cake contains 30 percent less cholesterol than our regular pound cake. Cholesterol lowered from 45 mg to 30 mg per serving. Contains 15 g fat per serving."

Source: Ref. 53.

For a food to be labeled "low fat," it also must be low in saturated fat and cholesterol. The specific criteria for "low-fat" foods are as follows:

Fat	Saturated fat	Cholesterol
≤3 g/standard serving	≤1 g/standard serving and no more than 15% of calories from saturated fat	≤20 mg/standard serving and 2 g or less of saturated fat

Meat, fish, and poultry can be labeled "lean" or "extra-lean" on the basis of the following criteria:

Lean meat and poultry: <10 g fat, <4.5 g saturated fat, and <95 mg cholesterol per 100 g

Extra lean meat, poultry, seafood, and game: <5 g fat, <2 g saturated fat, and <95 mg cholesterol per 100 g

For mixed dishes and meal-type products (frozen, canned, or shelf-stable entrees and dinners), there are no single values for fat, saturated fat, or cholesterol that can be applied to all products because the weight of an individual serving of these products varies. The definitions for "low fat," "low saturated fat," and "low cholesterol" are based on 100 g of mixed dishes and meal-type products. Table 19 shows the maximum amount of fat, saturated fat, and cholesterol allowable per weight of a serving of a mixed dish or meal-type product for it to qualify as "low fat," "low saturated fat," and/or "low cholesterol."

Table 19 Maximum Amount of Fat, Saturated Fat, and Cholesterol per Serving of Mixed Dishes and Meal-Type Products to Be Classified as Low Fat, Low Saturated Fat, and/or Low Cholesterol

Weight of serving (oz)	Fat (g)	Saturated fat (g)	Cholesterol (mg)
6	5	2	35
7	6	2	40
8	7	2	45
9	8	3	50
10	9	3	55
11	9	3	60
12	10	3	70
13	11	4	75

Source: Ref. 53.

Table 20 Nutrient Criteria for Health Claims

	Food product[a]	Main dish product[b]	Meal product[b]
Total fat, g	13	19.5	26
Saturated fat, g	4	6	8
Cholesterol, mg	60	90	120
Sodium, mg	480	720	960

[a]Per Reference Amount, per label serving size, and per 50 g when Reference Amount is 30 g or less *or* 2 tbsp or less.
[b]Per labeled serving.
Source: Ref. 53.

A health claim can be made on the nutrition label that describes the relationship between saturated fat and cholesterol *and* coronary heart disease. (Note: Other nutrient-disease relationships also can be acknowledged.) To make a health claim related to coronary heart disease, a serving of food must meet the specific nutrient criteria presented in Table 20.

In summary, the nutrition label will benefit patients who are following Step I and Step II diets. These labels will enable individuals to make food choices that are based either on the nutrient profile of the food or, more simply, on the nutrient or health claims on the label that meet federal guidelines. Patients, therefore, should find this information helpful in the implementation of their dietary therapy for an elevated blood cholesterol level. Of even greater significance is that the Nutrition Labeling and Education Act will enable all consumers to make more informed food choices and, thus, not only be more knowledgeable about good nutritional practices but actually plan and follow diets that meet contemporary dietary recommendations.

IMPLEMENTATION OF STEP I AND STEP II DIETS

The recommended approach to dietary therapy for patients with an elevated LDL cholesterol level is described in detail in the ATP II report (1). A summary of the implementation of Step I and Step II diets is presented in Figures 6 and 7. For patients without CHD, after starting on a Step I diet, the serum total cholesterol level should be measured at 4–6 weeks and at 3 months (Fig. 6). The patient's adherence to the diet should be assessed after 3 months. If the total cholesterol goal is achieved (<240 mg/dl with fewer than two risk factors or <200 mg/dl with two risk factors), then LDL cholesterol is measured to confirm that the goal has been met. If the LDL cholesterol goal is achieved, long-term monitoring can begin. If the blood cholesterol response is not satisfactory, the patient should be referred to a registered dietitian and, on the basis of his or her judgment, continue on a Step I diet or progress to a Step II diet. The patient should be evaluated again after 4–6

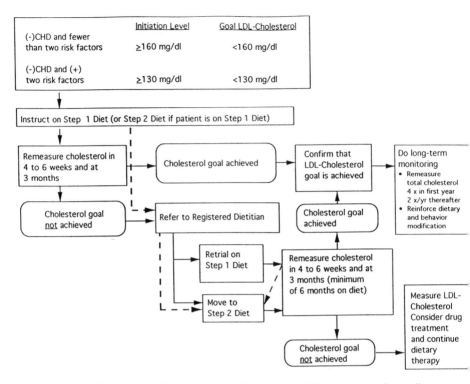

Figure 6   Algorithm of dietary treatment for patients without coronary heart disease. (Adapted from Ref. 1.)

weeks and at 3 months. Patients who do not achieve the LDL cholesterol goal on a Step I diet then should be instructed on a Step II diet.

When implementing the ATP II guidelines, it is likely that some patients already will be following a Step I diet since it is recommended for the population-at-large. These patients should be instructed initially on a Step II diet. Alternatively, other patients will be consuming a typical American diet or one that is very high in fat and saturated fat and, therefore, should be instructed initially on a Step I diet.

Patients with established CHD or other atherosclerotic disease should be referred to a registered dietitian and immediately begin a Step II diet (Fig. 7). A maximal approach to dietary therapy is important for these patients. However, a prolonged period of dietary therapy is not necessary before adding drug therapy. Patients on drug therapy should continue to follow a Step II diet.

Not only is it important to monitor a patient's response to diet therapy (i.e.

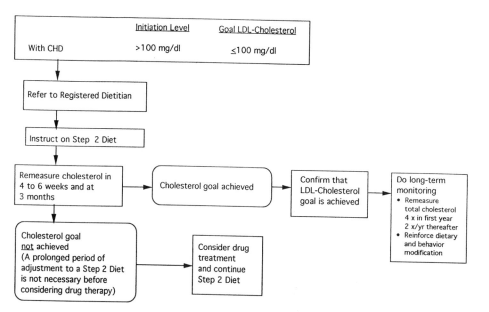

Figure 7 Algorithm of dietary treatment for patients with coronary heart disease. (Algorithm was drawn from information presented in Ref. 1.)

serum total and LDL cholesterol), it is also important to continually monitor dietary behaviors. Both short-term and long-term monitoring are recommended since recidivism is common, especially without frequent follow-up.

For patients without, as well as with, established CHD, achieving and maintaining a desirable body weight is a key aspect of implementing the Step I and Step II diets. For patients who must lose weight, dietary therapy will be more challenging. These patients will benefit from intensive educational efforts and frequent monitoring and follow-up.

ROLE OF PHYSICIANS, REGISTERED DIETITIANS, AND OTHER HEALTH PROFESSIONALS

Registered dietitians are nutrition experts who should play a key role in developing and implementing nutrition care plans for patients who require dietary therapy for an elevated LDL cholesterol level. The sheer number of individuals in the United States who require dietary therapy greatly exceeds the capacity of the dietetic profession to meet this demand. Therefore, other health professionals such as physicians, nurses, physicians' assistants, and health educators must assume a major role in implementing the ATP II guidelines. Table 21 outlines different ways

Table 21 Roles of Physicians, Registered Dietitians, and Other Health Professionals in the Implementation of the Step I and Step II Diets

Physician	Registered dietitian	Health professionals (nurses, physician assistants, health educators)
Assess CHD risk status and informs patient	Performs in-depth assessment of current diet and dietary habits	Assess current diet
Assess current diet and dietary habits	Develops individualized and intensive nutrition care plan	Plays an important role in patient education including initial dietary counseling
Organizes and informs patient of the therapeutic plan	Teaches patients and significant others the principles of Step I and Step II diets	Reinforces dietary advice
Emphasizes the importance of dietary therapy	Closely monitors the effectiveness of the dietary therapy plan	Monitors adherence to the Step I diet
Provides dietary guidance	Serves as a resource to patients regarding implementation of the therapeutic diet	Reinforces the treatment plan and emphasize the importance of dietary therapy
Monitors the effectiveness of the treatment plan		Refers patients to registered dietitians and other health professionals for individualized dietary therapy
Reinforces the importance of dietary therapy		
Refers patients to registered dietitians and other health professionals for individualized dietary therapy		

members of the health care team might share the responsibility of delivering dietary therapy to patients with elevated LDL cholesterol levels. The expertise of a registered dietitian would benefit patients who are having difficulties understanding and following a Step I diet. Moreover, because the Step II diet requires greater reductions in saturated fat and cholesterol intake, registered dietitians can play an important role in helping patients manage their prescribed diet. Ideally, registered dietitians should be involved from the outset of the treatment of patients with elevated LDL cholesterol levels. Since this is not always possible, registered dietitians and other qualified nutritionists must assume a leadership role in teaching other health professionals about the delivery of nutrition care services.

COST OF A BLOOD CHOLESTEROL–LOWERING DIET

Central to the adoption of a blood cholesterol–lowering diet is stressing that food costs need not be a barrier to dietary change. Several studies (both theoretical and actual cost analyses) have shown clearly that the cost of a diet lower in fat, saturated fat, and cholesterol is not higher than a typical American diet (54–57). In fact, theoretical estimates (54,55) have shown that such food costs are similar (54) or even lower (55) when compared to those of a typical American diet. While these studies were conducted in the mid- to late 1970s when there were far fewer fat-modified products on the market compared to the present, recent studies have supported this observation (56,57). Two contemporary intervention studies (57) have shown that a fat-modified/blood cholesterol–lowering diet costs the same (56) or even less (57) than a typical American diet. In the study conducted by Futhey et al. (56), food costs per 1000 kcal were similar when a small group of women ($n = 20$) in the Women's Intervention Nutrition Study (WINS) followed a diet that provided 17% of calories from fat versus 37% of calories from fat. Data collected on 109 hypercholesterolemic subjects from a rural area in the Assisting Primary Care Physicians with Lipid-Lowering Interventions (APPLI) study indicated that food costs of a blood cholesterol–lowering diet were as much as $1.67 less per day than their baseline diet (e.g., typical American diet) (57). Taken together, these studies indicate that food costs should not prohibit adoption of a blood cholesterol–lowering diet. Moreover, food costs may be less and, therefore, encourage patients to adopt this diet.

DIETARY TREATMENT OF COEXISTING DISORDERS

Patients with hypercholesterolemia frequently have coexisting disorders such as hypertriglyceridemia, hypertension, or diabetes. The treatment of hyper-cholesterolemic patients with multiple conditions increases the complexity of the therapy. The primary therapy still is dietary therapy. The Step I and Step II diets are recommended, with additional nutrient modifications. Restriction of certain nutrients (e.g., sodium, refined sugar, alcohol) makes these therapies challenging for both patients and dietitians.

Hypertriglyceridemia

Diet for the treatment of elevated triglycerides emphasizes weight loss, limited alcohol consumption, and a decrease in simple carbohydrate intake. For patients who consume alcohol, the intake should be very limited, at least on a trial basis. Some patients are sensitive to simple carbohydrates, whereas some are sensitive to an increase in the total carbohydrate content of the diet. In these patients, a high-carbohydrate diet will increase triglycerides further and decrease HDL cholesterol levels. In such cases, a Step I diet that provides ~30% of calories from fat and emphasizes complex carbohydrates is recommended initially (58). A diet that provides more than 55% of calories from carbohydrate will increase triglycerides in some patients. For these patients, some consideration is being given to increasing the total fat content of the diet (30–35% of calories or even higher) while keeping saturated fat at recommended levels. Some patients with familial hypertriglyceridemia (i.e., triglyceride levels >1000 mg/dl), on the other hand, require a reduction in total fat intake to 10–20% of calories. Thus, dietary therapy for patients with hypertriglyceridemia must be individualized.

Hypertension

The dietary guidelines for treating hypertension include reducing excess body weight, decreasing sodium intake to less than 2300 mg/day, and limiting alcohol intake (59). Weight reduction helps lower blood pressure in many hypertensive patients who are more than 10% above their ideal weight. A decrease in blood pressure often occurs with a loss of as little as 10 pounds (60). Weight reduction also enhances the effectiveness of antihypertensive medications (61). Regular aerobic physical activity may reduce blood pressure in addition to contributing to weight loss (62).

The major sources of sodium in the American diet are salt and other sodium-containing compounds added to food in processing, during preparation, at the table, or in restaurants. Sodium also is naturally present in food and water. Ready-to-eat cereal, bread, luncheon meat, cheese, condiments, and canned soup are common processed foods that contribute significant amounts of sodium to the diet. Restaurant food presents a special challenge to the person trying to limit sodium. For example, a 2½-cup serving of spaghetti with meat sauce contains about 1500 mg of sodium.

Excessive alcohol intake can increase blood pressure in addition to causing resistance to antihypertensive therapy (63). If alcoholic beverages are consumed, intake should be limited to no more than 1 ounce of alcohol per day, which is equivalent to 24 ounces of beer, 8 ounces of wine, or 2 ounces of 100-proof whiskey.

Several other dietary factors also may affect blood pressure (64). A potassium deficiency may increase blood pressure, and a high dietary intake of potassium may protect against developing hypertension. Potassium is easily obtainable from dietary sources such as bananas, cantaloupe, honeydew melon, nectarines, orange juice, potatoes, and skim milk. Potassium supplements may be necessary to prevent hypokalemia during diuretic therapy.

Calcium deficiency has been associated with an increased prevalence of hypertension in some studies (65,66). An increased intake of calcium has been found to lower blood pressure in some, but not all, hypertensive patients (reviewed in Ref. 64). At present, health experts recommend a daily calcium intake that meets the RDA (800–1200 mg) (59,64). Diets that provide amounts of calcium in excess of the RDA generally are not recommended to lower blood pressure (59). Dairy products are the best sources of calcium. Low-fat and fat-free dairy products provide as much calcium as their high-fat counterparts. Some dairy products, however, are rich sources of sodium (e.g., cheeses). Other sources of calcium include green leafy vegetables and fish with bones such as sardines and salmon.

Some evidence suggests that a low intake of magnesium is associated with higher blood pressure (67). This evidence is not strong enough to justify recommending an increased magnesium intake for the purpose of lowering blood pressure, although any diet should provide an adequate amount of magnesium (59). Magnesium is found in a wide variety of foods, including nuts, legumes, and green vegetables.

Caffeine can acutely raise blood pressure in people who do not habitually consume it (59). However, regular caffeine consumption appears to decrease this effect. Presently, there is no general recommendation to limit caffeine as a way of lowering blood pressure, although decreasing caffeine intake may be appropriate for specific caffeine-sensitive patients. The principal sources of caffeine are coffee, tea, chocolate, and some soft drinks (both regular and diet).

Diabetes

Dietary guidelines for the treatment of non–insulin-dependent diabetes mellitus (NIDDM) and insulin-dependent diabetes mellitus (IDDM) recommend a calorie intake appropriate for the patient to achieve and maintain a reasonable weight (68). This is a major concern since 60–90% of patients with NIDDM are overweight or obese. Weight reduction in NIDDM leads to reduction in hyperglycemia, hyperlipidemia, and hypertension. Improvement in glucose tolerance with weight reduction in NIDDM may eliminate or reduce the need for oral hypoglycemic agents or insulin.

The ratio of carbohydrate to fat in the diet will vary among individuals with diabetes. For individuals with normal lipid levels, ≤30% of calories from fat (and less than 10% of calories from saturated fat) is recommended (68). For individuals with hypertriglyceridemia a higher fat diet (that is low in saturated fat) may be recommended. The new guidelines place less emphasis on the type of carbohydrate in the diet and do not contraindicate sucrose, within reason. Fiber intake recommendations for persons with diabetes are now the same as those for persons without diabetes. A high intake of fiber, particularly soluble fiber, may lower total cholesterol and LDL cholesterol. Foods high in soluble fiber include legumes, oat bran and oatmeal, barley, and fruits containing pectin (see Table 3).

Recommendations for sodium intake are the same as those for the general population for individuals without hypertension (i.e., 2400–3000 mg/day). For patients with hypertension, sodium should be limited to 2400 mg/day.

STRATEGIES FOR IMPLEMENTING THE STEP I DIET ON A POPULATIONWIDE BASIS

Since the Step I diet is recommended for the entire population over 2 years of age (69), intervention approaches that target both groups and individuals are essential. Change is possible at many different levels; it occurs at the individual level and within social networks, organizations, communities, states and societies. Many intervention programs designed to change eating patterns have been conducted in various settings such as schools, worksites, families, fitness centers, and restaurants for different audiences such as youth, adults, workers, and restaurant and fitness center patrons. Figure 8 shows the multilayered levels of intervention for application of blood cholesterol–lowering strategies and the specific targets of change. Widespread intervention and education efforts at different levels both support and reinforce changes made at the individual level. Thus, family, friends, and the community can play an important role in helping individuals make dietary and other lifestyle changes.

At all levels, interventions can range from minimal to extensive and require vastly different resources. Interventions that target individuals include individual or group counseling, classes, personalized telephone and computer-assisted approaches with nutritionists, printed materials, home videotapes, special television programming, and computer-based assessment and feedback. Social-network-level interventions use similar strategies but target families and larger social networks such as circles of friends to support changing dietary behaviors. Information often is disseminated by influential persons. At the organizational level, interventions target larger groups and often entire groups that are affiliated with the organization. Frequently, classes are held or information is disseminated, and sometimes policies are established that affect various health practices. Examples include monitoring foods available for purchase on the premises and establishing cigarette smoking policies. Community-, state-, and national-level interventions affect entire populations of individuals. Legislation such as the Nutrition Labeling

Individual-Level Targets

Men/Women
Youth
Elderly
Minorities
Low income

Network-Level Targets

Extended social network
Peer groups
Families

Organization-Level Targets

Worksites
Restaurants
Supermarkets/Grocery stores
Schools and child care Centers
Cafeterias
Shopping Malls/Centers
Health Care Clinics
Professional groups/Social clubs
Senior citizens congregate Feeding Sites

Community/State/National-Level Targets

Media
Social norms
Public opinion
Local, state, national legislation
Food production/Food industry
Institutions of higher education
Cooperative Extension

Figure 8   A multilayered view of cholesterol-lowering strategies: levels of intervention and corresponding targets of change. (From Ref. 69.)

and Education Act is intended to have a significant impact on the food-consumption practices of the nation. Nutrition-oriented state health agency programs, communitywide multimedia campaigns, and grocery store nutrition programs are examples of interventions at the state and community levels that can have a significant impact on food behaviors.

The population-based approach to the implementation of dietary changes for blood cholesterol lowering is complementary to the patient-based approach (medical model) for treating individuals with elevated LDL cholesterol levels. Both are important in reducing the risk of CHD by lowering the average blood cholesterol level of the U.S. population. The intervention strategies for implementing dietary changes of patients versus groups of individuals (e.g., populations) are different. Whereas the implementation of these approaches has played a role in reducing the average blood cholesterol level in the United States, approximately 30% of the population still has blood cholesterol levels greater than 200 mg/dl. In addition, while food-consumption practices have improved in recent years, additional changes are needed for individuals and the population as a whole to adhere to contemporary dietary recommendations. Thus, both the population-based approach for reducing blood cholesterol levels and the patient approach for treating elevated LDL cholesterol levels are important for reducing the risk of CHD in the United States.

IMPLEMENTATION OF THE STEP I AND STEP II DIETS FOR CHILDREN

The Report of the Expert Panel on Blood Cholesterol Levels in Children and Adolescents (70) presents evidence that atherosclerosis begins in childhood and that lowering blood cholesterol levels in children and adolescents will favorably affect CHD risk status. The population approach is intended to lower the average blood cholesterol levels of children and adolescents through recommended changes in nutrient intake and eating patterns. The dietary recommendations for all children and adolescents over the age of 2 are congruent with those issued by many governmental and professional organizations for health promotion, optimal growth and development, and the prevention of chronic diseases. They are consistent with recommendations of the Step I diet.

NCEP dietary recommendations for all children over 2 years of age are:

Nutritional adequacy should be achieved by eating a wide variety of foods.
Energy (calories) should be adequate to support growth and development and to reach or maintain desirable body weight.
The following pattern of nutrient intake is recommended:
saturated fatty acids—less than 10% of total calories
total fat—average of no more than 30% of total calories
dietary cholesterol—less than 300 mg/day

Implementation of these dietary recommendations for all children requires the support of schools, health professionals, government agencies, and the food industry (70). Schools should provide meals that meet the recommended nutrient pattern for the entire population and that are appealing to children. In addition, the curriculum should teach children how to plan and follow a healthy diet. Health professionals can plan an important role by including nutrition education in well-child visits. Government agencies should provide more nutrition education to children, their caregivers, and their teachers. The food industry also should continue to develop foods low in saturated fat, total fat, and cholesterol and escalate marketing campaigns to promote healthy eating patterns among children.

An individualized approach to cholesterol lowering is recommended for children and adolescents who are at greatest risk for having an elevated blood cholesterol level in adulthood. As with adults who have high blood cholesterol levels, dietary therapy is the primary treatment approach—the Step I and Step II diets are recommended. To ensure adequacy of energy and nutrient (vitamins and minerals) intake by children on a Step II diet, a registered dietitian or other qualified nutrition professional should be consulted.

To implement dietary change in children, the following strategies are recommended (71):

Health professionals:
Educate and involve the child and his or her family about the dietary recommendations
Involve the child and the family in implementing these recommendations both in and outside the home
Encourage family support
Caregivers
Provide regular meals and snacks throughout the day
Provide a variety of acceptable food choices
Be aware of foods the child likes and dislikes and make dietary changes accordingly
Repeatedly offer new foods (increasing exposure increases acceptance)
Make dietary changes slowly; avoid extreme dietary changes
Inform teachers and other care providers of the child's needs
Create an environment in which the child does not feel different (children can participate in special celebrations with the provision of acceptable food substitutions or careful diet planning)
Understand the child will not adhere to a Step I or Step II diet every day
Have patience; dietary changes take time for everyone.

Diet is important for the prevention and treatment of high blood cholesterol levels in persons of all ages. The population approach and the individualized approach both are important in establishing good nutritional practices in childhood that are followed for a lifetime. A healthy diet that meets the recommendations of the Step I diet will decrease total and LDL cholesterol levels in childhood, adolescence, and into adulthood and, thus, reduce the risk and incidence of coronary heart disease. In addition, application of a Step I diet early in life will teach children good nutritional practices that can be followed for a lifetime as well as positively influence taste and food preferences.

SUMMARY

Implementing dietary change is essential for meeting the recommendations of the Step I and Step II diets. Diet is the key to lowering the blood cholesterol level of the U.S. population. Moreover, diet is the cornerstone of therapy for an elevated LDL cholesterol level. Thus, it is important that both patients and the public understand the principles of these diets and apply them effectively.

Successful application of the ATP II guidelines is, in large part, linked to implementing long-term dietary changes that lower blood cholesterol levels and facilitate achieving and maintaining a desirable body weight. Different approaches can be used (singly or together) to implement dietary changes that achieve the recommendations of the Step I and Step II diets. Patients can count grams of total fat (and grams of saturated fat), use the exchange lists of the American Heart Association, and/or apply various fat-reduction techniques. All of these dietary strategies are relatively easy to implement. Inherent in an effective dietary-therapy regimen are accurate initial and ongoing assessments of patients' diets. This is important in establishing the initial diet prescription and the subsequent treatment plan. The population-based approach for implementing dietary changes targets interventions to many audiences (groups and individuals) in different settings such as schools, worksites, households, restaurants, and supermarkets.

Because of the number of individuals who require dietary therapy for elevated blood cholesterol levels, registered dietitians and other health professionals must appreciate the significant effort that is required both on a patient and population basis to implement the dietary changes recommended by the National Cholesterol Education Program. Not only will achieving these recommended dietary changes benefit individuals at risk for CHD, including children from high-risk families, but on a population basis lowering blood cholesterol levels will significantly reduce morbidity and mortality from the leading cause of death in the United States.

RESOURCES

Resources for Professionals

Dietary Treatment of Hypercholesterolemia. A Handbook for Counselors (1988)
Contains easy-to-use information for use in counseling with the patient with hypercholesterolemia. Available through the American Heart Association.

Dietary Treatment for High Blood Pressure and High Blood Cholesterol. A Handbook for Counselors (1994)

Contains information for use in counseling patient in the Step I or Step II diets for hypercholesterolemia in combination with a 2-gram sodium diet. Available through the American Heart Association.

NCEP Report of the Expert Panel on Blood Cholesterol Levels in Children and Adolescents (Pediatrics 1992; 89(suppl):525–584)

Recommends two complementary strategies to lower blood cholesterol in children and adolescents. The population approach aims to lower average population levels of blood cholesterol; it encourages the adoption of eating patterns low in saturated fat, total fat, and cholesterol. The individualized approach aims to identify and treat children and adolescents whose high blood cholesterol levels during childhood put them at increased risk for high blood cholesterol and heart disease as adults. As part of this approach, the report gives recommendations for screening and treatment. Sample menus and nutrient information to help foster lower fat eating are included. 119 pages (#2732). Available from the National Institutes of Health, National Heart, Lung, and Blood Institute.

Prevalence of High Blood Cholesterol Among U.S. Adults (JAMA 1993; 269:3009–3014)

Presents the results obtained by applying the guidelines of NCEP's Expert Panel on the Detection, Evaluation, and Treatment of Cholesterol in Adults (Adult Treatment Panel or ATP II) to the most recent nationally representative serum total cholesterol and lipoprotein data (NHANES III). Statistics and data are presented in both table and text format. (#3096). Available from the National Institutes of Health, National Heart, Lung, and Blood Institute.

Public Awareness of Cholesterol: Data Fact Sheet (1990)

Presents the results of surveys conducted to measure public attitudes, knowledge, and practices regarding high blood cholesterol and to gain an understanding of what people know and what they still need to learn about cholesterol. 4 pages ($1). Available from the National Institutes of Health, National Heart, Lung, and Blood Institute.

Recommendations for Improving Cholesterol Measurement: Executive Summary (1990)

Provides guidelines and recommendations from the Laboratory Standardization Panel on Blood Cholesterol Measurement to implement the standardization of laboratory measurement and thus improve the reliability of cholesterol measurement in the United States. 16 pages (#2964A). Available from the National Institutes of Health, National Heart, Lung, and Blood Institute.

Second Report of the Expert Panel on Detection, Evaluation, and Treatment of High Blood Cholesterol in Adults ATP II (1994)

Offers practical guidelines for health professionals when detecting, evaluating, and treating patients with high blood cholesterol. Covers three subject areas: classification of blood cholesterol and patient evaluation, dietary treatment, and drug treatment. New recommendations in three important areas include cholesterol-lowering treatment in relation to risk status, HDL cholesterol, and an expanded approach to dietary therapy. (#3095). Available from the National Institutes of Health, National Heart, Lung, and Blood Institute.

Second Report of the Expert Panel on Detection, Evaluation, and Treatment of High Blood Cholesterol in Adults Executive Summary (JAMA 1993; 269:3015–3023)

Presents the Executive Summary of the guidelines recommended in the *Second Report of the Expert Panel on Detection, Evaluation, and Treatment of High Blood Cholesterol in Adults (ATP II)*. 28 pages. Available from the National Institutes of Health, National Heart, Lung, and Blood Institute.

Report of the Expert Panel on Population Strategies for Blood Cholesterol Reduction: Executive Summary (Arch Internal Med 1991; 151:1071–1084)

Contains the executive summary of the full report. Offers a set of recommendations designed to help healthy Americans age 2 and above lower their blood cholesterol levels. In addition to food consumption recommendations, this report identifies ways that health professionals, government agencies, and educators can help Americans lower their blood cholesterol levels. (#3047) 36 pages. Available from the National Institutes of Health, National Heart, Lung, and Blood Institute.

The Fifth Report of the Joint National Committee on Detection, Evaluation, and Treatment of High Blood Pressure (JNCV) (Arch Int Med 1993; 153:154–183)

Provides data from the third National Health and Nutrition Examination Survey (NHANES III) concerning the prevalence of hypertension. Also discusses the clinical evaluation and public health aspects of high blood pressure, treatment options, and concerns of special populations and situations. (#1088) 51 pages. Available from the National Institutes of Health, National Heart, Lung, and Blood Institute.

National High Blood Pressure Education Program; Working Group Report on Primary Prevention of Hypertension (Arch Int Med 1993; 153:186–208)

Examines the cause of high blood pressure and the approaches taken to prevent it. Also describes the influence of lifestyle factors on the development of hypertension. (#2669) 49 pages. Available from the National Institutes of Health, National Heart, Lung, and Blood Institute.

"CLIP" Food Log—Teaching & Assessment Tool, Cholesterol Lowering Intervention Program

The CLIP Food Log is easy to use and does not require lengthy calculations or computer analysis. Users tally food intake for one week; a quick review reveals the main source of cholesterol and saturated fat. Based on the MRFIT Food Record Rating system.

Nutrition Center/University of Pittsburgh
Graduate School of Public Health
Keystone Building, Suite 510
Pittsburgh PA 15213

A validated food frequency instrument for dietary assessment of fat intake is available from:

Fat Intake Scale
Northwest Lipid Research Clinic
465 Harborview Hall, 326 Ninth Avenue
Seattle, WA 98104

A food scoring tool to assess dietary adherence to a cholesterol-lowering diet is available from:

Dr. C. C. Tagney
Department of Clinical Nutrition
Rush-Presbyterian–St. Luke's Medical Center
1742 West Harrison 502SSH
Chicago, IL 60612

Resources for Patients

Dietary Treatment of Hypercholesterolemia: A Manual for Patients (1994)
Provides detailed information on the Step I and Step II diets for hyper-cholesterolemia. Available through the American Heart Association.

Dietary Treatment of High Blood Pressure and High Blood Cholesterol. A Manual for Patients (1994)
Provides detailed information on combining the Step I and Step II diets with a 2-gram sodium restriction. Available through the American Heart Association.

The Living Heart Brand Name Shopper's Guide, Revised and Updated
Lists more than 6100 products low in fat, saturated fat, and cholesterol and is a perfect companion to low-calorie and low-fat cookbooks. Send $16.95 ($14.95 [quality paperback] plus $2.00 for postage and handling) to:

Diet Modification Clinic
6565 Fannin, F770)
Houston, TX 77030

The Living Heart Guide to Eating Out
Contains values for fat, saturated fat, and cholesterol for more than 1600 American, ethnic, and fast foods plus 160 tips on choosing restaurant foods low in fat and sodium. Send $11.95 ($9.95 [quality paperback] plus $2.00 postage and handling to:

Diet Modification Clinic
6565 Fannin, F770
Houston, TX 77030

The Living Heart Diet

Provides information on heart disease and includes more than 500 recipes (with nutrient analysis). Send $15.50 ($12.50 [quality paperback] plus $3.00 for postage and handling) to:

Diet Modification Clinic
6565 Fannin, F770
Houston, TX 77030

American Heart Association Cookbook, 5th edition (1991)

Contains more than 550 recipes with emphasis on attaining reasonable weight and sodium modifications. Cost is $25.00 (hardcover), available at most bookstores.

American Heart Association Fat and Cholesterol Counter (1991)

Lists total fat, saturated fat, calories, cholesterol, and sodium in hundreds of foods. Cost is $3.50 (paperback), available at most bookstores.

American Heart Association Low-Salt Cookbook (1990)

Provides a complete guide to reducing fat and sodium in the diet. Cost is $19.95 (hardcover), available at most bookstores.

American Heart Association Low-Fat, Low-Cholesterol Cookbook (1989)

Provides a wealth of tips about grocery shopping and eating out and contains 200 new recipes developed for people who need to reduce their level of blood cholesterol. Cost is $18.95 (hardcover), available at most bookstores.

My New Weigh of Life Participant's Guide (1992)

A group-oriented program to make necessary food choices to reach desired body weight. Publication #AGRS-58. Cost is $7.00, available from the Pennsylvania State University Publications Distribution Center, 112 Ag Admin Bldg, University Park, PA 16802, (814) 865-6713.

Eating Smart for Your Heart

This is a 10-lesson nutrition education program for families of 4- to 10-year-old children with high blood lipid levels. The program includes talking books (illustrated story books with accompanying audiotapes) and an activity book for the child plus a Parent Guide with background information and suggested parent-child activities for the whole family. The lessons emphasize heart-smart eating to lower blood lipid levels. Available in three levels: Level 1 for 4- to 5-year-olds; Level 2 for 6- to 8-year-olds; and Level 3 for 9- to 10-year-olds. Price is $45.00/level plus $5.00 shipping and handling. Pennsylvania residents must add 6% sales tax or send a tax-exemption certificate. Send order to:

Penn State Nutrition Center
417 East Calder Way
University Park, PA 16801-5663
Ph: 814-865-6323
FAX: 814-865-5870
e-mail: etm101@psuvm.psu.edu

Parents' Guide. Cholesterol in Children. Healthy Eating Is a Family Affair (1992)
This booklet is designed for parents who want to encourage heart-healthy eating patterns in their families when a child has been found to have high blood cholesterol. It provides basic information about cholesterol and heart disease and the eating patterns recommended by the NCEP Expert Panel on Blood Cholesterol Levels in Children and Adolescents. This information is followed by practical tips on shopping for and planning meals, changing family habits, and making low-saturated fat, low-cholesterol foods appealing to children. (#3099). Available from the National Institutes of Health, National Heart, Lung, and Blood Institute.

Eating With Your Heart in Mind (1992)
This booklet is for children ages 7–10 with high blood cholesterol. Hands-on activities such as word games, puzzles, and connect-the-dots keep children's interest level high while they learn about eating in a low-saturated-fat, low-cholesterol way. (#3100). Available from the National Institutes of Health, National Heart, Lung, and Blood Institute.

Heart Health. . . . Your Choice (1992)
This booklet is for adolescents ages 11–14 with high blood cholesterol. Provides information on the relationship of high blood cholesterol to heart disease and how choosing foods low in saturated fat and cholesterol can reduce risk for future heart attacks and disease. Practical tips on snacking, fast foods, and physical activity are included. (#3101). Available from the National Institutes of Health, National Heart, Lung, and Blood Institute.

Healthy Heart Habits (1992)
This is designed for 15- to 18-year-olds with high blood cholesterol. Provides basic information about cholesterol and heart disease. Specific information on choosing foods low in saturated fat and cholesterol with emphasis on food labeling, the Step I diet, and physical activity is included. (#3102). Available from the National Institutes of Health, National Heart, Lung, and Blood Institute.

Step-by-Step Eating to Lower Your High Blood Cholesterol (1994)
This revised patient education booklet provides facts about high blood cholesterol and how it relates to heart disease. Contains general rules for Step I and Step II diets, physical activity, and weight loss to lower blood cholesterol, and tips for adopting heart-healthy eating habits and increasing physical activity. Includes tables listing the saturated fat and cholesterol content in selected foods, serving sizes for meat and cheese, and information on different types of physical activity. (#2920). Available from the National Institutes of Health, National Heart, Lung, and Blood Institute.

Eat Right to Lower Your High Blood Cholesterol (reprinted 1992)
This easy-to-read (5th-grade level) brochure was developed for patients with high blood cholesterol. The conversational text, presented in large typeface

with illustrations, explains the significance and importance of reducing high blood cholesterol and focuses on food choices and preparation—practical information that patients can apply to their own eating patterns. Charts list foods high and low in saturated fat and cholesterol. Menus illustrate an appropriate eating pattern. A tear-out shopping list of heart-healthy foods is provided to help patients when they shop. (#2972). Available from the National Institutes of Health, National Heart, Lung, and Blood Institute.

Eat Right to Help Lower Your High Blood Pressure (1992)

This easy-to-read (5th-grade level) brochure is intended for patients with high blood pressure. Focusing on practicality, the text provides shopping and menu ideas to help people develop low-salt, low-saturated-fat eating patterns. (#3289). Available from the National Institutes of Health, National Heart, Lung, and Blood Institute.

Facts About How to Prevent High Blood Pressure (1994)

This fact sheet explains what high blood pressure is, who is at risk, and what can be done to help prevent high blood pressure. Available from the National Institutes of Health, National Heart, Lung, and Blood Institute.

Facts About Blood Cholesterol (revised 1994)

This revised fact sheet reflects newly released NCEP ATP II as well as those recommendations targeting all healthy Americans to follow a low-saturated fat, low-cholesterol eating pattern to lower their blood cholesterol levels and thus reduce their risk of heart disease. The emphasis is on the benefits and how-to's of blood cholesterol reduction. Tips tell how to select and prepare foods lower in saturated fat and cholesterol, and a chart gives saturated fat, total fat, and cholesterol levels for basic foods. 20 pages (#2696). Available from the National Institutes of Health, National Heart, Lung, and Blood Institute.

Facts About Blood Cholesterol (1994 reproducible)

This is a black-and-white reproducible version of *Facts About Blood Cholesterol* listed above. (#2696a). Available from the National Institutes of Health, National Heart, Lung, and Blood Institute.

So You Have High Blood Cholesterol (1994)

This patient education booklet has been revised and describes the relationship of blood cholesterol to coronary heart disease. Provides the cutpoints of "high," "borderline," and "desirable" blood cholesterol levels of total and LDL cholesterol. Advice is offered for changing diet, becoming more physically active, and losing weight if necessary. Presents case studies on high-risk individuals to aid readers in understanding and making necessary lifestyle changes to reduce their risk of coronary heart disease. (#2922). Available from the National Institutes of Health, National Heart, Lung, and Blood Institute.

The Healthy Heart Handbook for Women (1992)

A 90-page spiral-bound resource that discusses each of the major risk factors and then targets preventative measures and special concerns of women.

Developed by the National Institutes of Health, National Heart, Lung, and Blood Institute.

Drawing the Line on Fat and Cholesterol: A Connect-the-Dots Approach to Better Eating, Health and Weight Control

A patented, connect-the-dots approach to keeping track of the intake of fat (both total and saturated) and cholesterol. Can be tailored to the Step I and Step II diets of the American Heart Association and National Cholesterol Education Program. Useful in cardiac rehabilitation programs and for clients needing to reduce fat and/or cholesterol intake. Book includes 30-day supply of connect-the-dots sheets. $10.95 + $3 S/H from:

Roberta Schwartz Wennik, M.S., R.D.
HealthPro
9120- 185th Place SW
Edmonds, WA

In addition, many pamphlets and brochures are available free from the American Heart Association. These include "AHA Diet: Eating Plan for Healthy Americans," "Recipes for Low-Fat, Low-Cholesterol Meals," "Nutritious Nibbles," and "Nutrition Labeling" (which teaches how to read food labels). Call the local AHA affiliate or 1-800-242-1793 to order or for further information.

Professional Groups

These professional groups can provide assistance to consumers or health professionals in locating a qualified nutritionist or health professional and/or providing resource materials for the management of elevated blood cholesterol levels.

Nutrition Hotline of The American Dietetic Association 1-800-366-1655
This service offers recorded nutrition messages in Spanish or English concerning current nutrition topics such as the new food label. It is also possible to speak to a registered dietitian or learn how to locate a local registered dietitian.

American Heart Association
7272 Greenville Avenue
Dallas, TX 75231 1-800-242-1793
Information that targets public and professional education as well as scientific publications is available for all areas of cardiovascular disease prevention.

National Heart Lung and Blood Institute Information Center
P.O. Box 30105
Bethesda, MD 20824-0105 1-301-251-1222
Professionals, patients, and the general public can receive a variety of National Cholesterol Education Program and National High Blood Pressure Education Program materials that target cardiovascular disease prevention in general, as well as strategies to reduce blood cholesterol, blood pressure, and other risk factors.

ACKNOWLEDGMENT

Supported, in part, by NIH grant HL 44177.

REFERENCES

1. Report of the Expert Panel on Detection, Evaluation, and Treatment of High Blood Cholesterol in Adults (ATP II). National Cholesterol Education Program, National Heart, Lung, and Blood Institute, U.S. Department of Health and Human Services, Bethesda, MD, 1994.
2. Sempos CT, Cleeman JI, Carroll MD, Johnson CL, Bachorik PS, Gordon DJ, Burt VL, Briefel RR, Brown CD, Lippel K, Rifkind BM. Prevalence of high blood cholesterol among US adults. An update based on guidelines from the Second Report of the National Cholesterol Education Program Adult Treatment Panel. JAMA 1993; 269:3009–3014.
3. Johnson CL, Rifkind BM, Sempos CT, Carroll MD, Bachorik PS, Briefel RR, Gordon DJ, Burt VL, Brown CD, Lippel K, Cleeman JI. Declining serum total cholesterol levels among US adults. The National Health and Nutrition Examination Surveys. JAMA 1993; 269:3002–3008.
4. Stephen, AM, Wald NJ. Trends in individual consumption of dietary fat in the United States, 1920-1984. Am J Clin Nutr 1990; 52:457–469.
5. CDC. Daily Dietary Fat and Total Food-Energy Intakes: National Health and Nutrition Examination Surveys III, Phase 1-U.S., 1988–1991. MMWR 1994; 43:120–124.
6. Life Sciences Research Office, Federation of American Societies for Experimental Biology. Nutrition Monitoring in the United States—an Update on Nutrition Monitoring. Prepared for the U.S. Dept. of Agriculture and the U.S. Dept. of Health and Human Services. Washington, DC: Public Health Services, 1989. DHHS publication 89-1255.
7. Unpublished data. Washington, DC: International Life Sciences Research Foundation, 1991.
8. Ornish D, Brown SE, Scherwitz LW, Billings JH, Armstrong WT, Ports TA, McLanahan SM, Kirkeeide RL, Brand RJ, Gould KL. Can lifestyle changes reverse coronary heart disease? The Lifestyle Heart Trial. Lancet 1990; 336:129–133.
9. Haskell WL, Spiller GA, Jensen CD, Ellis BK, Gates JE. Role of water-soluble dietary fiber in the management of elevated plasma cholesterol in healthy subjects. Am J Cardiol 1992; 69:433–439.
10. Whyte JL, McArthur R, Topping D, Nestel P. Oat bran lowers plasma cholesterol levels in mildly hypercholestrolemic men. J Am Diet Assoc 1992; 92:446–449.
11. Glore SR, Van Treeck D, Knehans AW, Guild M. Soluble fiber and serum lipids: a literature review. J Am Diet Assoc 1994; 94:425–436.
12. Raper N. Nutrient content of the U.S. Food Supply. Food Rev 1991; 14:13–27.
13. Putnam JJ. Food consumption, prices, and expenditures, 1966–87. U.S. Department of Agriculture, Economic Research Service, Statistical Bulletin No. 773. January, 1989. Washington, DC.
14. Rizek RL, Welsh SO, Marston RM, Jackson EM. Levels and sources of fat in the U.S.

food supply and in diets of individuals. In: Perkins EG, Visek WJ, eds. Dietary Fats and Health. Champaign, IL: American Oil Chemists' Society, 1983.

15. Light L. Eating Less Fat: A Progress Report on Improving America's Diet. Washington, DC: Institute for Science in Society, 1992.

16. Gorder DD, Dolecek TA, Coleman GG, Tillotson JL, Brown HB, Lenz-Litzow K, Bartsch GE, Grandits G. Dietary intake in the Multiple Risk Factor Intervention Trial (MRFIT): nutrient and food group changes over 6 years. J Am Diet Assoc 1986; 86: 744–751.

17. The Nutrition Monitoring Division, Human Nutrition Information Service. Nationwide Food Consumption Survey: CSF II. Continuing Survey of Food Intakes by Individuals—1986, Low-income women 19–50 years and their children 1–5 years, 4 days, 1986. Nutrition Today 1989; (Sept./Oct.):35–38.

18. Gorbach SL, Morrill-LaBrode A, Woods MN, Dwyer JT, Selles WD, Henderson M, Insull W Jr, Goldman S, Thompson D, Clifford C, Sheppard L. Changes in food patterns during a low-fat dietary intervention in women. J Am Diet Assoc 1990; 90: 802–809.

19. Thompson FE, Sowers MF, Frongillo EA Jr, Parpia BJ. Sources of fiber and fat in diets of US women aged 19 to 50: Implications for nutrition education and policy. Am J Public Health 1992; 82:695–702.

20. Frank GC, Berenson GS, Webber LS. Dietary studies and the relationship of diet to cardiovascular disease risk factor variables in 10-year-old children—The Bogalusa Heart Study. Am J Clin Nutr 1978; 31:328–340.

21. Kuczmarski RJ, Brewer ER, Cronin FJ, Dennis B, Graves K, Haynes S. Food choices among white adolescents: The Lipid Research Clinics Prevalence Study. Pediatr Res 1986; 20:309–315.

22. McPherson RS, Nichaman MZ, Kohl HW, Reed DB, Labarthe DR. Intake and food sources of dietary fat among schoolchildren in The Woodlands, Texas. Pediatrics 1990; 86:520–526.

23. Basch CE, Shea S, Zybert P. Food sources, dietary behavior, and the saturated fat intake of Latino children. Am J Public Health 1992; 82:810–815.

24. Ernst ND. Fatty acid composition of present day diets. In: Nelson GJ, ed. Health Effects of Dietary Fatty Acids. American Oil Chemists' Society, Champaign, IL, 1991:1–11.

25. Ainsworth BE, Jacobs DR Jr, Leon AS. Validity and reliability of self-reported physical activity status: the Lipid Research Clinics questionnaire. Med Sci Sports Exerc 1993; 25:92–98.

26. Stensland SH, Margolis S. Simplifying the calculation of body mass index for quick reference. J Am Diet Assoc 1990; 90:856.

27. Gans KM, Sundaram SG, McPhillips JB, Hixson ML, Linnan L, Carleton RA. Rate your plate: an eating pattern assessment and educational tool used at cholesterol screening and education programs. J Nutr Ed 1993; 25:29–36.

28. Dowdy A, Buck B, Retzlaff B. The NWLRC Fat Intake Scale: Further validation and utility with cholesterol-lowering diets. J Am Diet Assoc 1990; 90(suppl.): A-120.

29. Dietary Treatment of Hypercholesterolemia. A Handbook for Counselors. American Heart Association, Dallas, TX, 1988.

30. Kristal AR, White E, Shattuck AL, Curry S. Anderson GL, Fowler A, Urban N. Long-term maintenance of a low-fat diet: durability of fat-related dietary habits in the Women's Health Trial. J Am Diet Assoc 1992; 92:553–559.

31. Cleveland LE, Escobar AJ, Lutz SM, Welsh SO. Method for identifying differences between existing food intake patterns and patterns that meet nutrition recommendations. J Am Diet Assoc 1993; 93:556–563.

32. Sigman-Grant M, Zimmerman S, Kris-Etherton PM. Dietary approaches for reducing fat intake of preschool-age children. Pediatrics 1993; 91:955–960.

33. DeLeeuw ER, Windham CT, Lauritzen GC, Wyse BW. Developing menus to meet current dietary recommendations: implications and applications. J Nutr Ed 1992; 24(3):136–144.

34. Smith-Schneider LM, Sigman-Grant MJ, Kris-Etherton PM. Dietary fat reduction strategies. J Am Diet Assoc 1992; 92:34–38.

35. Lloyd HM, Paisley CM, Mela DJ. Changing to a low fat diet: attitudes and beliefs of UK consumers. Eur J Clin Nutr 1993; 47:361–373.

36. Sporny LA, Contento IR. Stages of Dietary Change with Respect to Fat Intake: Behavioral and Process-Related Correlates. Presented at the 1992 Society for Nutrition Education Annual Meeting Proceedings, Washington, DC, July 1992, p. 67.

37. Burrows ER, Henry HJ, Bowen DJ, Henderson MM. Nutritional applications of a clinical low fat dietary intervention to public health change. J Nutr Ed 1993; 25: 167–175.

38. Ursin G, Ziegler RG, Subar AF, Graubard BI, Haile RW, Hoover R. Dietary patterns associated with a low-fat diet in the National Health Examination follow-up study: identification of potential confounders for epidemiologic analyses. Am J Epidemiol 1993; 137:916–927.

39. Light L, Portnoy B, Blair JE, Smith JM, Tuckermanty E, Tenney J, Matthews O. Nutrition education in supermarkets. Family Commun Health 1989; 12:43–52.

40. Pennington JAT, Wisniowski LA, Logan BG. In-store nutrition information programs. J Nutr Ed 1988; 20:5–10.

41. Levy AS, Mathews O, Stephenson M, Tenney JE, Schucker RE. The impact of a nutrition information program of food purchases. Public Policy Market 1985; 4:1–13.

42. Russo JE, Staelin R, Nolan CA, Russell GJ, Metcalf BL. Nutrition information in the supermarket. J Consumer Res 1986; 13:48–70.

43. Jeffery RQ, Pirie PL, Rosenthal BS, Gerber WM, Murray DM. Nutrition education in supermarkets: an unsuccessful attempt to influence knowledge and product sales. J Behav Med 1982; 5:189–200.

44. Ernst ND, Wu M, Frommer P, Katz E, Mathews O, Moskowitz J, Pinsky JL, Pohl S, Schreiber GB, Sondik E, Tenney J, Wilbur C, Zifferblatt S. Nutrition education at the point of purchase: the Foods for Health project evaluated. Prev Med 1986; 15:60–73.

45. Soriano E, Dozier DM. Selling nutrition and heart-healthy diet behavior at the point-of-purchase. J Appl Nutr 1978; 30:56–65.

46. Mullis R, Hunt MK, Foster M, Haefeld L, Lansing D, Snyder P, Price P. Environmental support of healthful behavior: The Shop Smart for Your Heart grocery program. J Nutr Ed 1987; 19:225–228.

47. Shannon B, Mullis RM, Pirie PL, Pheley AM. Promoting better nutrition in the grocery store using a game format: the Shop Smart Game Project. J Nutr Ed 1990; 22:183–188.

48. Dougherty MF, Wittsten AB, Guarino MA. Promoting low-fat foods in the supermarket using various methods, including videocassettes. J Am Diet Assoc 1990; 90: 1106–1108.

49. Mullis R, Pheley A, Snyder P, Price K. Final Report: Point of Purchase Videotapes in the Produce Department. A collaborative effort of the United Fresh Fruit and Vegetable Association and the University of Minnesota School of Public Health. Minneapolis: University of Minnesota School of Public Health, 1988.

50. Scott JA. Point-of-purchase nutrition education. Aust J Nutr Diet 1992; 49:4–9.

51. Scott JA, Pollard CM. Supermarket sleuth: evaluation of a novel nutrition education strategy. Aust J Nutr Diet 1992; 49:11–17.

52. Glanz K, Mullis RM. Environmental interventions to promote healthy eating: a review of models, programs, and evidence. Health Ed Q 1988; 15:395–415.

53. Browne, MG. Label Facts for Healthful Eating. Educator's Resource Guide. National Food Processors Association. Dayton, OH: The Mazer Corporation, 1993.

54. Berwick DM, Cretin S, Keeler EB. Cholesterol, Children and Heart Disease, An Analysis of Alternatives. New York: Oxford University Press, 1980.

55. Scott LW, Foreyt JP, Young J, Reeves RS, O'Malley MP, Gotto AM Jr. Are low-cholesterol diets expensive? J Am Diet Assoc 1979; 74:558–561.

56. Futhey KL, Buzzard IM, Asp EH. Cost analysis of low fat intervention diets. Presented at the 1993 Society for Nutrition Education Annual Meeting Proceedings, Minneapolis, MN, July 1993, p. 76.

57. Shaul JA, Rader JM, Jenkins PL, Mitchell DC, Shannon BM, Pearson TA. Does a cholesterol-lowering diet cost more? Circulation 1993; 88(suppl):abstract #2067.

58. NIH Consensus Development Panel on Triglyceride, High-Density Lipoprotein, and Coronary Heart Disease. Triglyceride, high-density lipoprotein, and coronary heart disease. JAMA 1993; 269(4):505–510.

59. The Fifth Report on the Joint National Committee on Detection, Evaluation, and Treathment of High Blood Pressure (JNCV). National High Blood Pressure Education Program, National Heart, Lung, and Blood Institute, National Institutes of Health, Bethesda, MD, 1992.

60. Stamler R, Stamler J, Grimm R, Gosch F, Elmer P, Dyer A, Berman R, Fishman J, Van Heel N, Civinelli J, McDonald A. Nutritional therapy for high blood pressure. Final report of a four-year randomized controlled trial—the Hypertension Control Program. JAMA 1987; 257:1484–1491.

61. Little P, Girling G, Hasler A, Trafford A. A controlled trial of a low sodium, low fat, high fibre diet in treated hypertensive patients: effect on antihypertensive drug requirement in clinical practice. J Human Hypertens 1991; 5:175–181.

62. Arroll B, Beaglehole R. Does physical activity lower blood pressure: a critical review of the clinical trials. J Clin Epidemiol 1992; 45:439–447.

63. World Hypertension League. Alcohol and hypertension—implications for management: a consensus statement by the World Hypertension League. J Human Hypertens 1991; 5:227–232.

64. National High Blood Pressure Education Program. Working Group Report on Primary

Prevention of Hypertension. National High Blood Pressure Education Program, National Heart, Lung, and Blood Institute, National Institutes of Health, Bethesda, MD, NIH Publication No. 93–2669.

65. McCarron DA, Morris CD, Henry HJ, Stanton JL. Blood pressure and nutrient intake in the United States. Science 1984; 224:1392–1398.

66. Witteman, JCM, Willett WC, Stampfer MJ, Colditz GA, Sacks FM, Speizer FE, Rosner B, Hennekens CH. A prospective study of nutritional factors and hypertension among US women. Circulation 1989; 80:1320–1327.

67. Stitt FW, Clayton DG, Crawford MD, Morris JN. Clinical and biochemical indicators of cardiovascular disease among men living in hard and soft water areas. Lancet 1973; 1:122–126.

68. Nutrition recommendations and principles for people with diabetes mellitus. J Am Diet Assoc 1994; 94:504–506.

69. Report of the Expert Panel on Population Strategies for Blood Cholesterol Reduction. National Cholesterol Education Program, National Heart, Lung, and Blood Institute, U.S. Department of Health and Human Services, Bethesda, MD, 1990.

70. Report of the Expert Panel on Blood Cholesterol Levels in Children and Adolescents. National Cholesterol Education Program, National Heart, Lung, and Blood Institute, U.S. Department of Health and Human Services, Bethesda, MD, 1991.

71. Satter E. How to Get Your Kid to Eat. . . . But Not Too Much. Palo Alto, CA: Bull Publishing Company, 1987.

72. Anderson JW, Bridges SR. Dietary fiber content of selected foods. Am J Clin Nutr 1988; 47:440–447.

73. Pennington JAT. Bowes & Church's Food Values of Portions Commonly Used. 16th ed. Philadelphia: J.P. Lippincott Co., 1994.

74. Grant JP. Handbook of Total Parenteral Nutrition. Philadelphia: W.B. Saunders, 1980:15.

75. Human Nutrition Information Service (USDA). Report of the Dietary Guidelines for Americans. Hyattsville, MD: U.S. Government Printing Office 1990:24.

76. Reduced-fat, low-calorie eating more popular than ever, Calorie Control Comment 1993; 15(1):3.

77. Cahn HS. A major error in nomograms for estimating body mass index. Am J Clin Nutr 1991; 54:435–437.

78. Bray GA. Definitions, measurements, and classifications of the syndromes of obesity. Int J Obes 1978; 2:99–113.

12

Drug Therapy

Henry N. Ginsberg

Columbia University College of Physicians and Surgeons, New York, New York

INTRODUCTION

The treatment of hypercholesterolemia has been a matter of intense interest to the medical community since reports from the Framingham Heart Study over two decades ago indicated that dyslipidemias, including elevated serum total and low-density lipoprotein (LDL) cholesterol, represent major independent risk factors in the development of cardiovascular disease (1). Evidence for the long-term clinical benefits of diet and pharmacotherapy derives from primary and secondary intervention trials demonstrating diminutions in cardiovascular morbidity and mortality with reductions in serum cholesterol (2–8). Although these studies have been reviewed in other chapters of this book, it would be useful to note aspects of some of them that speak to the issue of the efficacy of drug therapy for cholesterol lowering. The Lipid Research Clinics Coronary Primary Prevention Trial (LRC-CPPT) documented a reduction in coronary artery disease (CAD) risk of 19% associated with declines in total serum and LDL cholesterol of 8 and 12%, respectively. However, in the subgroup of patients taking all of their medication (approximately six packets of cholestyramine per day), there was a 50% reduction in CAD events associated with reductions in LDL cholesterol of 35% and in total cholesterol of 25% (3). Thus, although critics of cholesterol-lowering trials have suggested that the overall LRC data reflect the lack of efficacy of intervention trials, the data indicate that marked reductions in CAD incidence are possible when significant reductions in plasma total and LDL cholesterol are achieved.

Similarly, although early regression trials such as the NHLBI Type II Coronary Intervention Study (9) demonstrated only modest differences between controls and cholestyramine-treated patients, more potent combination drug therapy resulted in significant differences based on computerized analyses of reputed angiograms in both the study conducted by Brown et al. (10,11) and those carried out by Blankenhorn and colleagues (12–14). In the former study, a group of 146 men with CAD and high serum levels of apoprotein B were treated with either a 3-hydroxy-3-methylglutaryl–coenzyme A (HMG-CoA) reductase inhibitor or niacin, together with the bile-acid resin colestipol. Marked reductions in total and LDL cholesterol and significant increases in HDL cholesterol were achieved (9–11). Similarly, in the Cholesterol Lowering Atherosclerosis Study (CLAS), therapy with colestipol, niacin, and diet dramatically improved serum lipoprotein profiles, with marked lowering of LDL cholesterol and triglyceride levels and significant increases in high-density lipoprotein (HDL) cholesterol concentrations (12–14).

This brief overview indicates that when significant alterations in the lipid profile are required, they can be achieved by the appropriate use of one or more pharmacological agents. More important, when these significant improvements in lipid and lipoprotein levels are achieved, they are associated with significant reductions in both CAD events and in the progression of CAD lesions. As a logical consequence of these findings, the National Cholesterol Education Program (NCEP) recently proposed updated guidelines for the management of hypercholesterolemia in high-risk subjects (16). The second report of the NCEP Adult Treatment Panel (ATP II), like the 1988 report (ATP I), underscores the role of an initial trial of diet to lower LDL cholesterol in patients whose levels exceed desirable levels. Simultaneously, however, the new guidelines continue the aggressive stance toward pharmacological therapy in patients with multiple risk factors for CAD and take an even more aggressive approach to individuals with established CAD. In setting goals for LDL cholesterol of less than 130 mg/dl for individuals with multiple risk factors and less than 100 mg/dl for patients with preexisting CAD, the Panel has acknowledged the necessity of combined diet and drug therapy for a significant number of Americans. While there is debate about the overall efficacy of such an approach to primary prevention, there is little debate about an aggressive approach in secondary intervention. In any event, the overall success of any intervention will be determined by the appropriateness of the therapeutic approach taken.

NONPHARMACOLOGICAL THERAPY

Dietary intervention is addressed in detail in other chapters, but a few points relevant to its use together with pharmacological treatments may be helpful. An important issue is the length of time that one should wait after instituting diet treatment before moving on to diet plus drug. First, the physician or health provider should decide how likely it is that the patient will ultimately require pharmacological intervention. This decision should be based on the entire risk profile of the patient as well as the goals and desires of the patient and his or her family. As noted above, secondary intervention, with its goals of reaching an LDL cholesterol level of less than 100 mg/dl, will almost always require drug treatment together with diet; in this instance, a few months of diet therapy alone might be utilized to allow the patient to learn the diet and see the benefits of that aspect of treatment. Whether only the Step I diet or the Step I followed by the Step II diet should precede drug therapy is, I believe, a decision best made by the physician and the patient. In primary prevention, the Step I diet might be maintained for more than 6 months before considering addition of drugs. In this case, advancing the patient to the Step II diet prior to initiation of pharmacotherapy seems very appropriate. A key point to remember is that diet is maintained even if drug therapy becomes necessary.

DRUG THERAPY: OVERVIEW

By one estimate (17) 52 million Americans aged 20 years or older require dietary therapy under the ATP II guidelines. Of these, 12.7 million (approximately one-fourth) would fail to attain cholesterol targets with dietary therapy alone and would be potential candidates for pharmacotherapy. Approximately 4 million (one third) of these patients have established CAD. It is clear that this represents a very large burden for the health care system. It is, therefore, imperative that drug therapy be as efficacious as possible. As a first step toward this goal, the physician must be able to categorize the patient so that the most appropriate class of pharmacological agents can be chosen. Although several classification systems exist, a practical starting point can be to simply review the plasma lipid profile, which includes the total cholesterol and triglyceride levels and the HDL cholesterol concentration. From these values, the LDL cholesterol level can be estimated* using the following formula:

$$\text{LDL cholesterol} = \text{total cholesterol} - \left(\frac{\text{triglyceride}}{5} + \text{HDL cholesterol} \right)$$

With this information, the physician can divide his patients into those with elevated LDL cholesterol and normal triglyceride levels; elevated LDL cholesterol and elevated triglyceride levels; and normal LDL and elevated triglyceride levels. HDL cholesterol levels may be normal in any of these three groups, although they

---

*The estimation of LDL is only valid for fasting plasma triglyceride values less than 400 mg/dl. For triglyceride levels greater than 400 mg/dl, consultation with a specialist is recommended.

will usually be low in the two groups with elevated triglyceride concentrations. The term "elevated" refers to levels greater than the 90th percentile for LDL cholesterol and triglycerides; it does not refer to the LDL targets that would apply depending on risk factors and the absence or presence of CAD. Categorizing patients by their LDL cholesterol and triglyceride levels will give the health care provider some insight into the pathophysiological process underlying the patient's hyperlipidemia and the likelihood of responsiveness to different drugs. In the discussion of drug therapy, the efficacy of each single agent and combination of agents will be presented within the context of these different types of dyslipidemia.

MONOTHERAPY FOR HYPERCHOLESTEROLEMIA

There classes of lipid-lowering agents are recommended by the NCEP ATP II as first-line therapy against hypercholesterolemia: the bile-acid sequestrants or binding resins, niacin, and the HMG-CoA reductase inhibitors. The fibric acid derivative gemfibrozil is suggested by the NCEP ATP II as a second-line agent for hypercholesterolemia. Fibric acid derivatives are more effective as triglyceride-lowering agents. Finally, probucol, which has only modest LDL-lowering activity but significantly reduces HDL cholesterol, is also not recommended as a major cholesterol-lowering agent. Probucol may have utility in the extremely rare case of homozygous familial hypercholesterolemia, where its use has been associated with reduction in xanthomas.

Bile-Acid Resins

Cholestyramine and colestipol, the currently available bile-acid resins, have been in use as lipid-lowering agents for almost three decades. These drugs interfere with reabsorption of bile acids in the intestine, and this in turn results in a compensatory increase in bile-acid synthesis from cholesterol in hepatocytes. Reduction in intracellular cholesterol induces an up-regulation of LDL-receptor expression and receptor-mediated clearance of LDL cholesterol from the circulation. The bile acid sequestrants are primary agents in the treatment of patients with elevated levels of LDL cholesterol and normal triglycerides. Sequestrants produce dose-dependent decreases on the order of 15–25% in total cholesterol and 20–35% in LDL cholesterol (18). Modest increases in HDL cholesterol are also achieved by these agents. Bile-acid resins are recommended particularly for young adult men and premenopausal women with moderate cholesterol elevations (16).

The efficacy and safety of cholestyramine and colestipol have been evaluated in numerous controlled clinical trials, including the LRC-CPPT (2,3). In this large multicenter, placebo-controlled, primary prevention study, reductions of 13.4% and 20.3% in total and LDL cholesterol, respectively, were achieved on the average dose of 16 g/day of cholestyramine. The bile-acid resins also possess excellent safety profiles. The fact that cholestyramine and colestipol are not absorbed systemically may account for their relative lack of nongastrointestinal (GI) side effects. In the LRC-CPPT trial (2,3), the only adverse events that occurred more frequently in the cholestyramine group than the placebo group were GI-related, principally constipation and heartburn. These effects are dose related, manageable, and tend to diminish with time. In a number of patients, however, the adverse GI effects associated with bile-acid resins present significant compliance problems during long-term use. This is particularly true in patients who have pre-existing gastrointestinal problems such as chronic constipation or irritable bowel syndrome. The bile-acid sequestrants are contraindicated in patients with significant bowel diseases such as ileitis or ulcerative colitis. In addition, administration requires patient education; the resins are delivered by a powder vehicle and must be mixed with liquid. An unpleasant, gritty texture may result, diminishing palatability for many patients.

Although these aesthetic problems and the gastrointestinal side effects tend to diminish the usefulness of the resins, their safety profile supports their use, at least in low doses, whenever possible. Additionally, many of the problems listed above can be avoided or diminished if patients start with very small doses of the resins, increase the dose slowly, and are advised to maintain the dose just below the point where side effects begin to significantly affect lifestyle. It has become clear that tolerance of two or three packets of either resin, cholestyramine or colestipol, is high, and that reductions of 10–20% in total cholesterol (as a result of LDL lowering) can be achieved with these levels of adherence. In the near future, there should be several new preparations of bile acid sequestrants available in tablet or pill form; these preparations will at least avoid the aesthetic problems inherent in the powdered forms.

A physiological limitation of the resins' efficacy derives from the ability of the liver to increase endogenous cholesterol synthesis to meet the need for bile acid synthesis. This "downside" of the mechanism of action of the resins is actually the basis for their synergistic interaction with the reductase inhibitors, making this combination extremely effective for reducing LDL cholesterol concentrations (see below). A further limitation of the sequestrants is their tendency to raise triglyceride levels through compensatory increases in hepatic synthesis of very low-density lipoprotein (VLDL), which contains triglyceride as a major core component. This phenomenon essentially limits the use of the resins as single agents to individuals with normal triglyceride levels. The resins can be used successfully in combination with niacin, fibrates, or reductase inhibitors in patients with hypertriglyceridemia and elevated LDL cholesterol levels (see below) (16–18).

Both cholestyramine and colestipol are anion-exchange resins and prevent or delay absorption of other drugs from the GI tract. Compromised absorption has been reported on concomitant administration with thiazide diuretics, β-adrenergic blockers, thyroid hormone supplements, anticoagulants, cardiac glycosides, and fat-soluble vitamins (16). As a result of these findings and the lack of data for most drug classes, it is recommended that resins be used 1–2 hours after or 4 hours before any other medications are ingested. This clearly becomes a problem in patients taking other regularly scheduled medications.

Resins may be the least cost-effective pharmacological method of normalizing serum lipid profiles. In an analysis of common cholesterol-lowering agents, cholestyramine proved least efficient in terms of cost per percent reduction in LDL cholesterol: average cost was $347 per percent reduction over 5 years (21). Sequestrants also proved least cost effective in elevating HDL cholesterol.

Niacin

This is another agent with a very long record of efficacy. In addition, favorable responses to niacin can be demonstrated along virtually the entire lipid and lipoprotein profile. Niacin diminishes both total and LDL cholesterol in the range of 15–25%. It is particularly potent as a triglyceride-lowering agent, reducing VLDL levels 25–35%. However, niacin receives most interest because of its ability to significantly affect HDL cholesterol levels, which can be raised as much as 15–25% (18). Thus, on the basis of proven activity against the three major lipoproteins—VLDL, LDL, and HDL—niacin would appear to be an optimal agent.

Niacin's probable mechanism of action involves inhibition of intracellular lipolysis and, therefore, release of free fatty acids from adipocytes. Based on recent studies in cultured liver cells, decreased free fatty acid delivery to the liver would result, in turn, in diminished hepatic output of apoprotein B–containing lipoproteins such as VLDL and LDL; hence niacin's significant efficacy as an agent to lower both plasma triglycerides and LDL cholesterol levels. Niacin's efficacy as an HDL-raising agent probably also stems from its effects on apoprotein B–containing lipoproteins. Fewer free fatty acids arriving at the liver results in lowered VLDL triglyceride levels, which, in turn, lead to less exchange between HDL cholesteryl ester and VLDL triglyceride. Lower plasma free fatty acid levels might, according to recent data, also directly reduce core lipid exchange between VLDL and HDL. Of note, niacin is the only major lipid-lowering agent that diminishes Lp(a). The basis of this activity is as yet undefined.

Efficacy of monotherapy with niacin was confirmed in the long-term, large-scale Coronary Drug Project, in which this agent was the only drug of the five initially tested to significantly reduce definite myocardial infarction (8). A longer-term follow-up of that study (15-year total) showed a significant 11% decrease in all-cause mortality among patients randomized to niacin in this secondary prevention study (6). Niacin has also been used successfully in combination therapy in three studies of regression of coronary artery disease. In all of those studies, increased HDL cholesterol levels appeared to play an important role in the outcomes. In all of these studies, short-acting or "regular" niacin was used. Niacin is also available in several slow-release preparations that gained favor in the 1980s because of the reduced incidence of flushing associated with these preparations. However, recent studies have suggested both less efficacy and greater potential toxicity associated with the slow-release preparation, and they are not recommended by most lipid specialists.

Like the bile-acid resins, niacin exhibits a solid safety record, having been in use as a lipid-lowering agent for almost 30 years. However, unlike the resins, niacin is a systemic agent, and its use can be associated with a variety of bothersome, as well as potentially significant, systemic side effects that can limit patient acceptability. For instance, it frequently causes uncomfortable and potentially dose-limiting cutaneous flushing. Pruritus can be associated with the flushing, but both tend to subside after several weeks in most patients and may be minimized by initiating therapy at lower doses. In fact, a typical schedule for initiating therapy would be to use 100 mg bid for several days (or a week), followed by doubling of the dose every 4–6 days. Stabilization of the dose in the range of 1–2 g/day may be useful to allow a new lipid steady state to be reached. Whenever flushing or other complaints become significant, reduction in dose to the previous level is usually adequate. Further increases can then be attempted after several more days at the lower dose. Flushing may also be minimized or reduced by premedication with aspirin or ibuprofen, because the vasodilator reactions that are the basis of this system are probably prostaglandin mediated. Of note, initiation of therapy at low doses avoids possible episodes of orthostatic hypotension associated with vasodilatation. As noted earlier, flushing is markedly reduced or eliminated in most patients when the slow-release niacin preparations are used. However, these preparations appear to have a greater potential for systemic, particularly hepatic, toxicity (see below).

Less common adverse effects of niacin include elevations of liver enzymes (transaminitis), GI distress, altered glucose disposal, and elevated serum uric acid with or without gouty arthritis. Although transaminitis may occur in as many as 3–5% of patients on full doses of niacin (>2 g/day), actual hepatotoxicity or hepatitis is uncommon. If hepatitis occurs, it does so more commonly upon administration of the slow-release preparation or very high doses (>3 g/day) of regular niacin (22). To limit GI upset, it is usually recommended that niacin be taken with meals. Myopathy, including rhabdomyolysis, has been reported when niacin is used alone or in concert with certain other agents (see below) (23).

Contraindications to niacin use include a history of jaundice or active hepatic disease, gallbladder disease, gouty arthritis or hyperuricemia, and peptic ulcer disease. Niacin is generally contraindicated in individuals with diabetes mellitus (see below). In any event, monitoring of uric acid, glucose, and liver function is advised in all patients receiving niacin (16).

In the analysis of lipid-lowering agents by Schulman and colleagues (21), niacin emerged as the most cost-effective alternative in terms of cost per percent LDL cholesterol reduction and HDL cholesterol elevation. Yet the study also reported that estimated additional costs incurred by medication intolerance (e.g., initiation of new lipid-lowering agent, use of secondary medications) can raise the cost of niacin by as much as 60%.

HMG-CoA Reductase Inhibitors

Consistent with the increasing evidence supporting their efficacy profiles, high patient tolerability, and favorable safety experience during the past decade, HMG-CoA reductase inhibitors are now included by the NCEP ATP II among first-line alternatives for treatment of hypercholesterolemia (16). The category includes three members: lovastatin, simvastatin, and pravastatin.

HMG-CoA reductase inhibitors directly inhibit the rate-limiting step in hepatic cholesterol biosynthesis: the conversion of HMG-CoA to mevalonate. A consequent rise in LDL-receptor expression on hepatocytes enhances receptor-mediated clearance of LDL cholesterol from the circulation. Additionally, several studies in humans indicate that inhibition of reductase is associated with reduced secretion of apoprotein B–containing lipoproteins by the liver. Thus, both reduced entry of VLDL and LDL into plasma and increased efficiency of LDL removal from plasma can explain the marked reduction in LDL cholesterol levels seen in most patients receiving reductase inhibitors. At usually recommended therapeutic doses, the HMG-CoA reductase inhibitors decrease total cholesterol in the range of 15–30% and LDL cholesterol 20–40%. Larger reductions can be achieved with higher doses. Interestingly, treatment with reductase inhibitors often produces reductions in triglycerides of 10–20%; this may be the result of reduced secretion of VLDL by the liver. HDL cholesterol levels rise about 5–10% during reductase therapy (18,24–26). Although declines in apoprotein B levels have also been demonstrated (27), the HMG-CoA reductase inhibitors appear to exert minimal effects on apolipoprotein A-I, A-II, and Lp(a) (27–29).

The HMG-CoA reductase inhibitors are exceedingly well tolerated and approximately equivalent with respect to safety profiles during monotherapy. In comparison with other lipid-lowering agents, HMG-CoA reductase inhibitors are relatively free of significant adverse effects, which have been reported by fewer than 5% of patients in controlled clinical trials (24–26, 30–32). The most common side effects are mild GI disturbances (e.g., nausea, abdominal pain, diarrhea, constipation, flatulence), which rarely warrant discontinuation of therapy. Headache, fatigue, pruritus, and myalgia are other minor side effects that seldom prompt treatment termination. Mild, transient elevations in liver enzymes have been reported with all of the HMG-CoA reductase inhibitors. Elevations in serum aminotransferases to more than three times upper limits of normal have occurred in less than 2% of patients in controlled clinical trials. Therapy should be discontinued when elevations of this magnitude occur. Significant elevations in hepatic enzymes appear to be dose related and thus more common in patients with severe familial hypercholesterolemia (FH), who require higher doses to achieve a therapeutic effect. Monitoring of hepatic aminotransferase levels according to manufacturers' labeling is recommended for all HMG-CoA reductase inhibitors. The recommended starting doses for lovastatin, pravastatin, and simvastatin are 20, 20, and 5–10 mg/day, respectively. Maximal doses for each are four times the starting dose. As described in more detail later, use of reductase inhibitors in combination with resins allows the physician to use lower doses of both, thereby reducing the potential for adverse effects.

A rare but potentially serious adverse effect of HMG-CoA reductase inhibitors is myopathy, with muscle symptoms and serum creatine phosphokinase (CPK) elevations more than 10 times the upper limit of normal (33). In the EXCEL study of more than 8000 patients followed for 48 weeks, myositis occurred in only five lovastatin-treated patients (30). In large controlled clinical trials, one definite case has been documented with pravastatin (34) and two cases of questionable origin were reported with simvastatin (25). Other isolated case reports of myopathy on simvastatin, with and without pronounced rises in CPK, have also been published (35). It should be noted that the early concern that lovastatin use might be associated with the development of corneal opacities has not been confirmed with longer follow-up of much larger numbers of patients on each of the reductase inhibitors.

Finally, with respect to cost, analysis by Schulman and colleagues (21) indicated that an HMG-CoA reductase inhibitor was a cost-effective alternative for regulating serum lipids. Further, once-daily dosing is frequently possible with the HMG-CoA reductase inhibitors, a schedule generally more convenient than the multiple-daily dosing regimens typical of bile-acid resins and niacin.

MONOTHERAPY FOR HYPERTRIGLYCERIDEMIA

In patients with normal LDL cholesterol levels and elevated plasma triglycerides, the rationale for drug therapy is much less clearly founded than it is for elevated LDL cholesterol levels. The reasons for uncertainty regarding treatment of hypertriglyceridemia include the lack of consistent data that plasma fasting triglycerides are predictive of CAD incidence and the lack of data from intervention trials where lowering of triglyceride levels has been achieved in the absence of changes in other lipoproteins. Hence, in the Helsinki Heart Study, triglyceride reductions were typically associated with concomitant reductions in LDL cholesterol and increases in HDL cholesterol levels. Similarly, in the Stockholm trial, HDL rose concomitantly with falls in plasma triglyceride levels. The ATP II recommendations for evaluating and treating hypertriglyceridemia focus on the associated LDL and HDL concentrations as guidelines for therapy. Thus, the patient's overall risk profile can be used to set goals for LDL cholesterol, using a reduced HDL level (commonly associated with hypertriglyceridemia) as a concomitant major risk factor. Of course, when triglyceride levels are greater than 500 mg/dl, the risk of developing pancreatitis increases and a direct focus on triglyceride-lowering is recommended.

Niacin

Niacin should be the first drug used when treating someone with hypertriglyceridemia. The positive and negative characteristics of niacin were outlined earlier in this chapter, but it is relevant to note that since LDL will be the target of intervention in most patients with elevated plasma triglyceride levels (ATP II), the potency of niacin against both triglycerides and LDL cholesterol makes it the first choice. Niacin's ability to raise HDL cholesterol further strengthens its first-place ranking.

Fibrates

The fibric acid derivatives available in the United States are clofibrate and gemfibrozil. Clofibrate, first used 3 decades ago, is rarely used now and will not be discussed. Gemfibrozil was the drug used in the Helsinki Heart Study.

The mechanism of action of gemfibrozil, and the fibrates in general, is only partially understood. This class of drugs appears to stimulate the enzyme lipoprotein lipase, and this is probably the basis for much of their potent triglyceride-lowering effects. However, some studies indicated that treatment with fibrates also reduces VLDL triglyceride entry into plasma. Fibrate stimulation of peroxisomal fatty acid oxidation may be important in this aspect of their triglyceride-lowering actions. In any event, gemfibrozil treatment is associated with 25–40% reductions in plasma triglyceride concentrations. Postprandial triglyceride levels, which are linked to fasting levels, are also reduced during treatment with gemfibrozil.

The ability of gemfibrozil, and the fibrates in general, to raise HDL cholesterol derives mainly from their triglyceride-lowering effects. Thus, reducing one of the substrates for the cholesteryl-ester transfer protein–mediated exchange of HDL cholesteryl ester for VLDL (and chylomicron) triglyceride results in increased HDL cholesterol levels. Apo A-I levels also rise during treatment with fibrates, and although this may be a concomitant of reduced triglyceride–cholesteryl ester exchange and stabilization of the HDL particle, some studies have suggested that apo A-I synthesis is increased during fibrate treatment.

Fibrates are not outstanding LDL cholesterol–lowering agents. In patients with isolated elevations of LDL cholesterol, gemfibrozil induces about a 10–15% reduction in LDL levels (16). It is for that reason that AP II did not list gemfibrozil as a primary agent for lowering LDL levels. Further, in hypertriglyceridemic subjects, when plasma triglycerides fall, LDL cholesterol levels may fall, remain unchanged, or rise. Although the latter outcome seems undesirable, the physiological regulation of LDL levels is quite complex in subjects with hypertriglyceridemia and interpretation of changes in LDL cholesterol is not straightforward. In the Helsinki Heart Study, subgroup analysis suggested that equal benefit, in terms of reduced CAD events, was achieved independent of whether LDL fell, remained unchanged, or increased slightly. The physician must be aware of the possibility of significant increases in LDL cholesterol that can accompany otherwise potentially beneficial falls in triglycerides and increases in HDL cholesterol during fibrate therapy. Such rises may require a change to another drug or addition of a second agent (see below).

The safety record of fibrates has been a subject of controversy. In the short-term, these drugs are extremely well tolerated, with mild GI distress, in the form of epigastric pain, being the major side effect. Transaminitis can occur in 2–3% of subjects but does not usually require cessation of treatment. Rarely, a hepatitis-like picture can occur. Myopathy with myositis is also a rare occurrence with the fibrates alone or in combination. Idiosyncratic granulocytopenia has also been reported. The long-term safety issues have been the source of controversy; in the WHO study, long-term use of clofibrate was associated with increased incidence of gallbladder and GI disease, including GI malignancies (4,7). Overall mortality in that study was increased in the clofibrate group compared to placebo, even though ischemic heart disease events were reduced by 20%. In the Helsinki Heart Study, gallbladder disease was somewhat increased in the gemfibrozil group, but GI malignancy was not different from placebo (5). All of the fibrates appear to make the bile more lithogenic, and long-term use is probably associated with a twofold increase in gallstone formation.

Overall, the lack of potency against LDL cholesterol leaves gemfibrozil as an alternative to niacin in the treatment of severe hypertriglyceridemia. Its use in moderate hypertriglyceridemia will mainly be in combination with LDL-lowering drugs, again as an alterative to niacin.

COMBINATION THERAPY

As noted several times in the preceding discussions, proper use of combinations of agents should enable the health care provider to achieve much greater changes in plasma lipids while simultaneously avoiding some of the adverse effects associated with high doses of any single agent. The use of combination therapy will be a mainstay of secondary intervention regimes, where LDL levels below 100 mg/dl are, together with optimal triglyceride and HDL cholesterol levels, the goal. Combination therapy is not without shortcomings, however. Potential problems include specific side effects and increased cost. Thus, the potent combinations reviewed below must be used judiciously.

In choosing the "right" combination, the physician should use the categories of dyslipidemia described above for monotherapy. Thus, for isolated elevations of LDL cholesterol, resins plus niacin, resins plus reductase inhibitors, or niacin plus reductase inhibitors are the logical choices. For combinations of elevated LDL cholesterol concentrations and triglyceride levels, particularly in the setting of a significantly reduced HDL cholesterol, resins and niacin or resins and gemfibrozil have been commonly used. With the availability of the reductase inhibitors, many physicians have used these more potent LDL-lowering agents to replace the resins in these combinations with niacin or gemfibrozil.

Bile-Acid Sequestrants and Niacin

Historically, the most common combination regimen has been a bile-acid resin and niacin, which together can decrease LDL cholesterol 45–55% (36). This combination was used with extreme efficacy in the CLAS trial, as one of the two combination regimens in the study by Brown et al., and in the study of patients with familial hypercholesterolemia by Kane et al. (37). The latter was the only angiographic study to date that has included females. Because of niacin's triglyceride-lowering capabilities, it can more than offset the commonly observed rise in triglycerides that occurs when bile acid sequestrants are used alone. In the CLAS trial, triglyceride concentrations fell about 35% on this combination. Additionally, HDL cholesterol concentrations increase between 15 and 25% with continued use of resins and niacin, mainly as a result of niacin's action on HDL metabolism. Thus, the combination of bile acid sequestrants and niacin is very potent against all types of lipid disorders and, at least theoretically, should be the first choice among the available combinations.

Unfortunately, although no additional side effects are unique to this combination, the resins and niacin both have a number of bothersome side effects that make this a difficult combination for many patients (see above). In addition, the two must be taken at different times; niacin is best taken with meals, while the resins can be taken at any other time. In subjects who have CAD and are receiving several cardiac medications, timing of the resin can become a problem. Despite these difficulties, this combination should be attempted in most patients with combined hyperlipidemia as well as in patients with severe, isolated elevations of LDL cholesterol.

Bile-Acid Sequestrants and HMG-CoA Reductase Inhibitors

Another extremely effective combination for achieving marked reductions in LDL cholesterol is that of an HMG-CoA reductase inhibitor and a bile-acid sequestrant. These two drugs exhibit highly complementary mechanisms of action in combination therapy for severe hypercholesterolemia. By disrupting the enterohepatic circulation of bile acids, the sequestrants stimulate conversion of hepatic cholesterol to bile acids. This phenomenon is in turn associated with both an up-regulation of hepatic LDL-receptor expression and an increase in endogenous cholesterol synthesis. The latter effect tends to limit the efficacy of the bile-acid sequestrants. Addition of a reductase inhibitor blocks this compensatory mechanism and further stimulates receptor-mediated LDL cholesterol clearance. Further, this combination is particularly attractive because the HMG-CoA reductase inhibitor acts systemically, whereas the resin is nonsystemic; systemic drug interactions are thus minimized (38,39).

When used in combination with sequestrants, HMG-CoA reductase inhibitors reduce LDL cholesterol approximately 30–60% in various studies. Most trials have involved the combination of lovastatin with cholestyramine or colestipol in patients with FCHL (39–42). The efficacy of such combination therapy has been maintained for as long as 5 years. Vega and Grundy (43) observed a 48% decline in LDL cholesterol among 10 patients with primary hypercholesterolemia treated with lovastatin-colestipol. In the few reports on simvastatin-resin combinations, similar declines in total and LDL cholesterol have been attained. Simvastatin (40 mg/day) with cholestyramine (12 g/day) significantly diminished total and LDL cholesterol (44). Use of the combination in 66 patients with type IIa hypercholesterolemia maintained significant reductions in LDL cholesterol of 45% over a year. Combination therapy with pravastatin and cholestyramine was evaluated in a multicenter trial that enrolled 248 patients with LDL cholesterol levels above the 90th percentile (45). As against placebo, there was a 51% decrease in LDL cholesterol levels after 8 weeks of treatment with pravastatin (40 mg/day) and cholestyramine (24 g/day). Similar results were achieved in another trial in which addition of cholestyramine (24 g/day) to pravastatin therapy also raised HDL cholesterol 11–18% after 4 weeks in 33 patients with primary hypercholesterolemia (46).

Combination therapy with HMG-CoA reductase inhibitors and resins has also achieved regression of atherosclerotic lesions in patients with established CAD. In 146 such men with apo B levels of >125 mg/dl who were followed for 2½ years, regression in any of nine proximal coronary segments was significantly more frequent (32%) and progression less frequent (21%) in groups treated with lovastatin-colestipol (vs. control). In contrast, of the control group of patients treated with placebo or resin alone, 46% exhibited lesion progression and only 11% regression (10). Associated changes in the lipoprotein profile included a 46% reduction in LDL cholesterol and a 15% increase in HDL cholesterol among the lovastatin-colestipol group. Similar favorable findings were reported by Kane and colleagues (37), who demonstrated lesion regression in 72 patients with heterozygous FH treated with various combination lipid-lowering regimens, including an HMG-CoA reductase inhibitor and bile-acid resin.

In all of the foregoing trials, the HMG-CoA reductase inhibitor–bile-acid resin combination proved safe and generally well tolerated. Sequestrant-induced GI disturbances may compromise compliance slightly compared with that observed with the HMG-CoA reductase inhibitor alone. Mild, transient, and asymptomatic rises in liver enzymes and CPK have also been reported.

Niacin and HMG-CoA Reductase Inhibitors

Another potentially attractive combination, especially for patients who cannot tolerate sequestrants, is niacin and a HMG-CoA reductase inhibitor. These agents also act by complementary mechanisms to reduce serum LDL cholesterol. Niacin and the reductase inhibitors both appear to reduce hepatic secretion of VLDL, whereas HMG-CoA reductase inhibitors also enhance receptor-mediated clearance of the metabolic byproduct of VLDL, which is LDL. Because niacin reduces VLDL output, combination therapy with an HMG-CoA reductase inhibitor is also appropriate for patients with concurrent elevations of LDL cholesterol and triglyceride. As noted previously, niacin is the most potent HDL-raising agent, and when used together with a reductase inhibitor can increase HDL levels 25–30%.

In eight patients with heterozygous FH treated with high-dose (80 mg/day) lovastatin and niacin (3 g/day), Illingworth and Bacon (36) reported a mean decline of 49% in LDL cholesterol, an increase in HDL cholesterol of 25%, and a decrease in triglyceride of 43%, findings generally consistent with those of other investigators. Similarly, pravastatin (40 mg/day) and niacin (2 g/day) reduced LDL cholesterol 43%, as against 16% for niacin monotherapy, in 158 hypercholesterolemic patients (47).

The HMG-CoA reductase inhibitor–niacin combination is also generally well tolerated. Niacin-induced cutaneous flushing may compromise patient compliance but is often readily manageable. Rises in hepatic transaminases are occasionally encountered but are usually mild, transient, and asymptomatic. More serious and possibly dose-related hepatotoxic events might be anticipated in rare instances, based on case reports of transaminase elevations and fulminant hepatic failure with high-dose or sustained-release niacin monotherapy (22,23,48,49).

On the other hand, myopathy, occasionally with rhabdomyolysis and renal failure, has been reported when lovastatin is administered along with niacin, gemfibrozil, erythromycin, and cyclosporine (50–52). Severe myopathy has been reported in organ-transplant patients, whose multiple-drug (e.g., cyclosporine) regimens may predispose them to liver dysfunction and, hence, elevated systemic HMG-CoA reductase inhibitor levels. To date, myopathy during niacin combination therapy has not been reported with either pravastatin or simvastatin.

Bile-Acid Sequestrants and Gemfibrozil

This is another good combination for individuals with combined hyperlipidemia. It is not as effective as resin and niacin, but may be better tolerated. Because of the limited effects of gemfibrozil on LDL cholesterol, it is recommended for individuals whose combined hyperlipidemia is characterized mainly by hypertriglyceridemia. For the same reason, bile-acid sequestrants and gemfibrozil are not the combination of choice for isolated elevations of LDL cholesterol that require more than monotherapy. Actually, a frequent indication for this combination is a rise in LDL cholesterol during therapy of isolated hypertriglyceridemia in an individual who cannot tolerate niacin. In such cases, addition of a resin will reverse the gemfibrozil-associated increase in LDL levels without otherwise affecting the efficacy of the fibrate. Similarly, a significant rise in triglycerides during monotherapy with a resin can be reversed by the addition of gemfibrozil. Once again, niacin would be the drug of choice (possibly as monotherapy) in this type of patient.

The combination of bile-acid sequestrant and gemfibrozil can achieve LDL cholesterol reductions of 25–30%, triglyceride reductions of 30–40%, and increases in HDL cholesterol of 15–20%. Compliance with this regimen is usually limited by the same gastrointestinal complaints that accompany either drug alone; there does not seem to be an increase in such problems with combination treatment, however. There are no adverse reactions unique to this combination. As with all drugs taken in combination with bile-acid sequestrants, intake should be separated by several hours.

Gemfibrozil and HMG-CoA Reductase Inhibitors

The use of gemfibrozil with a reductase inhibitor derives its potential potency from the synergism of their actions. Fibrates tend to lower triglycerides and elevate HDL cholesterol mainly by stimulating lipoprotein lipase and promoting VLDL clearance from plasma. HMG-CoA reductase inhibitors complement this action by augmenting receptor-mediated clearance of LDL, which is generated from VLDL. Additionally, both drugs reduce the input of VLDL into the circulation. Consistent with the effects of fibrates used alone, declines in triglyceride and increases in HDL cholesterol during combination therapy with HMG-CoA reductase inhibitors are substantial; reductions in total and LDL cholesterol are less pronounced than with other combination regimens, however (36, 53, 54). Hence, the major use for the combination of gemfibrozil and a reductase inhibitor is in the treatment of combined hyperlipidemia. For instance, in 80 patients with mixed hyperlipidemia, total cholesterol reductions of 22%, LDL cholesterol reductions of 35%, and triglyceride reductions of 35% were achieved with lovastatin and gemfibrozil (55). In patients with FCHL, institution of simvastatin-gemfibrozil therapy reduced LDL cholesterol by 39% in an 18-month open study (56). Triglyceride fell by 50%, and HDL cholesterol rose by 16%. Similar results were obtained with a combination of pravastatin and gemfibrozil, which lowered LDL cholesterol by 37.1% and raised HDL cholesterol by 16.8% (57). However, Illingworth and Bacon reported no additional significant diminution in LDL cholesterol after addition of gemfibrozil to lovastatin therapy for heterozygous FH (53). Several of the other combinations listed above are much more efficacious for FH.

The principal limitation of the HMG-CoA reductase inhibitor–fibrate regimen is the possibility of myopathy, which has been reported when gemfibrozil is combined with lovastatin (33, 55). In some cases, myopathy has progressed to severe rhabdomyolysis and renal failure. Incidences of myopathy on lovastatin-gemfibrozil range from 5 to 8% (42,69). Myopathy has not been encountered with administration of HMG-CoA reductase inhibitors other than lovastatin during combined therapy with gemfibrozil. Further long-term trials are warranted, however, to evaluate the safety of individual HMG-CoA reductase inhibitors in combination with fibric acid derivatives.

CONCLUSION

Extensive primary and secondary prevention studies have elucidated a strong trend toward diminished CAD morbidity and mortality with reductions in LDL cholesterol. In addition newer studies indicate that increases in HDL cholesterol, and possibly reductions in plasma triglycerides, also impact favorably on CAD. Angiographic studies have provided an anatomical rationale for the efficacy of lipid-lowering therapy. For the approximately 25% of hypercholesterolemic patients who fail to respond to diet alone, the clinical challenge is to tailor monotherapy or combination lipid-lowering regimens to individual needs. Among the factors influencing selection are the nature and severity of lipoprotein disturbances (e.g., phenotype, secondary causes) as well as the safety, tolerability, and cost effectiveness of each approach. The NCEP ATP II guidelines recommend as first-line pharmacotherapeutic agents the bile-acid resins, niacin, and HMG-CoA reductase inhibitors. When lipid disorders resist low-fat diet and monotherapy, combination therapy is effective and generally well tolerated, with the notable exceptions of case reports involving myopathy on binary regimens with lovastatin and either niacin or gemfibrozil. Further long-term clinical experience is necessary to determine whether the biopharmaceutical profiles of individual HMG-CoA reductase inhibitors translate into distinct safety benefits during combination lipid-lowering therapy.

REFERENCES

1. Castelli WP, Wilson PWF, Levy D, Anderson K. Serum lipids and the risk of coronary artery disease. Atherosclerosis Rev 1990; 21:7–19.
2. Lipid Research Clinics Program. The Lipid Research Clinics Coronary Primary Prevention Trial results: I. Reduction in incidence of coronary heart disease. JAMA 1984; 251:351–364.

3. Lipid Research Clinics Program. The Lipid Research Clinics Coronary Primary Prevention Trial results: II. The relationship of reduction in incidence of coronary heart disease to cholesterol lowering. JAMA 1984; 251:365–374.

4. Oliver MF, Heady JA, Morris JN. Report from the committee of principal investigators: a cooperative trial in the primary prevention of ischaemic heart disease using clofibrate. Br J Heart Dis 1978; 40:1069–1118.

5. Frick MH, Elo O, Haapa K, et al. Helsinki Heart Study: primary-prevention trial with gemfibrozil in middle-aged men with dyslipidemia. N Engl J Med 1987; 317:1237–1245.

6. Canner PL, Berge KG, Wenger NK, et al. Fifteen year mortality in Coronary Drug Project patients: long-term benefit with niacin. J Am Coll Cardiol 1986; 8:1245–1255.

7. Committee of Principal Investigators. WHO Clofibrate Trial: a cooperative trial in the primary prevention of ischaemic heart disease using clofibrate. Br Heart J 1978; 40:1069–1118.

8. Coronary Drug Project Research Group. Clofibrate and niacin in coronary heart disease. JAMA 1975; 231:360–381.

9. Levy RI, Brensike JF, Epstein SE, et al. The influence of changes in lipid values induced by cholestyramine and diet on progression of coronary artery disease: results of the NHLBI type II coronary intervention study. Circulation 1984; 69:325–337.

10. Brown G, Albers JJ, Fisher LD. Regression of coronary artery disease as a result of intensive lipid-lowering therapy in men with high levels of apolipoprotein B. N Engl J Med 1990; 323:1289–1298.

11. Brown BG, Zhao XQ, Sacco DE. Lipid lowering and plaque regression: new insights into prevention of plaque disruption and clinical events in coronary disease. Circulation 1993; 87:1781–1791.

12. Blankenhorn DH, Nessim SA, Johnson RL. Beneficial effects of combined colestipol-niacin therapy on coronary atherosclerosis and coronary venous bypass grafts. JAMA 1987; 257:3233–3240.

13. Cashin-Hemphill L, Mack WJ, Pogoda JM, et al. Beneficial effects of colestipol-niacin on coronary atherosclerosis: a 4-year follow-up. JAMA 1990; 264:3013–3017.

14. Blankenhorn DH, Selzer RH, Crawford DW, et al. Beneficial effects of colestipol-niacin therapy on the common carotid artery: two- and four-year reduction of intima-media thickness measured by ultrasound. Circulation 1993; 88:20–28.

15. Watts GF, Lewis B, Brunt JNH, et al. Effects on coronary artery disease of lipid-lowering diet, or diet plus cholestyramine, in the St Thomas' Atherosclerosis Regression Study (STARS). Lancet 1992; 339:563–569.

16. Expert Panel on Detection, Evaluation, and Treatment of High Blood Cholesterol in Adults. Summary of the second report of the National Cholesterol Education Program (NCEP) Expert Panel on Detection, Evaluation, and Treatment of High Blood Cholesterol in Adults (Adult Treatment Panel II). JAMA 1993; 269:3015–3023.

17. Sempos CT, Cleeman JI, Carroll MD, et al. Prevalence of high blood cholesterol among US adults: an update based on guidelines from the second report of the National Cholesterol Education Program Adult Treatment Panel. JAMA 1993; 269:3009–3014.

18. Ginsberg HN, Arad Y, Goldberg IJ. Pathophysiology and therapy of hyperlipidemia. In: Antonaccio M, ed. Cardiovascular Pharmacology. 3d ed. New York: Raven Press, 1990:485–513.

19. Gurakar A, Hoeg JM, Kostner G, Papadopoulus NM, Brewer HB. Levels of Lp(a) decline with neomycin and niacin treatment. Atherosclerosis 1985; 57:293–301.

20. Carlson LA, Hamsten A, Asplund A. Pronounced lowering of serum levels of lipoprotein Lp(a) in hyperlipidaemic subjects treated with nicotinic acid. J Intern Med 1989; 226:271–276.

21. Schulman KA, Kinosian B, Jacobson TA, et al. Reducing high blood cholesterol level with drugs: cost-effectiveness of pharmacologic management. JAMA 1990; 264: 3025–3033.

22. Henkin Y, Oberman A, Hurst DC. Niacin revisited: clinical observations on an important but underutilized drug. Am J Med 1991; 91:239–246.

23. Litin SC, Anderson CF. Nicotinic acid-associated myopathy: a report of three cases. Am J Med 1989; 86:481–483.

24. Bradford RH, Shear CL, Chremos AN, et al. Expanded clinical evaluation of lovastatin (EXCEL) study results: I. Efficacy in modifying plasma lipoproteins and adverse event profile in 8245 patients with moderate hypercholesterolemia. Arch Intern Med 1991; 151:43–49.

25. Boccuzzi SJ, Bocanegra TS, Walker JF, Shapiro DR, Keegan ME. Long-term safety and efficacy profile of simvastatin. Am J Cardiol 1991; 68:1127–1131.

26. Pravastatin Multicenter Study Group II. Comparative efficacy and safety of pravastatin and cholestyramine alone and combined in patients with hypercholesterolemia. Arch Intern Med 1993; 153:1321–1329.

27. Nozaki S, Vega GL, Haddox RJ, Dolan ET, Grundy SM. Influence of lovastatin on concentrations and composition of lipoprotein subfractions. Atherosclerosis 1990; 84:101–110.

28. Kostner G, Gavish D, Leopold B, et al. HMG-CoA reductase inhibitors lower LDL cholesterol without reducing Lp(a) levels. Circulation 1989; 80:1313–1319.

29. Scanu A, Lawn R, Berg K. Lipoprotein(a) and atherosclerosis. Ann Intern Med 1991; 115:209–218.

30. Dujovne CA, Chremos AN, Pool JL, et al. Expanded clinical evaluation of lovastatin (EXCEL) study results: IV. Additional perspectives on the tolerability of lovastatin. Am J Med 1991; 91(suppl 1B):25S–30S.

31. Newman TJ, Kassler-Taub KB, Gelarden RT. Safety of pravastatin in long-term clinical trials conducted in the United States. J Drug Devel 1990; 3(suppl I): 275–281.

32. Bradford RH, Downton M, Chremos AN, et al. Efficacy and tolerability of lovastatin in 3390 women with moderate hypercholesterolemia. Ann Intern Med 1993; 118: 850–855.

33. Pierce LR, Wysowski DK, Gross TP. Myopathy and rhabdomyolysis associated with lovastatin-gemfibrozil combination therapy. JAMA 1990; 264:71–75.

34. McGovern ME, Mellies MJ. Long-term experience with pravastatin in clinical research trials. Clin Ther 1993; 15:57–64.

35. England JF, Viles A, Walsh JC. Muscle and liver side effects associated with simvastatin therapy (abstr). Austr/N Zeal J Med 1991; (suppl 2):512.

36. Illingworth DR, Bacon S. Treatment of heterozygous familial hypercholesterolemia with lipid-lowering drugs. Arteriosclerosis 1989; 9(suppl I):121–134.

37. Kane JP, Malloy MJ, Ports TA, Phillips NR, Diehl JC, Havel RJ. Regression of coronary atherosclerosis during treatment of familial hypercholesterolemia with combined drug regimens. JAMA 1990; 264:3007–3012.

38. Erkelens DW. Combination drug therapy with HMG CoA reductase inhibitors and bile acid sequestrants for hypercholesterolemia. Cardiology 1990; 77(suppl 4):33–38.

39. Mabuchi H, Sakai T, Sakai Y. Reduction of serum cholesterol in heterozygous patients with familial hypercholesterolemia: additive effects of compactin and cholestyramine. N Engl J Med 1983; 308:609–613.

40. Lees AM, Stein SW, Lees RS. Therapy of hypercholesterolemia with mevinolin and other lipid-lowering drugs (abstr). Arteriosclerosis 1986; 6:544a.

41. Illingworth DR. Mevinolin plus colestipol in therapy for severe heterozygous familial hypercholesterolemia. Ann Intern Med 1984; 101:598–604.

42. Grundy SM, Vega GL, Bilheimer DW. Influence of combined therapy with mevinolin and interruption of bile acid reabsorption on low density lipoprotein metabolism in heterozygous familial hypercholesterolemia. Ann Int Med 1985; 103; 339–343.

43. Vega GL, Grundy SM. Treatment of primary moderate hypercholesterolemia with lovastatin (mevinolin) and colestipol. JAMA 1987; 257:33–38.

44. Emmerich J, Aubert I, Bauduceau B. Efficacy and safety of simvastatin (alone or in association with cholestyramine): a 1-year study in 66 patients with type II hyperlipoproteinaemia. Eur Heart J 1990; 11(2):149–155.

45. Betteridge DJ. Clinical efficacy of pravastatin. In: La Rosa JC, ed. New Advances in the Control of Lipid Metabolism: Focus on Pravastatin. London: Royal Society of Medicine Services, Ltd., 1989.

46. Pan HY, DeVault AR, Swites BJ. Pharmacokinetics and pharmacodynamics of pravastatin alone and with cholestyramine in hypercholesterolaemia. Clin Pharmacol Ther 1990; 48:201–207.

47. Davignon J, Roederer G, Hayden M. Comparative efficacy and safety of pravastatin, nicotinic acid, and the two combined in patients with hypercholesterolemia. Program and abstracts of the 9th International Symposium on Atherosclerosis. Rosemont, Ill: International Atherosclerosis Society, 1991.

48. Henkin Y, Johnson KC, Segrest JP. Rechallenge with crystalline niacin after drug-induced hepatitis from sustained-release niacin. JAMA 1990; 264:241–243.

49. Mullin GE, Greenson JK, Mitchell MC. Fulminant hepatic failure after ingestion of sustained-release nicotinic acid. Ann Intern Med 1989; 111:253–255.

50. Norman JD, Illingworth DR, Munson J. Myolysis and acute renal failure in a heart-transplant recipient receiving lovastatin (letter). N Engl J Med 1988; 318:146–147.

51. Reaven P, Witztum JL. Lovastatin, nicotinic acid, and rhabdomyolysis (letter). Ann Intern Med 1988; 109:597–598.

52. Corpier CL, Jones PH, Suki WN, et al. Rhabdomyolysis and renal injury with lovastatin use: report of two cases in cardiac transplant recipients. JAMA 1988; 260:239–241.

53. Illingworth DR, Bacon S. Influence of lovastatin plus gemfibrozil on plasma lipids and lipoproteins in patients with heterozygous familial hypercholesterolaemia. Circulation 1989; 79:590–596.

54. Glueck CJ, Oakes N, Speirs J. Gemfibrozil-lovastatin therapy for primary hyper-cholesterolemia. Am J Cardiol 1992; 70:1–9.
55. Wirebaugh SR, Shapiro ML, McIntyre TH. A retrospective review of the use of lipid-lowering agents in combination, specifically gemfibrozil and lovastatin. Pharmaco-therapy 1992; 12:445–450.
56. Da Col PG, Fonda M, Fisicaro M. Tolerability and efficacy of combination therapy with simvastatin plus gemfibrozil in type IIb refractory familial combined hyper-lipidemia. Curr Ther Res 1993; 53:473–483.
57. Wiklund O, Angelin B, Bergman M. Pravastatin and gemfibrozil alone and in combination for the treatment of hypercholesterolemia. Am J Med 1993; 94:13–20.

13

Measuring Cholesterol and Other Lipids and Lipoproteins in Cardiovascular Risk Assessment

G. Russell Warnick

Pacific Biometrics, Inc., Pacific Biometrics Research Foundation, Seattle, Washington

INTRODUCTION

Cholesterol has become one of the most common important analytes measured in the clinical laboratory. International efforts to reduce the impact of coronary heart disease (CHD) on public health have focused attention on improving the reliability and convenience of the cholesterol and related assays. Current practice guiding laboratory measurement of total serum cholesterol, triglycerides, HDL cholesterol, and LDL cholesterol derives from recommendations of expert panels convened by the National Cholesterol Education Program (NCEP). Decision cutpoints and recommended workup sequences for adults are provided by the Expert Panel on Detection, Evaluation, and Treatment of High Blood Cholesterol in Adults (1,2), recently updated from the original recommendations (3). Guidelines for children were given by the Expert Panel on Blood Cholesterol Levels in Children and Adolescents (4). Expert laboratory panels have developed performance goals and detailed recommendations for reliable measurement of the lipid and lipoprotein analytes (5–9). Measurements of the lipids and lipoproteins are reviewed here in the context of NCEP clinical and laboratory guidelines.

Current clinical guidelines for case-finding (1) advise measuring the high-density lipoprotein (HDL) cholesterol together with total cholesterol as the first step in the workup. NCEP decision cutpoints are summarized in Table 1. Adults with elevated cholesterol, ≥ 240 mg/dl, low HDL cholesterol, considered a risk

Table 1   Treatment Decision Cutpoints

Total cholesterol		
Desirable blood cholesterol	<200 mg/dl	
Borderline-high blood cholesterol	200–239 mg/dl	
High blood cholesterol	≥240 mg/dl	
LDL cholesterol		
Dietary therapy		
	Initiation level	*LDL goal*
Without CHD and >2 risk factors	≥160 mg/dl	<160 mg/dl
Without CHD and ≥2 risk factors	≥130 mg/dl	<130 mg/dl
With CHD	>100 mg/dl	≤100 mg/dl
LDL cholesterol		
Drug treatment		
	Initiation level	*LDL goal*
Without CHD <2 risk factors	≥190 mg/dl	<160 mg/dl
Without CHD and ≥2 risk factors	≥160 mg/dl	<130 mg/dl
With CHD	≥130 mg/dl	≤100 mg/dl
HDL cholesterol		
Low HDL cholesterol	<35 mg/dl	
Protective HDL cholesterol	>60 mg/dl	
Triglycerides		
Desirable	<200 mg/dl	
Borderline	200–399 mg/dl	
Elevated	400–1000 mg/dl	
Severely elevated/pancreatitis	>1000 mg/dl	

factor when ≤35 mg/dl, or CHD require analysis of the lipoproteins to determine the low-density lipoprotein (LDL) cholesterol, the primary analyte for treatment decisions. Patients with borderline elevated cholesterol from 200 to 239 mg/dl and with two or more risk factors also require quantitation of LDL cholesterol. The risk factors considered in the classification scheme are age (males ≥45 years and females ≥55), family history of premature CHD, smoking, hypertension, and diabetes. HDL cholesterol of 60 mg/dl or higher is considered protective and subtracts one from the total number of risk factors.

Treatment decisions are based on the LDL cholesterol level. For patients without symptoms of CHD, LDL cholesterol values of ≥160 mg/dl require treatment. Borderline cases with LDL cholesterol from 130 to 159 and two or more risk factors are also candidates for treatment. For adults with symptoms of CHD, an LDL cholesterol of 100 mg/dl or higher is considered appropriate for treatment. Since the lipid and lipoprotein values are subject to natural biological fluctuations, it is recommended that one confirm each of the values by repeat measurement at least a week later before making treatment decisions.

For children and adolescents a similar classification scheme is followed, but with lower cutpoints for total and LDL cholesterol. The cutpoints for borderline and high cholesterol are 170 and 200 mg/dl, respectively, while those for LDL cholesterol are 110 and 130 mg/dl. Selective measurement of cholesterol is recommended for children whose parents have elevated cholesterol or premature heart disease.

The clinical recommendations from the NCEP panels direct the practice in routine clinical laboratories to measurements of total, HDL, and LDL cholesterol and triglycerides. The triglycerides are primarily associated with chylomicrons, very-low density lipoproteins (VLDL), and intermediate-density lipoproteins (IDL) thought to be atherogenic, but the association of triglycerides with risk of CHD in epidemiological studies is ambiguous. A recent NIH-sponsored Consensus Conference (10) concluded that evidence to date is insufficient to justify aggressive intervention and recommended treatment of elevated triglycerides primarily by the dietary regimens suggested for disorders of total and HDL cholesterol. The NCEP has supported cutpoints for triglycerides of 200, 400, and 1000 mg/dl (1,2). Below 200 is desirable, 200–399 is considered borderline-high, 400–1000 is high, and >1000 is very high. Treatment is justified for any patient with triglyceride levels of ≥400. Triglyceride values of >1000 mg/dl are considered severely elevated, with risk of pancreatitis likely. In routine practice triglycerides have generally been measured as part of a lipoprotein panel to facilitate estimating LDL cholesterol. Some protein components of the lipoproteins, the apolipoproteins Lp(a), apo B, and apo A-I, have been of moderate diagnostic interest in clinical laboratories and more so in lipid specialty and research laboratories but have not been incorporated into the NCEP guidelines. Similarly, measurement of other lipoprotein fractions and lipids is of interest primarily in specialty laboratories and for research studies.

Accuracy is especially important in the analysis of the lipids and lipoproteins. Since these analytes are risk factors, not diagnostic factors, decision cutpoints cannot be easily established by an individual laboratory or manufacturer as is done for other diagnostic analytes. Decision points have been set by the NCEP expert panels based on population distributions and coronary disease risk relationships established in large long-term epidemiological studies. Standardization of the lipid analytical methods in the participating research laboratories made results comparable among laboratories and over time. For other research laboratories to generate comparable results and for routine clinical laboratories to obtain reliable classification of patients using the national decision cutpoints, the methods must be standardized to the same accuracy base of the national studies, a process that will be described subsequently.

CHOLESTEROL MEASUREMENT

Total serum cholesterol has traditionally been the first measurement in the lipid workup, and the lipoproteins are generally quantitated based on their cholesterol content. Early analytical methods (reviewed in Ref. 11) used strong acids such as sulfuric and acetic and chemicals such as acetic anhydride or ferric chloride, which produce a characteristic color with cholesterol. Since these reactions are relatively nonspecific, partial or full extraction by organic solvents was used to improve specificity. The current cholesterol reference method still employs hexane extraction after hydrolysis with alcoholic KOH coupled with Liebermann-Burchard color reagent, composed of sulfuric and acetic acids and acetic anhydride (12,13). The reference method is tedious but gives good agreement with the "gold standard" method developed and applied at the U.S. National Institute for Standards and Technology, the so-called Definitive Method employing isotope-dilution mass spectrometry (14).

In recent years virtually all routine cholesterol measurements and most research analyses have been modernized to enzymic reagents. The development of enzymic reagents has revolutionized the practice of laboratory medicine. The specificity of the enzymes allows reasonably accurate quantitation without the necessity for extraction or other pretreatment. Coupled with highly automated microprocessor-controlled robotic instruments, enzymic reagents have made the measurements highly automated and efficient as well as more precise and accurate. Cholesterol, and often triglycerides, are included in routine test panels on automated batch and discrete chemistry analyzers. Since the lipoproteins, HDL and LDL, require pretreatment steps that are not as easily automated, these analytes have not benefited as much from automation, although newer methods are now being developed to improve these assays. A new class of compact analysis system developed for physician office and home use will be briefly reviewed subsequently.

Although other enzymic reaction sequences have been described, only one sequence is in common use for assaying cholesterol (15,16). Cholesteryl ester hydrolase enzyme cleaves the fatty acid residue from the cholesteryl ester producing unesterified or free cholesterol. Cholesterol is reacted by cholesterol oxidase enzyme to produce hydrogen peroxide, which feeds into a common color sequence using horseradish peroxidase to couple two colorless chemicals, forming a colored compound that can be monitored at about 500 nm. Early commercial versions were not perfected, exhibiting, for example, incomplete hydrolysis of the esters (17), but the reagents have progressed considerably in recent years, and most, calibrated appropriately, can be expected to give reliable results. This reaction sequence, even though universally applied directly to serum without pretreatment can, nevertheless, be subject to interference. The most common is from reducing substances such as vitamin C and bilirubin, which can interfere with the peroxidase-catalyzed color reaction (18).

It is essential for accuracy that the method be calibrated or traceable to the cholesterol reference method. In the case of cholesterol, the reference system is quite advanced and complete, having served as a model for standardization of laboratory analytes (6,19). The Definitive Method at the National Institute for Standards and Technology provides the accuracy target but is too expensive and complicated for frequent use (14). The Reference Method developed and applied at the Centers for Disease Control (CDC), calibrated by an approved primary reference standard to the Definitive Method, provides a transferable, practical reference link (13). The Reference Method is now accessible in a network of standardized laboratories, the Cholesterol Reference Method Laboratory Network established in the United States and some European and Asian countries, which extends standardization to manufacturers and clinical laboratories (19). The network was established primarily to provide direct accuracy comparisons using fresh native serum specimens, necessary for reliable accuracy transfer because of analyte-matrix interaction problems on processed reference materials (20–22).

In the early stages of cholesterol standardization directed to diagnostic manufacturers and routine laboratories, commercial lyophilized or freeze-dried materials were used. These materials, made in large quantities often with spiking or artificial addition of analytes, were assayed by the Definitive and/or Reference Methods and distributed widely for accuracy transfer. Subsequently, biases were observed in enzymic assays on native specimens from patients. Even though such manufactured reference materials are convenient, stable, and amenable to shipment at ambient temperatures, the manufacturing process, especially spiking and lyophilization, altered the measurement properties in enzymic assays. These enzymic assays, sensitive to the nature of the analyte and the specimen matrix, were altered such that results were not representative of results on patient specimens. In order to achieve reliable feedback on accuracy and transfer of the accuracy base, direct comparisons with the Reference Method on actual patient specimens were determined to be necessary (19).

The CDC Cholesterol Network program not only offers reliable accuracy comparisons but has developed a formal certification program whereby laboratories and manufacturers can document traceability to the National Reference System for Cholesterol (19). Laboratories performing cholesterol analysis can select a commercial method that has been certified by the Network (Table 2). Certification does not assure all aspects of quality in a reagent but primarily assures that the accuracy is traceable to the National Reference System for Cholesterol within accepted limits and that the precision is acceptable. Certification is most efficient through manufacturers, but individual laboratories desiring to confirm the performance of their system by completing a certification protocol can

Table 2 Cholesterol Diagnostic Products Traceable to the
National Reference System for Cholesterol[a]

Analytical systems	Reagents
Baxter Diagnostics	EM Diagnostic Systems
Beckman Instruments	Medical Analysis Systems
Boehringer Mannheim Diagnostics	Reagents Applications
Cholestech	
DuPont	
Eastman Kodak	
Intrumentation Laboratory	
Olympus America	
Roche Diagnostic Systems	
Wako Diagnostics	

[a]Further information regarding the manufacturer's certification program or
the list of the certified clinical laboratories can be obtained from Mary M.
Kimberly of the CDC at (404) 488-4126.

contact the CDC. This Network program will eventually be expanded to include
triglycerides, HDL cholesterol, and LDL cholesterol. At the present time, standardization of these analytes is much less complete than for cholesterol with
reasonably well-validated reference methods but no definitive methods. The CDC,
in conjunction with the National Heart Lung and Blood Institute, has conducted a
Lipoprotein Standardization Program covering HDL and triglycerides in addition
to cholesterol, which is available to NIH-funded research laboratories (19). There
is currently no formal standardization program for LDL cholesterol, which is of
concern considering that LDL cholesterol is the major decision analyte in the
NCEP guidelines.

The NCEP laboratory panels (6–9) have established requisite analytical performance goals based on clinical needs for routine measurements (Table 3). For
analysis of total cholesterol the performance goal for total error is 8.9%. That is,
the overall error should be such that each individual cholesterol measurement falls
within plus or minus 8.9% of the Reference Method value. Actually, a statistical
nuance is that since the goals are based on 95% certainty, 95 of 100 measurements
should fall within the total error limit. One can assay a specimen many times and
calculate the mean to determine the usual value or the central tendency. The scatter
or random variation around the mean is described by the standard deviation; an
interval of ± 1 standard deviation interval around the mean includes by definition
two thirds of the observations. In the laboratory, since the scatter or imprecision is
often proportional to the level, random variation is usually specified in relative
terms as CV, the coefficient of variation or relative standard deviation, the standard

Table 3 NCEP Analytical Performance Goals

	Precision CV	Bias	Total error
Cholesterol	3%	±3%	±8.9%
HDL, 1993			
≥42 mg/dl	6%	±10%	±21.8%
<42 mg/dl	SD ≤ 2.5		
HDL, 1998			
≥42 mg/dl	4%	±5%	±12.8%
<42 mg/dl	1.7 mg/dl		
LDL	4%	±4%	±11.8%
Triglycerides	5%	±5%	±14.8%

deviation divided by the mean. Overall accuracy or systematic error is described as bias, the difference between the mean and the true value. Bias is primarily a function of the method's calibration and may vary by concentration. Of greatest concern in this context is bias at the NCEP decision cutpoints. The bias and CV targets presented in Table 2 are representative of performance, that will meet the NCEP total error goals.

Acceptable analytical performance derives from following established principles of laboratory quality assurance. Aspects of quality assurance include selection of reliable methods, reagents, and equipment; use of appropriate applications described in detailed protocols; provision for well-trained and conscientious staff; and vigorous follow-up of performance problems. An important element in the conventional laboratory is analysis of quality-control materials, which should preferably closely emulate actual patient specimens. For the present, most suitable are quality-control pools that are prepared from freshly collected patient serum, aliquoted into securely sealed vials, quick-frozen, and stored at −70°C. Pools of fresh frozen serum are essential for monitoring lipoprotein separation and analysis and preferable for monitoring cholesterol and other lipid measurements. Since commercial pools often undergo matrix alterations, which change their analysis characteristics, results may not represent results on patients. At least two pools should be analyzed, preferably with levels at or near decision points for each analyte.

Precision is a prerequisite for accuracy; a method may have no overall systematic error or bias but if imprecise will still be inaccurate on individual measurements. With the shift to modern automated analyzers, analytic variation has in general become of less concern than biological and other sources of preanalytical variation. Cholesterol levels are affected by many factors, which can be categorized into biological, clinical, and sampling sources as summarized in Table 4 (23). Changes in lifestyle practices affecting usual diet, exercise, weight,

Table 4 Sources of Preanalytical Variation

Biological	Clinical	Sampling
Diet	Myocardial infarction	Fasting status
Obesity	Stroke	Posture
Smoking	Hypertension	Anticoagulent
Exercise	Nephrosis	Analyte instabilities
Alcohol intake	Diabetes	
	Infections	
	Transplantation drugs	
	Other diseases	
	Pregnancy	
	Environment	

and smoking patterns can result in fluctuations in the observed cholesterol values and the distribution of the lipoproteins. Similarly the presence of clinical conditions, various diseases, or the medications used in their treatment affect the amount of cholesterol circulating in the lipoproteins. Conditions during blood collection such as fasting status, posture, the choice of anticoagulant in the collection tube and prolonged storage can alter the measured cholesterol or the distribution of cholesterol in the lipoproteins. Typical observed variation over a year for total cholesterol averages approximately 6.1% CV. Thus, in the average patient measurements made over a year would fall two thirds of the time within plus or minus 6.1% of the mean cholesterol and 95% of the time within twice this range. Some patients may exhibit substantially more biological variation. Thus preanalytical variation is generally relatively large in relation to the usual analytical variation, typically less than 3% CV, and must be considered in interpreting cholesterol results. Some factors such as posture and blood collection can be standardized to minimize the variation. The NCEP guidelines (1) recommend making decisions based on the average of two or three measurements to factor out the effect of both preanalytical and analytical sources. The use of stepped cutpoints in the workup also reduces the practical effect of variation.

Serum, usually collected in serum separator vacuum tubes with clotting enhancers, for convenience reasons has been the fluid of choice for cholesterol measurement in the routine laboratory. EDTA plasma has traditionally been the choice in lipid research laboratories, especially for lipoprotein separations, because the anticoagulant is thought to enhance stability. EDTA does have potential disadvantages, which discourage routine use. Micro-clots that form during storage plug the sampling probes on the modern chemistry analyzers. EDTA osmotically draws water from red cells, diluting the plasma constituents, and the dilution effect can vary, depending on such factors as fill volume and extent of mixing.

Since the NCEP cutpoints are based on serum values, measurements made on EDTA plasma require correction by the factor 1.03. Tradeoffs must be considered in selecting the appropriate fluid for a particular study; the consensus seems to be moving toward serum.

HDL CHOLESTEROL

The measurement of HDL cholesterol has assumed greater importance in the latest NCEP treatment guidelines. Previously HDL cholesterol was measured as a risk factor but otherwise was not considered in treatment decisions. Following recommendations of an NIH-sponsored consensus panel (10), the NCEP guidelines now include HDL cholesterol measurement with total cholesterol in the first medical workup. Since the risk associated with HDL cholesterol is expressed over a relatively small concentration range, accuracy in the measurement is especially important. This analyte is considered a high priority for standardization.

In both routine clinical and research laboratories for diagnostic purposes, HDL is separated almost exclusively by chemical precipitation. This is usually accomplished with polyanions such as heparin used together with divalent cations such as manganese (24,25). The apo B of VLDL and LDL is rich in positively charged amino acids, which form complexes with the polyanions; the divalent cations neutralize the charged groups on the lipoproteins, rendering them insoluble. In practice the HDL cholesterol measurement is a two-step procedure. The precipitation reagent is added to serum or plasma to aggregate non-HDL lipoproteins, which are sedimented by centrifugation. Early methods used centrifugation forces around 1500 × g requiring lengthy centrifugation times of 10–30 minutes. Newer methods use high speed centrifuges with forces of 10,000–15,000 × g, decreasing centrifugation times to 3–5 minutes. The HDL is then quantified as the cholesterol in the supernate, with analysis commonly performed using one of the enzymic assays modified for the lower HDL cholesterol range.

The earliest common precipitation method used heparin in combination with manganese to precipitate the apo B–containing lipoproteins (26–28). Because manganese produced interference with enzymic assays, alternative reagents were developed (29). Sodium phosphotungstate (30,31) with magnesium became common in routine use but because of its sensitivity to reaction conditions and greater variability is being replaced by dextran sulfate (a synthetic heparin) with magnesium (32). The earliest dextran sulfate methods used material of 500 kDa, but 50 kDa, considered more specific, is now becoming more common (33). Polyethylene glycol also precipitates lipoproteins, but because 100-fold higher concentrations than the polyanions are required with consequent larger dilutions and highly viscous reagents, which are difficult to pipette precisely, this reagent is less common (34–36). HDL cholesterol can also be measured electrophoretically; after separation on gels such as agarose, lipoprotein bands are visualized with lipophilic dyes or enzymic cholesterol reagents. Early electrophoretic methods were considered to be imprecise (37), but a new automated system has been demonstrated to give precise and accurate quantitation (38). This method of lipoprotein quantitation will be described in more detail in connection with measurement of LDL cholesterol. Numerous commercial versions of these various precipitation reagents are available. The various methods may give quite different results, since there has been no general standardization program available.

A significant problem with HDL precipitation methods is interference by elevated triglyceride levels, which prevent sedimentation of the precipitate (39). When VLDL and chylomicrons, rich in triglyceride, are present, the low density of the aggregated lipoproteins may prevent their sedimenting or even cause floating during centrifugation. This incomplete sedimentation, indicated by cloudiness, turbidity, or particulate matter floating in the supernate, results in overestimation of HDL cholesterol. High-speed centrifugation will reduce the proportion of turbid supernates. Predilution of the specimen also promotes clearing but may lead to errors in the cholesterol analysis. Turbid supernates may also be cleared by ultrafiltration, a method that works well but is tedious and inefficient.

NCEP analytical performance targets for HDL cholesterol measurement (8) (Table 3) specify that by 1993 bias will be within 10% of the reference method and CV within 6% (SD within 2.5 mg/dl for HDL less than 42 mg/dl). This is a challenge for many laboratories: about half of the specialty laboratories participating in the CDC-NHLBI Lipoprotein Standardization Program do not meet this target (40). In addition, the 1993 target falls far short of meeting the clinical need. To determine HDL cholesterol with requisite reliability, within 10% bias and 95% certainty, and including analytical variation at the 1993 goal combined with the average biological 8.4% over a year, measurement of eight replicate samples would be required. For measurements meeting the 1998 ideal goal, four replicate samples are required. Since even this number is considered impractical, the NCEP recommendation specifies decisions based on three replicate specimens. Improvement must occur in HDL measurement to provide reliable results for diagnostic purposes.

The accepted reference method for HDL cholesterol is a three-step procedure developed at CDC involving ultracentrifugation to remove VLDL, heparin-manganese precipitation to remove LDL, and analysis of supernatant cholesterol by the Abell-Kendall assay (13). This method is tedious and expensive and will compromise the ability to do the necessary accuracy comparisons using fresh specimens. The need for a simpler direct precipitation method has been recognized and the CDC Network laboratory group is attempting to select and validate an equivalent reference method, most likely using direct heparin-manganese or dextran sulfate (50 kDa) precipitation with Abell-Kendall cholesterol analysis.

The diagnostics industry recognizes the need to improve not only reliability but also efficiency in the HDL measurement. New methods in development will streamline and possibly automate the preliminary HDL-separation step. In the near future, technology should offer improvements in the HDL method that will help to bring laboratory performance in line with analytical goals.

LDL CHOLESTEROL

LDL cholesterol as the validated atherogenic lipoprotein is the primary basis for treatment decisions in the NCEP clinical guidelines (1). The common research method for accurate LDL cholesterol quantitation and the basis for the reference method is designated beta-quantification, beta referring to the electrophoretic term for LDL. The beta-quantification technique involves a combination of ultracentrifugation and chemical precipitation (28,41). Ultracentrifugation of serum at the native density of 1.006 g/liter is used to float VLDL and any chylomicrons for separation. The fractions are recovered by slicing the tube between the fractions and pipetting. Ultracentrifugation is preferred for the VLDL separation because other methods, such as precipitation, are not highly specific for VLDL and subject to interference from chylomicrons. Ultracentrifugation is a robust but tedious method that can give reliable results provided technique is meticulous.

In a separate step chemical precipitation as described above is used to separate HDL from either the whole serum or the infranate obtained from ultracentrifugation. Compared to ultracentrifugation the precipitation step is efficient and convenient as well as relatively robust for HDL separation. Cholesterol is quantitated in serum and in the three lipoprotein fractions by enzymic or other assay methods. LDL cholesterol is calculated as the difference between cholesterol measured in the infranate and in the HDL fraction. VLDL cholesterol is usually calculated as the difference between that in whole serum and in the infranate fraction. Calculation of VLDL cholesterol by difference is considered more reliable than direct measurement in the top fraction because quantitative recovery of the lipoproteins in the top is more difficult than in the bottom. Recovery and analysis of the top is nevertheless recommended; comparison of the sum of the top and bottom fractions to the total serum value gives a useful recovery check on technique. Losses in recovery of the bottom can have a substantial effect on reported values for LDL and VLDL, a problem often present but unappreciated. For example, a 5% recovery loss in the bottom fraction can easily result in a 7% negative error in LDL cholesterol and a 100% positive error in the VLDL cholesterol, both calculated by difference. The ultracentrifugation step makes beta-quantification tedious and inaccessible for routine diagnostic purposes.

A simpler technique for LDL cholesterol quantitation, common in both routine and research laboratories, that bypasses ultracentrifugation is the so-called Friedewald calculation or derived beta-quantification (42). HDL is separated by precipitation and its cholesterol assayed; cholesterol and triglycerides are measured in the serum. VLDL cholesterol is estimated as triglycerides divided by 5, an approximation that works reasonably well in most normolipemic specimens. The presence of elevated triglycerides (400 mg/dl is the accepted limit) chylomicrons, and beta-VLDL characteristic of the rare type III hyperlipoproteinemia preclude this estimation. The estimated VLDL cholesterol and measured HDL cholesterol are subtracted from total serum cholesterol to estimate or derive LDL cholesterol.

The Friedewald estimation has been used almost universally in estimating LDL cholesterol in routine clinical practice. Investigations in lipid specialty laboratories have suggested that the method is reliable for patient classification provided the underlying measurements are made with appropriate accuracy and precision (43,44). There is considerable concern about the reliability in routine laboratories, since the error in estimated LDL cholesterol includes the cumulative error in each of the underlying measurements used in the calculations—cholesterol, triglycerides, and HDL cholesterol. The NCEP laboratory expert panel reviewed performance data and concluded that the level of analytical performance required to derive LDL cholesterol accurately enough to meet clinical needs was beyond the capability of most routine laboratories. In order to meet the requisite 4% CV NCEP goal for LDL cholesterol (Table 3), a laboratory would be required to achieve half the NCEP goals for each of the underlying measurements. The NCEP panel concluded that alternative methods are needed for routine diagnostic use, preferably ones that directly separate LDL for cholesterol quantitation (9).

In response to the NCEP request, several direct LDL cholesterol methods have recently been developed or refined for general use (45). One commercial method (Genzyme Corporation) that is FDA approved and compatible as a pretreatment step with a variety of analytical systems employs immunochemical separation. A mixture of antibodies specific to epitopes on the apolipoproteins of VLDL and HDL is immobilized on latex beads. Specimen is added to the beads in a microfiltration device, which is mixed and subjected to centrifugation. VLDL and HDL are retained by the filter while LDL passes through. Cholesterol in the LDL filtrate is assayed by enzymic reagent using any of the automated chemistry analyzers. This method has been found to correlate with beta-quantification on both fasting and nonfasting specimens (46).

Nuclear magnetic resonance (NMR) spectroscopy has been used for research investigations and has shown potential for routine direct quantitation of lipoproteins (47,48). Analytical quantitation by NMR takes advantage of characteristic energy shifts for protons in either methyl or methylene groups in the lipoproteins to estimate lipoprotein cholesterol concentrations by computer analysis of the signals from the resonance envelope. NMR has potential for highly automated high-volume analysis. The disadvantages of NMR for routine analysis are that the equipment is expensive and specialized skills are required, making this method more suitable for high-volume settings such as referral laboratories.

Electrophoretic methods (reviewed in Refs. 49,50) have a long history of use in qualitative and quantitative analysis of lipoproteins. Electrophoresis allows separation and quantitation of major lipoprotein classes and provides a visual display useful in detecting unusual or variant patterns. Agarose has been the most common media for separation of whole lipoproteins, providing a clear background and convenience (51–54). Electrophoretic methods have, in general, been considered more useful for qualitative analysis but less than desirable for lipoprotein quantitation because of poor precision and large systematic biases compared to other methods (37). Recent improvements to a commercial automated electrophoretic system suggest, however, that electrophoretic quantitation can be precise and accurate. Evaluations demonstrate good separation of the major lipoprotein classes, including Lp(a), and precise and accurate quantitation of HDL and LDL cholesterol in comparison with the reference methods (38).

Reliable quantification of LDL cholesterol, critical in the NCEP program, will require both general implementation of improved measurement technology and a standardization program. Industry has responded to the challenge to develop better technology. Experience with the newer methods will determine their reliability and cost-effectiveness for routine diagnostic use. The responsible institutions and professional societies should consider it a priority to develop and implement programs for standardization of this important analyte.

TRIGLYCERIDES

Much of the current demand for routine triglyceride measurement stems from their use in the estimation of LDL cholesterol. As direct methods become available—alternatives to the Friedewald estimation—both the need for measurement and the requirements for accuracy and precision can be expected to decrease. Triglycerides will be measured initially with LDL cholesterol to detect elevations. Patients for whom treatment of elevated triglyceride is appropriate will be monitored. For the majority of patients, regular measurements of triglycerides will be less important.

Virtually all routine and most research measurements of triglycerides employ enzymic reagents. Several reaction sequences are available, all having in common the use of lipase enzymes to cleave fatty acids from the glycerol backbone. The freed glycerol, in turn, participates in any one of several enzymic sequences. One of the more common earlier reactions ending in a ultraviolet-absorbing product used glycerol kinase and pyruvate kinase culminating in the conversion of NADH to NAD+ with an associated decrease in absorbency monitored at 340 nm (55). This reaction is relatively susceptible to interference and side reactions. The UV endpoint is also less convenient for modern analyzers, so this and other UV sequences are gradually being replaced by the second sequence, which feeds into the same peroxidase color reaction described for cholesterol (56). The intermediate steps involve glycerol kinase and glycerol phosphate oxidase.

All enzymic triglyceride reaction sequences react with endogenous free glycerol, a universal and significant source of interference (57,58). In most specimens, the endogenous free glycerol contributes to a 10–20 mg/dl overestimation of triglycerides. A few specimens, about 20%, will have higher glycerol with levels increased in certain conditions such as diabetes and liver disease. Few routine laboratories—but most research laboratories—incorporate some type of correction for endogenous free glycerol. The most common correction, designated double cuvette blanking, is to perform a second parallel measurement without the lipase enzyme to quantitate only the free glycerol blank, which is subtracted from the total glycerol measurement of the complete reaction sequence to determine a net or blank-corrected glycerol (59). Another approach, designated the single cuvette blank, is to begin with the lipase free reagent and, after a brief incubation, take a blank reading that includes only the free glycerol. The lipase is then added as a second separate reagent and, after additional incubation, a total reading is taken, which, corrected for the blank by the instrument, represent net triglycerides (60). A commercial variation on this is to react the free glycerol to a colorless product and then add a key ingredient of the color reaction with the lipase. Since these blank corrections are expensive and perhaps unjustified by the accuracy requirements in routine practice, a convenient and easily implemented interim alternative, designated calibration blanking, is to simply adjust calibrator set points to compensate for the average free glycerol content of specimens. This approach is reasonably reliable because the free glycerol levels are relatively low and uniform in most specimens. Most specimens will be corrected appropriately; a few will be undercorrelated but will be more accurate than uncorrected. This method also fits well with the CDC network approach to standardization. In comparison analyses of fresh specimens, the Network laboratory can simply provide blank corrected values for adjustment of triglyceride calibrator set points.

The accepted Reference Method for triglyceride assay, designated the chromotropic acid method, involves hydrolysis, extraction, and color reaction with chromotropic acid (61). The assay is tedious, not well characterized, and not applied except at the CDC. The CDC Network group is attempting to develop a reliable but convenient enzymic assay for use in accuracy transfers as an equivalent Reference Method. Accuracy in the triglyceride measurement is not as critical because the cutpoints are wide-spaced and because the physiological variation is so large (CV $\sim$ 25–30%) that analytical variation is relatively insignificant.

COMPACT ANALYZERS

A major area of development in recent years has been in compact analysis systems for use in point of care testing, at the patient's bedside, in the physician's office, at wellness centers, and even in the home (62). First-generation systems, introduced in the 1980s, were relatively large and measured cholesterol and triglycerides as well as other common analytes, usually separately and sequentially. HDL separations involving off-line pretreatment steps were subsequently developed. Second-generation systems became smaller and more sophisticated, offering separation of HDL and analysis of cholesterol and triglycerides simultaneously from fingerstick blood in an integrated system. Noninstrumented systems with a thermometer-like reading are available for total cholesterol and are in development for HDL cholesterol. This new technology offers the capability for measuring lipids and lipoproteins reliably outside the conventional laboratory.

APOLIPOPROTEINS

The apolipoproteins that have generated some interest for diagnostic purposes deserve brief mention in this review (63). Apolipoprotein B-100 (apo B) is the major protein of LDL and VLDL; in most individuals the bulk of apo B resides in LDL and is an indicator of LDL level. Since apo B can be measured directly in serum by immunoassay methods, which are steadily improving in reliability, some have advocated replacing the LDL cholesterol measurement (64), which requires at a minimum a pretreatment step. Studies of CHD risk prediction, to date primarily cross-sectional (65) with a few prospective studies (66), suggest that apo B is indeed a good predictor and may identify some patients at increased risk in spite of normal LDL cholesterol (67). Others take a more conservative view, citing deficiencies in assay methodology and particularly standardization as well as the lack of intervention data (68). Considerations are similar but not as compelling for apo A-I, the major protein of HDL. These apolipoprotein measurements are common in search laboratories and supplement the conventional lipoprotein measurements in patient management in lipid specialty centers but have not yet been accepted or recommended by any organization such as NCEP for use in routine practice.

The Lp(a) lipoprotein (69,70), often referred to as "lipoprotein little a," is an interesting variant of LDL containing an additional protein unit with homology to plasminogen, which confers thrombotic properties. Because Lp(a) is more likely to be taken up by peripheral cells than LDL, it is highly atherogenic. Because of its similarity to plasminogen, it binds to fibrin clots and to cells but is not converted to plasmin as is plasminogen, thereby interfering with clot lysis. Thus Lp(a) contributes to both the chronic and acute events leading to a myocardial infarct. As such, Lp(a) is a powerful predictor of CHD risk independent of LDL cholesterol and apo B. As reliable methods for quantitation become available and standardized, measurement of Lp(a) will likely become common in CHD risk assessment. Since Lp(a) is genetically heterogeneous and levels and risk correlate with the isoform size, qualitative assessment of isoform distribution may also be important.

The clinical guidelines of the NCEP drive the laboratory measurement of the lipid and lipoprotein risk factors for CHD. Measurements of primary interest are total serum cholesterol, HDL cholesterol, LDL cholesterol, and triglycerides. Methods for cholesterol measurement are advanced and reasonably well standardized. Improvements in reliability and convenience as well as general standardization are needed for HDL and especially LDL cholesterol. Methods for triglycerides are adequate, and standardization is less critical.

ACKNOWLEDGMENTS

The author wishes to express appreciation to Sherry Hamman and Nicolette Roberge for their assistance in the preparation of this manuscript.

REFERENCES

1. National Cholesterol Education Program. Second Report of the Expert Panel on Detection, Evaluation, and Treatment of High Blood Cholesterol in Adults (Adult Treatment Panel II). NCEP, 1993.
2. National Cholesterol Education Program. Summary of the Second Report of the National Cholesterol Education Program (NCEP) Expert Panel on Detection, Evaluation, and Treatment of High Blood Cholesterol in Adults (Adult Treatment Panel II). JAMA 1993; 60(23):3015.
3. National Cholesterol Education Program. Report of the National Cholesterol Education Program Expert Panel on Detection, Evaluation and Treatment of High Blood Cholesterol in Adults. Arch Intern Med 1988; 148:36–69.
4. National Cholesterol Education Program. Report of the Expert Panel on Blood Cholesterol Levels in Children and Adolescents. National Institutes of Health Publication No. 91–2732, 1991.
5. National Cholesterol Education Program Laboratory Standardization Panel. Current status of blood cholesterol measurements in clinical laboratories in the United States. Clin Chem 1988; 34:193–201.
6. Report from the Laboratory Standardization Panel of the National Cholesterol Education Program. Recommendations for Improving Cholesterol Measurement. Bethesda, MD: National Institutes of Health, 1990. NIH publication no. 90-2964.
7. National Cholesterol Education Program Lipoprotein Measurement Working Group. Recommendations for Triglyceride Measurement. NIH Publication. In press.
8. National Cholesterol Education Program Lipoprotein Measurement Working Group. Recommendations for Measurement of High Density Lipoprotein Cholesterol. NIH Publication. In press.
9. National Cholesterol Education Program Lipoprotein Measurement Working Group. Recommendations for Measurement of Low Density Lipoprotein Cholesterol. NIH Publication. In press.
10. NIH Consensus Development Panel on Triglyceride, High-Density Lipoprotein, and

Coronary Heart Disease. Triglyceride, high-density lipoprotein,and coronary heart disease. JAMA 1993; 269:505–510.

11. Zak B. Cholesterol methodologies: a review. Clin Chem 1977; 23(7):1201–1214.

12. Abell LL, Levy BB, Brody BB, Kendall FC. A simplified method for the estimation of total cholesterol in serum and demonstration of its specificity. J Biol Chem 1952; 195:357–366.

13. Duncan IW, Mather A, Cooper GR. The Procedure for the Proposed Cholesterol Reference Method. Atlanta: Division of Environmental Health Laboratory Sciences, CEH, Centers for Disease Control, 1982.

14. Cohen A, Hertz HS, Mandel J, et al. Total serum cholesterol by isotope dilution/mass spectrometry: a candidate definitive method. Clin Chem 1980; 26:854–860.

15. Richmond W. Preparation and properties of a cholesterol oxidase from *Nocardia* sp. and its application to the enzymatic assay of total cholesterol in serum. Clin Chem 1973; 19:1350.

16. Allain CC, Poon LS, Chang CSG, et al. Enzymatic determination of total serum cholesterol. Clin Chem 1974; 20:470.

17. Wiebe DA, Bernert Jr JT. Influence of incomplete cholesteryl ester hydrolysis on enzymic measurements of cholesterol. Clin Chem 1984; 30(3):352–356.

18. McGowan MW, Artiss JD, Zak B. Spectrophotometric study on minimizing bilirubin interference in an enzyme reagent mediated cholesterol reaction. Microchem J 1982; 27:564–573.

19. Myers GL, Henderson LO, Cooper GR, Hassemer DJ. Standardization of lipid and lipoprotein measurements. In: Rifai N, Warnick GR, eds. Methods for Clinical Laboratory Measurements of Lipid and Lipoprotein Risk Factors. Washington, DC: AACC Press, 1991:101–125.

20. Greenberg N, Li ZM, Bower GN. National Reference System for Cholesterol (NRS-CHOL): problems with transfer of accuracy with matrix materials. Clin Chem 1988; 24:1230–1231.

21. Kroll MH, Chesler R, Elin RJ. Effect of lyophilization on results of five enzymatic methods for cholesterol. Clin Chem 1989; 35:1523–1526.

22. Eckfeldt JH, Copeland KR. Accuracy verification and identification of matrix effects. Arch Pathol Lab Med 1983; 117:381–386.

23. Cooper GR, Myers GL, Smith SJ, Schlant RC. Blood and lipid measurements: variations and practical utility. JAMA 1992; 267:1652–1660.

24. Burstein M, Legmann P. Lipoprotein precipitation. In: Clarkson TB, Kritchevsky D, Pollak OJ, eds. Monographs on Atherosclerosis. Vol. II. New York: S. Karger, 1982:1.

25. Levin SJ, High-Density Lipoprotein Cholesterol: Review of Methods. The American Society of Clinical Pathologists. Check sample 1989; Core Chemistry, No. PTS 89–2(PTS-35):5(2).

26. Burstein M, Samaille J. Sur un dosage rapide du cholesterol lie aux α- et aux β-lipoproteines du serum. Clin Chim Acta 1960; 5:609.

27. Fredrickson DS, Levy RI, Lindgren FT. A comparison of heritable abnormal lipoprotein patterns as defined by two different techniques. J Clin Invest 1968; 47:2446–2457.

28. Manual of laboratory operations, lipid research clinics program, lipid and lipoprotein analysis. Washington, DC: NIH, U.S. Dept. of Health and Human Services, 1983.

29. Steele BW, Koehler DF, Azar MM, Blaszkowski TP, Kuba K, Dempsey ME. Enzymatic determinations of cholesterol in high-density-lipoprotein fractions prepared by a precipitation technique. Clin Chem 1976; 22;98–101.

30. Burstein M, Scholnick HR. Lipoprotein-polyanion-metal interactions. Adv Lipid Res 1973; 11:68–105.

31. Lopes-Virella MF, Stone P, Ellis S, Colwell JA. Cholesterol determination in high-density lipoproteins separated by three different methods. Clin Chem 1977; 23: 882–884.

32. Warnick GR, Cheung MC, Albers JJ. Comparison of current methods for high-density lipoprotein cholesterol quantitation. Clin Chem 1979; 25:596–604.

33. Warnick GR, Benderson J, Albers JJ, et al. Dextran sulfate-Mg^{2+} precipitation procedure for quantitation of high-density-lipoprotein cholesterol. Clin Chem 1982; 28:1379–1388.

34. Viikari J. Precipitation of plasma lipoproteins by PEG-6000 and its evaluation with electrophoresis and ultracentrifugation. Scand J Clin Lab Invest 1976; 36:265–268.

35. Allen JK, Hensley WJ, Nicholls AV, Whitfield JB. An enzymic and centrifugal method for estimating high-density lipoprotein cholesterol. Clin Chem 1979; 25: 325–327.

36. Demacker PNM, Vos-Janssen HE, Hijmans AGM, Van't Laar A, Jansen AP. Measurement of high-density lipoprotein cholesterol in serum: comparison of six isolation methods combined with enzymic cholesterol analysis. Clin Chem 1980; 26:1780–1786.

37. Warnick GR, Nguyen T, Bergelin RO, Wahl PW, Albers JJ. Lipoprotein quantification: an electrophoretic method compared with the lipid research clinics method. Clin Chem 1982; 28:2116–20.

38. Warnick GR, Leary ET, Goetsch J. Electrophoretic quantification of LDL-cholesterol using the Helena REP (abstr). Clin Chem 1993; 39:1122.

39. Warnick GR, Albers JJ, Bachorik PS, et al. Multi-laboratory evaluation of an untrafiltration procedure for high density lipoprotein cholesterol quantification in turbid heparin-manganese supernates. J Lipid Res 1981; 22:1015–1019.

40. Meyers GL. Communication with author.

41. Belcher JD, McNamara JR, Grinstead GF, Rifai N, Warnick GR, Bachorik P, Frantz Jr. I. Measurement of low density lipoprotein cholesterol concentration. In: Rifai N, Warnick GR, eds. Methods for Clinical Laboratory Measurement of Lipid and Lipoprotein Risk Factors. Washington, DC: AACC Press, 1991:75–86.

42. Friedewald WT, Levy RI, Fredrickson DS. Estimation of the concentration of low-density lipoprotein cholesterol in plasma, without use of the preparative ultra-centrifuge. Clin Chem 1972; 18:499–502.

43. Warnick GR, Knopp RH, Fitzpatrick V, Branson L. Estimating low-density lipoprotein cholesterol by the Freidewald equation is adequate for classifying patients on the basis of nationally recommended cutpoints. Clin Chem 1990; 36:15–19.

44. McNamara JR, Cohn JS, Wilson PWF, Schaefer EJ. Calculated values for low-density

lipoprotein cholesterol in the assessment of lipid abnormalities and coronary disease risk. Clin Chem 1990; 36:36–42.

45. Rifai N, Warnick GR, McNamara JR, Belcher JD, Grinstead GF, Frantz Jr. ID. Measurement of low-density-lipoprotein cholesterol in serum: a status report. Clin Chem 1992; 38:150.

46. Leary ET, Tjersland G, Warnick GR. Evaluation of the Genzyme immunoseparation regent for direct quantitation of LDL cholesterol. Clin Chem 1993; 39:1124.

47. Otvos JD, Jeyarajah EJ, Bennett DW. Quantification of plasma lipoproteins by proton nuclear magnetic resonance spectroscopy. Clin Chem 1991; 37:377–386.

48. Otvos JD, Jeyarajah EJ, Bennett DW, Krauss RM. Development of a proton nuclear magnetic resonance spectroscopic method for determining plasma lipoprotein concentrations and subspecies distributions from a single, rapid measurement. Clin Chem 1992; 38:1632–1638.

49. Lewis LA, Opplt JJ. CRC Handbook of Electrophoresis. Vol. 1. Boca Raton, FL: CRC Press, Inc., 1980.

50. Lewis LA, Opplt JJ. CRC Handbook of Electrophoresis. Vol. 2. Boca Raton, FL: CRC Press, Inc., 1980.

51. Noble RP. Electrophoretic separation of plasma lipoproteins in agarose gel. J Lipid Res 1968; 9:693.

52. Lindgren FT, Silvers J, Jutagir R, et al. A comparison of simplified methods for lipoprotein quantitation using the analytic ultracentrifuge as a standard. Lipids 1977; 12:278.

53. Conlon D, Blankstein LA, Pasakarnis PA. Quantitative determination of high-density lipoprotein cholesterol by agarose gel electrophoresis updated. Clin Chem 1979; 24:227.

54. Papadopoulos NM. Hyperlipoproteinemia phenotype determination by agarose gel electrophoresis updated. Clin Chem 1978; 24:227–229.

55. Bucolo G, Yabut J, Chang TY. Mechanized enzymatic determination of triglycerides in serum. Clin Chem 1975; 21:420–424.

56. McGowan M, Artiss J, Strandbergh DR, Zak B. A peroxidase-coupled method for the colorimetric determination of serum triglycerides. Clin Chem 1983; 29:538–542.

57. Stinshoff K, Weisshaar D, Staehler F, Hesse D, Gruber W., Steier E. Relation between concentrations of free glycerol and triglycerides in human sera. Clin Chem 1977; 23:1029–1032.

58. Cole TG. Glycerol blanking in triglyceride assays: is it necessary? Clin Chem 1990; 36:1267.

59. Warnick G. Enzymatic methods for quantification of lipoprotein lipids. In: Albers JJ, Segrest JP, eds. Methods in Enzymology. Orlando, FL: Academic Press, 1986:101.

60. Sullivan DR, Kruijswijk Z, West CE, Kohlmeier M, Katan MB. Determination of serum triglycerides by an accurate enzymatic method not affected by free glycerol. Clin Chem 1985; 31:1227–1228.

61. Lofland HB, J. A semiautomated procedure for the determination of triglycerides in serum. Anal Biochem 1964; 9:393–400.

62. Warnick GR, Compact analysis systems for cholesterol, triglycerides and high-density lipoprotein cholesterol. Cur Opin Lipidol 1991; 2:343–348.

63. Labeur C, Shepherd J, Rosseneu M. Immunological assays of apolipoproteins in plasma: methods and instrumentation. Clin Chem 1990; 36:591–597.
64. Sniderman AD. Is it time to measure apolipoprotein B? Arteriosclerosis 1990; 10: 665–667.
65. Genest JJ, Bard JM, Fruchart JC, Ordovas JM, Wilson PWF, Schaefer EJ. Plasma apolipoproteins (a), A-I, A-II, B, E, and C-III containing particles in men with premature coronary artery disease. Atherosclerosis 1991; 90:149–158.
66. Stalmpfer MJ, Sacks FM, Salvini S, Willett WC, Hennekens CH. A prospective study of cholesterol, apolipoproteins, and the risk of myocardial infarction. N Engl J Med 1991; 325:373–381.
67. Albers JJ, Brunzell JD, Knopp RH. Apoprotein measurements and their clinical application. In: Rifkind BM, Lippel K, eds. Clinics in Laboratory Medicine. Vol. 9, No. 1. Philadelphia: W.B. Saunders Company, 1989:137–152.
68. Vega GL, Grundy SM. Does measurement of apolipoprotein B have a place in cholesterol management? Arteriosclerosis 1990; 10:558–671.
69. Utermann G. The mysteries of lipoprotein(a). Science 1989; 246:904–910.
70. Loscalzo, J. Lipoprotein (a) a unique risk factor for atherothrombotic disease. Arteriosclerosis 1990; 10:672–679.

14

The Cost-Effectiveness of Programs to Lower Serum Cholesterol

David J. Cohen

Harvard Medical School and Beth Israel Hospital, Boston, Massachusetts

Lee Goldman

Harvard Medical School, Harvard School of Public Health, and Brigham and Women's Hospital, Boston, Massachusetts

Milton C. Weinstein

Harvard School of Public Health and Brigham and Women's Hospital, Boston, Massachusetts

The Adult Treatment Panel of the National Cholesterol Education Program (NCEP) first published recommendations for screening and treating hypercholesterolemia in 1987 (1). Since that time, the public health and economic implications of these and other recommendations have been the subject of considerable analysis and debate. There are several important reasons for these concerns. First, the economic burden of coronary heart disease (CHD) in Western society is considerable. Approximately 1.5 million Americans suffer a heart attack each year (2). The American Heart Association estimates that the direct costs of CHD were between $41.5 and $56 billion in 1990, not including indirect costs due to lost time and earning potential (2,3). Second, hypercholesterolemia is prevalent in Western cultures. Based on the NCEP guidelines, approximately 60 million American adults are estimated to be candidates for dietary treatment, while nearly 10 million Americans may be candidates for pharmacological therapy (4,5). Clearly, the costs of implementing nutritional and drug therapy over such a large population would be considerable. Finally, while the benefits of cholesterol reduction are well established for patients with known CHD and for middle-aged men with moderate-to-severe hypercholesterolemia, there is no definitive proof that these benefits can be extended to other populations.

As U.S. health care costs continue to rise, the pressure to control costs is increasing (6,7). Consequently, responsible decisions regarding the implementation of health care programs must consider both the costs and likely effectiveness of these programs. Before undertaking an ambitious treatment program such as widespread screening and treatment of hypercholesterolemia, the overall cost-effectiveness of these interventions must be assessed. The techniques of cost-effectiveness analysis can be used to provide a framework for estimating both the costs and the benefits of any medical technology or program and for comparing the program with the available alternatives (8,9).

THE COST OF PROGRAMS TO SCREEN AND TREAT HYPERCHOLESTEROLEMIA

Thorough economic evaluation of any medical treatment or practice must begin by determining the costs of the program. The net cost of any medical treatment can be calculated as (8):

$$\Delta C_{Net} = \Delta C_{Rx} + \Delta C_{SE} - \Delta C_{Morb} + \Delta C_{RxLE}$$

where ΔC_{Rx} is the direct medical and health care costs of the treatment (drugs, laboratory tests, physician services, etc.) and ΔC_{SE} represents the additional direct medical care costs incurred because of adverse side effects of treatment. In the context of cholesterol reduction, for example, this could include the costs of additional cholecystectomies performed in patients treated with lipid-lowering drugs and the costs of diagnosing and treating patients with hepatitis caused by nicotinic acid or lovastatin. The third component (ΔC_{Morb}) includes direct medical costs that are saved when morbidity is avoided because of the treatment under consideration. Thus, ΔC_{Morb} would include any cost savings because myocardial infarctions or revascularization procedures were prevented or delayed by cholesterol lowering. The final component (ΔC_{RxLE}) represents the costs of treating other diseases that would not have occurred if lives had not been prolonged by the treatment under consideration. Few economic evaluations of cholesterol lowering have incorporated this component of cost, however.

When considering the costs of drug treatment for hypercholesterolemia, both the direct costs of treatment and the induced costs of monitoring to assess the efficacy of treatment and to detect any adverse side effects are substantial (Table 1). The annual costs of drug treatment for an individual patient have been estimated to range from $353/yr for generic niacin to $1766/yr for lovastatin at 40 mg/day (10). In general, nonpharmacological costs, such as the costs of blood tests and physician visits, account for 30–40% of the overall cost of treatment, but for the least expensive drugs, they can represent as much as 75% of the total.

Although the annual per capita cost of treating hypercholesterolemia is not particularly alarming, two factors make the overall economic burden of treatment

Table 1 Estimated Annual Direct and Induced Costs of Cholesterol-Lowering Drug Treatment

Drug/Daily dose	Wholesale drug cost ($)[a]	Monitoring costs ($)[b]	Total cost (1993 $)
Cholestyramine 16 g	1094	339	1433
Colestipol 20 g	952	339	1291
Gemfibrozil 1200 mg	722	356	1078
Lovastatin 20 mg	691	384	1075
Lovastatin 40 mg	1382	384	1766
Niacin (generic) 3 g	77	276	353
Probucol 1 g	739	362	1101
Pravastatin 20 mg	598	384	982
Simvastatin 10 mg	623	384	1007

[a]From Drug Topics Redbook 1992 and February 1993 update.
[b]Includes induced costs for monitoring and treatment of drug side effects. Derived from Ref. 10 and converted to 1993 dollars based on the Medical Care Consumer Price Index.

much more concerning. First is the duration of treatment. Since cholesterol lowering for primary prevention may require life-long treatment, the total treatment cost for any individual increases by a factor of 40–50, a lifetime total of $14,000–$70,000 depending on the particular agent used. Even if future costs are discounted at 5% per year (see below), the present value of the lifetime cost of cholesterol lowering treatment is expected to be about $7000–$35,000.

The second factor is the prevalence of hypercholesterolemia and other CHD risk factors in Western cultures. Using prevalence data from the National Health and Nutrition Survey, Sempos et al. estimated that approximately 36% of the current U.S. population (63.8 million individuals) may require at least dietary intervention and careful follow-up for treatment of hypercholesterolemia based on the current NCEP Adult Treatment Panel guidelines (4). The number of individuals who would require even more costly pharmacological treatment depends on the effectiveness of dietary intervention. Based on the distribution of risk factors in the Framingham Offspring Study and assuming that dietary interventions produce, on average, a 10% reduction in total serum cholesterol, Wilson and colleagues estimated that 7% of the U.S. adult male population (approximately 5.7 million) and 4% of the U.S. adult female population (3.8 million) would be candidates for pharmacological lipid-lowering therapy (5). Clearly, the costs of drug treatment for a population of this size would be considerable—ranging from $2.75 to $17 billion per year depending on the specific drugs used.

Another "hidden" cost of cholesterol-lowering treatment is the cost of screening—those resources required to identify individuals who would benefit from either dietary or drug therapy. Garber and colleagues estimated the cost of screening tests for the U.S. Medicare population alone to be $143 million (11). In a more comprehensive analysis based on the recommendations of the Canadian Consensus Conference on Cholesterol Lowering (12), Grover and colleagues estimated the cost of comprehensive cholesterol screening for all adults in Canada to be between $463 and $561 million in the first year—approximately 1% of the total Canadian health care budget (13).

Clearly the costs of implementing populationwide cholesterol screening and intervention programs are substantial. Given current pressures on the health care system to control costs, it is therefore imperative that such expensive programs be founded on solid evidence that reducing serum cholesterol levels improves or extends life.

THE RELATIONSHIP BETWEEN SERUM CHOLESTEROL AND CORONARY HEART DISEASE: EVIDENCE FOR THE EFFECTIVENESS OF CHOLESTEROL REDUCTION

Epidemiological Data

There is a broad range of scientific evidence to support the relationship between serum cholesterol and coronary heart disease (1,3,14). Cross-cultural and migrant epidemiological data demonstrate a consistent relationship between cholesterol and CHD (15–17). Epidemiological data from prospective cohorts such as the Framingham Heart Study, the Multiple Risk Factor Intervention Trial, and the Lipid Research Clinics Program Follow-up Study also demonstrate that the serum cholesterol level (or the LDL cholesterol level) is an independent risk factor for the development of CHD and CHD mortality (18–20). These studies and others also demonstrate that the HDL cholesterol level is an independent factor that protects against both CHD and cardiovascular mortality (21–25).

Clinical Trials: Primary Prevention of Coronary Heart Disease

Clinical trials have demonstrated that reducing serum cholesterol levels in middle-aged men with moderate elevations of serum cholesterol can reduce the incidence of CHD over a 5- to 7-year follow-up period (26,27). In addition to providing the best evidence of the overall effectiveness of cholesterol reduction for primary prevention, these intervention studies have implications for the effectiveness of cholesterol-lowering treatment for other populations. In the Lipid Research Clinics Primary Prevention Trial, treatment with cholestyramine reduced cholesterol levels by an average of 8.5% compared with placebo and resulted in a 19% reduction in definite CHD incidence (from 9.8% to 8.1%) over 7 years of follow-up (28). This 2% decline in coronary heart disease risk for every 1% decrease in serum cholesterol level was essentially the same reduction that would have been predicted from the Framingham Heart Study's risk equations (29). Meta-analysis of numerous smaller cholesterol-lowering intervention trials has demonstrated the same relationship between cholesterol reduction and the degree of subsequent risk reduction (30). Because many of the epidemiological analyses have been based on a single cholesterol level, however, it has been postulated that regression-dilution bias may lead to a substantial underestimate of the true relationship; a 1% change in cholesterol may actually correspond to a 3% change in coronary risk (31,32). The similarity between the extent of CHD risk reduction and the predictions of the Framingham Heart Study suggests that the effectiveness of cholesterol reduction in other populations (e.g., women, elderly individuals) might be estimated based on the Framingham risk equations. This assumption (albeit unproven) underlies much of the current support for cholesterol reduction in these populations and has been used extensively to analyze the potential cost-effectiveness of such interventions (33–36).

Several other consistent findings of these primary prevention trials have important implications for the cost-effectiveness of cholesterol lowering for primary coronary prevention. As expected, there is a time lag between the initiation of treatment to lower cholesterol and the resulting reduction in coronary risk. In both the Lipid Research Clinics Primary Prevention Trial and the Helsinki Heart Study, this time lag was approximately 2 years (26,27). Consequently, most analyses of the cost-effectiveness of cholesterol-lowering have incorporated a 2-year delay between initiation of treatment, with its attendant costs and inconvenience, and the realization of any benefits.

It is important to note, however, that no trial of cholesterol lowering for primary coronary prevention has demonstrated an improvement in all-cause mortality. Although most studies have demonstrated a trend toward reduced mortality due to CHD (26,27,37), these benefits have generally been offset by minor increases in the rates of death from cancer (38,39), accidents, or violence (26,27). In a meta-analysis of cholesterol-lowering primary prevention trials, Muldoon and colleagues found a 15% reduction in CHD mortality (one-tailed $p = 0.06$) but statistically significant increases in the rates of death due to cancer (43%, 95% CI 8–90%) and death "not related to illness" (76%, 95% CI 19–158%) (40). In more than 24,000 men treated in these controlled trials, there was no evidence of improved overall survival. Most investigators have attributed these increases in noncardiovascular mortality to inadequate study power, chance, or inadequate duration of follow-up (3,41), but the consistency of this finding suggests that cholesterol lowering—especially by medications—may increase noncardiovascular mortality. This increase in noncardiovascular mortality in the primary prevention trials is also consistent with the higher levels of noncardiovascular mortality seen with lower cholesterol levels in epidemiological studies (42).

In assessing the potential benefit of cholesterol reduction for populations other than those enrolled in the primary prevention trials, it is critical to compare the overall risk of CHD, which will be reduced by cholesterol lowering, with the risk of death from other causes, which may be increased by cholesterol lowering. Compared with middle-aged men, all other age/sex groups either have a lower risk of developing CHD or have rates that are less dependent on cholesterol levels (e.g., older individuals) (43); in these groups, the risk-benefit ratios for cholesterol reduction may be less favorable. Thus, any extrapolation of the existing data to suggest that pharmacological treatment to lower serum cholesterol results in improved overall survival in individuals without known coronary disease—especially in individuals at lower risk than the subjects in these trials (i.e., middle-aged men)—remains unproven. Since drug treatment to reduce serum cholesterol levels is not without risk or cost, efforts to target treatment at populations at highest risk for developing coronary heart disease are likely to be most effective—and consequently, most cost-effective.

Cholesterol Reduction for Secondary Coronary Prevention

The serum cholesterol level remains an important predictor of recurrent coronary events in patients with established CHD (44). Moreover, because absolute coronary risk is so much greater for individuals with established CHD (44), a given difference (change) in serum cholesterol level is associated with a much greater difference (change) in absolute coronary event rates in individuals with known CHD compared with individuals without CHD. For example, in the Coronary Drug Project, the difference in 5-year CHD mortality between the highest and lowest cholesterol quintiles (among patients with established CHD) was 67 deaths per 1000 patients (45), while in the Multiple Risk Factor Intervention Trial, the difference in 6-year CHD mortality (among patients without preexisting CHD) was only 8 deaths per 1000 patients (46). Thus, the absolute risk of coronary events due to any given degree of cholesterol elevation is much higher in secondary than in primary prevention. Because of the high levels of both absolute and attributable coronary risk in individuals with elevated serum cholesterol and established CHD, cholesterol-lowering interventions for secondary coronary prevention are particularly attractive.

Secondary coronary prevention is also attractive because extensive cholesterol screening (i.e., "case-finding") is unnecessary, and the patient population may be highly motivated to be compliant. Finally, secondary prevention may be somewhat less expensive than primary prevention, because the duration of treatment for any individual will be shortened. However, secondary prevention cannot benefit those individuals in whom the first manifestation of coronary disease is fatal.

The benefit of cholesterol lowering for secondary coronary prevention is confirmed by the results of controlled clinical trials. Numerous studies have demonstrated that cholesterol reduction—either by diet or medication—after an initial myocardial infarction reduces the risk of recurrent myocardial infarction (45,47–49). Angiographic studies have also documented a reduced rate of progression or even regression of coronary atherosclerosis when aggressive cholesterol reduction was used as secondary prevention (50–53). Several studies including the nicotinic acid arm of the Coronary Drug Project, the Stockholm Ischemic

Heart Disease Study, and the Oslo Diet/Heart Intervention Study demonstrated significant reductions in both coronary and all-cause mortality with cholesterol reduction (47,48,54). Although other trials have not demonstrated a reduction in overall mortality, these studies were limited by a relatively brief (<8 year) follow-up period (49). Meta-analysis of eight secondary prevention trials demonstrated a relative risk of 0.88 for cardiovascular mortality and 0.91 for all-cause mortality in patients randomized to cholesterol-lowering treatment compared with control patients (49).

COST-EFFECTIVENESS ANALYSIS: METHODOLOGICAL CONSIDERATIONS

Cost-effectiveness analysis is a technique for identifying the net health benefits that can be derived by allocating a fixed amount of scarce health care resources (8). The analysis compares the cost and clinical outcomes of one program to those of alternative programs.

In contrast to benefit-cost analysis, in which both health benefits and medical care costs are measured in dollar terms, cost-effectiveness analysis has the advantage of measuring health benefits in natural units (9). For comparisons among programs to treat a single condition or set of conditions, disease-specific intermediate outcomes can be used as effectiveness measures. For example, in an analysis of alternate treatments for hypertension, a natural unit of effectiveness would be the change in diastolic blood pressure; for interventions to lower cholesterol, the change in serum cholesterol level would be an appropriate outcome measure. However, to compare alternative health care programs for treating different conditions or disease states, more general measures of overall health benefit are required. In general, health benefits include extensions of life or improvements in quality of life and are measured as years of life or "quality-adjusted life years" (QALYs). Quality-adjusted life years are derived by multiplying the time spent in a given health state by a weighting factor that represents the quality of life in the state. This weighting factor reflects the preferences of the target population regarding the relative desirability of time spent in various health states (8,9,55). Aspects of quality of life that are relevant to the evaluation of cholesterol-lowering interventions include ay decrement in quality of life due to restrictive diets or the need for drug treatment as well as improvements in quality of life by avoiding myocardial infarctions, revascularization procedures, or angina. To date, however, most economic analyses of cholesterol lowering have ignored quality-of-life considerations.

An incremental cost-effectiveness ratio is derived by dividing a program's incremental cost relative to the next most effective alternative by its incremental effectiveness (9). Programs that both reduce costs and improve clinical outcomes are favored on both grounds and should clearly be adopted, but most health care programs achieve clinical benefit only by increasing costs (56–58). Similarly, programs that are both more expensive and less effective than alternative programs for treating the same individuals are said to be "dominated" and should not be adopted. Finally, the remaining programs are available for adoption in order of increasing cost-effectiveness ratios until the available budget is exhausted. By this method, a decision maker can allocate the available health care resources so as to maximize the total health benefits that can be achieved.

An important limitation of cost-effectiveness analysis is that it does not specifically indicate whether any program or treatment under evaluation is "cost-effective." Although programs with low cost-effectiveness ratios are more favorable than programs with higher ratios, the threshold cost-effectiveness ratio above which a program should not be funded depends on both the alternative uses available for the same resources and the size of the overall health care budget. Nonetheless, within the U.S. health care system there is a growing consensus about the desirability of cost-effectiveness ratios. Incremental cost-effectiveness ratios of less than $20,000 per quality-adjusted life year gained—such as those for coronary artery bypass grafting for left main coronary disease (58) or the treatment of severe hypertension (diastolic blood pressure > 105 mmHg) (59)—are generally viewed as quite favorable. Incremental cost-effectiveness ratios between $20,000 and $40,000 per additional QALY are also consistent with many other accepted treatments such as hemodialysis (60), treatment of mild hypertension with diuretics (61), and implantable defibrillator treatment for survivors of out-of-hospital cardiac arrest (62,63), while cost-effectiveness ratios of $40,000–$60,000/QALY are somewhat less attractive. Cost-effectiveness ratios above $100,000/QALY are higher than those of most accepted treatments and are therefore generally regarded as unattractive.

Several other specific features of cost-effectiveness analysis are also worthy of comment. The importance of *incremental* cost-effectiveness analysis cannot be overemphasized. As in general economics, the principle of diminishing marginal returns also applies to medical economics. Thus, the incremental benefits of any medical intervention generally fall as the intervention's scope is broadened (9). In medical cost-effectiveness analysis, this implies that the incremental costs and benefits of an intervention should be calculated relative to the next most intensive alternative, which is also presumably the next most effective as well. For example, since the overall benefits of cholesterol reduction for the primary prevention of CHD are likely to be increased by lowering the threshold for treatment of hypercholesterolemia from 300 to 240 mg/dl, it is important to calculate the costs and effectiveness of cholesterol-lowering treatment for individuals with cholesterol levels between 240 and 300 mg/dl as incremental to the costs and effectiveness of treating just those individuals with cholesterol levels above 300 mg/dl. Similarly, within a population of hypercholesterolemic individuals, universal cholesterol reduction for primary prevention is more intensive than restricting cholesterol reduction to those individuals with proven CHD (i.e., secondary prevention). Thus the incremental cost-effectiveness of cholesterol lowering for primary coronary prevention should be calculated relative to secondary coronary prevention in this population.

The principle of discounting also plays an important role in the economic evaluation of cholesterol-lowering interventions. The intrinsic value of money and health may vary over time. Future costs are generally less onerous than present costs, while future health benefits are generally less desirable than equivalent present benefits. Cost-effectiveness analyses generally incorporate these principles of time-preference by discounting both future costs and future benefits (8). Although there is no universally accepted discount rate, most analyses use a rate of 5% per year and explore the sensitivity of the results to modest variations in this rate. For preventive programs such as cholesterol lowering, costs are generally incurred many years before any expected benefits. Thus, discounting lowers the projected costs less than it lowers the expected benefits and results in less favorable cost-effectiveness ratios than analyses that do not incorporate discounting.

Cost-effectiveness analysis is a powerful analytical tool to guide the allocation of scarce resources. In general, cost-effectiveness analyses employ simulation models built from a set of assumptions and based on empirical evidence from a variety of sources. Such simulation models are no more or less accurate than the empirical evidence and assumptions on which they are based. Because the assumptions on which a model is based might prove incorrect, sensitivity analyses, in which the effect of varying one or more of the assumptions is explored, are critical to the interpretation of the analysis. If the results of the analysis are stable, even in the face of plausible variations in the underlying assumptions, then the conclusions of the analysis may be accepted with a reasonable degree of confidence. If the results are highly sensitive to variations in an assumption, however, the analysis should be viewed with caution and further empirical studies would be needed to reduce the uncertainty regarding the assumption.

COST-EFFECTIVENESS OF INTERVENTIONS TO LOWER SERUM CHOLESTROL

Over the past several years, a number of analyses of the cost-effectiveness of cholesterol-lowering interventions have been published in the medical literature (10,33–36,64,65). The major methodological and analytical features of these studies are summarized in Table 2. Although each analysis is unique, most of these studies share several common features. First, most analyses have assumed that the effect of cholesterol reduction for primary prevention of CHD can be modeled using the Framingham Heart Study's logistic risk equations to describe the relationship between the total cholesterol level and the incidence of CHD (29). The use of the Framingham risk equations is based on the finding that the relationship between the degree of cholesterol reduction and the observed reduction in CHD risk in the major intervention studies is similar to that predicted by the Fra-

Table 2 Overview of Studies of the Cost-Effectiveness of Cholesterol Reduction

Authors (Ref.)	Year	Alternatives considered	Population analyzed + risk factor stratification	Model assumptions	Incremental CEA?	Special considerations
Oster and Epstein (33)	1987	Cholestyramine	Total cholesterol level, age, sex, smoking, hypertension, diabetes	Framingham risk equations; survival after CHD based on Framingham results	For additional years of treatment	Examined incremental cost-effectiveness of varying duration of treatment; only study to include additional costs of treating conditions other than CHD in years of life extended by treatment.
Kinosian and Eisenberg (64)	1988	Cholestyramine, colestipol, or oat bran	Smoking status	Efficacy based on 7-year results from LRC-PPT	No	Only study to include indirect costs
Schulman et al. (10)	1990	Cholestyramine, colestipol, niacin, gemfibrozil, lovastatin, or probucol	No (not relevant)	Cost-effectiveness measured as cost per unit change in lipid subfraction or CHD risk	No	Considered effects of noncompliance and possibility of switching agents; analysis only designed to indicate which drugs are "efficient" treatments for hyperlipidemia

Study	Year	Treatment	Variables	Data source		Comments
Martens et al. (34)	1990	Simvastatin or cholestryamine	Age, cholesterol level	Framingham risk equations and life expectancies	No	None
Kristiansen et al. (65)	1991	"Drug treatment" or individual diet vs. populationwide diet	Norwegian males age 40–49 only	Tromso heart study MI rates	Yes	Cost-utility analysis—only study to consider negative impact of drug and diet therapy on quality of life
Hay et al. (35)	1991	Lovastatin	Total cholesterol level, age, sex, smoking status, blood pressure	Framingham risk equations and life expectancies	No	Assumed cholesterol reduction achieves only 90% of predicted benefit
Goldman et al. (36)	1991	Lovastatin (20, 40, or 80 mg/day)	Total cholesterol level, age, sex, smoking status, relative weight	Framingham risk equations, event rates for U.S. population with CHD	Yes	Only study to consider primary prevention as incremental to secondary prevention

Diet = Dietary therapy; LRC-PPT = Lipid Research Clinics Primary Prevention Trial; CHD = coronary heart disease.

mingham equations (30). Among other features, these equations assume that the effect of serum cholesterol as a risk factor is multiplicative with other risk factors and with the baseline risk of CHD. This approach is appropriate for middle-aged men and is generally considered the best option for estimating the potential effectiveness of cholesterol reduction in populations other than those in which primary prevention has been proven effective.

Second, all of these analyses have assumed that the only effects of cholesterol lowering are to reduce CHD incidence and mortality. No analyses to date have considered the possible impact of an increase in non-CHD mortality due to cholesterol reduction. Other similarities among the existing analyses include the assumption of a 2-year lag period between the initiation of cholesterol-lowering therapy and any effect on coronary heart disease incidence, the use of a 5% discount rate for both costs and health effects, and the use of cost per year of life gained as the main cost-effectiveness measure—ignoring quality-of-life considerations.

Despite these similarities, each analysis has important differences that preclude direct comparisons among the various studies (see Table 2). These differences include which therapeutic alternatives were compared, the particular population considered, stratification by the levels of risk factors other than serum cholesterol, consideration of secondary as well as primary coronary prevention, and inclusion of non-CHD treatment costs. Nonetheless, there are a number of consistent findings from which several important conclusions about the cost-effectiveness of cholesterol lowering can be drawn.

Cost-Effectiveness of Secondary Prevention of Coronary Heart Disease

The most cost-effective indication for cholesterol-lowering therapy is for secondary prevention in patients with existing CHD. Goldman et al. used the Coronary Heart Disease Policy Model—a simulation of the entire U.S. population between the ages of 35 and 85 (66)—to study the cost-effectiveness of lovastatin for secondary as well as primary coronary prevention (36). They found that lovastatin used for secondary prevention in individuals with preexisting coronary disease had highly favorable cost-effectiveness ratios (Table 3). For men ages 35–54 years with serum cholesterol levels above 250 mg/dl, treatment with lovastatin 20 mg/day was estimated to save both money and lives. For older men and women of all ages with serum cholesterol levels above 250 mg/dl, treatment with 20 mg/day lovastatin had an estimated cost effectiveness ratio of less than $20,000 per year of life saved (1989 dollars). Even with serum cholesterol levels under 250 mg/dl, the estimated cost-effectiveness ratios for secondary prevention with 20 mg/day lovastatin were less than $40,000 per year of life saved except in young women. Thus, using lovastatin for secondary coronary prevention appears to have highly

Table 3 Estimated Cost (in dollars) per Year of Life Saved for Lovastatin as Secondary Prevention

	Age (yr)				
	35–44	45–54	55–64	65–74	75–84
Pretreatment Cholesterol level ≥ 250 mg/dl					
Men					
Lovastatin 20 mg/day	[a]	[a]	1,600	10,000	19,000
Lovastatin 40 mg/day	14,000	86,000	17,000	27,000	38,000
Women					
Lovastatin 20 mg/day	4,500	3,500	8,100	12,000	15,000
Lovastatin 40 mg/day	49,000	30,000	29,000	30,000	29,000
Pretreatment Cholesterol level < 250 mg/dl					
Men					
Lovastatin 20 mg/day	38,000	16,000	17,000	25,000	30,000
Lovastatin 40 mg/day	120,000	57,000	48,000	53,000	58,000
Women					
Lovastatin 20 mg/day	210,000	73,000	36,000	30,000	23,000
Lovastatin 40 mg/day	310,000	150,000	81,000	62,000	45,000

[a]Therapy estimated to save both lives and money
Source: Adapted from Ref. 36.

favorable cost-effectiveness ratios in most individuals, even those with only modest elevations of serum cholesterol (e.g., 200–249 mg/dl).

That secondary coronary prevention is more cost-effective than primary prevention is not surprising. Since the absolute risk of recurrent coronary events and CHD mortality is 5–7 times higher for patients with preexisting CHD compared with individuals without known coronary heart disease (49,67,68), the potential for risk reduction for an individual is far greater in secondary than primary prevention. Moreover, the potential adverse effects of cholesterol lowering on noncardiovascular mortality will be less important in patients with preexisting CHD. The overall benefit of any cholesterol-lowering intervention represents the sum of its effects on cardiovascular and noncardiovascular mortality. To the extent that treatments to lower serum cholesterol may increase noncardiovascular mortality (42), cholesterol-lowering interventions will be most effective—and most cost-effective—in those individuals with the greatest absolute risk of mortality due to CHD. Even a small increase in the risk of noncoronary and noncardiovascular death may outweigh the benefit of cholesterol lowering in a population at low risk for developing CHD. By comparison, in patients with preexisting CHD, in whom more than 80% of deaths are due to recurrent coronary events (49), any absolute adverse effect of cholesterol lowering on noncardiovascular mortality is likely to be small compared with the absolute benefit that would be achieved by reducing cardiovascular mortality.

Effect of Other CHD Risk Factors on the Cost-Effectiveness of Primary Prevention

Numerous observational epidemiological studies have demonstrated that the risk of developing CHD is a multiplicative function of serum cholesterol and other independent risk factors (e.g., sex, age, cigarette smoking, hypertension, glucose intolerance) (29,69–71). The multiplicative nature of this risk function implies that the effect of any degree of cholesterol reduction on overall CHD risk is greater for individuals whose baseline risk is higher. Using a model of CHD incidence based on the Framingham multiple logistic risk equations, Taylor et al. demonstrated that in otherwise low-risk individuals (i.e., nonsmokers without hypertension), the expected gains in life expectancy from a 6.7% reduction in serum cholesterol, such as was observed in the Multiple Risk Factor Intervention Trial, would be only 3 days to 3 months depending on age and initial cholesterol level (72). In contrast, those individuals whose blood pressure and cigarette consumption was at the 90th percentile had estimated gains in life expectancy that were six times greater. Similarly, they found that for any particular combination of other risk factors, the gains in life expectancy were significantly greater at higher baseline cholesterol levels.

Thus, it is not surprising that targeting cholesterol reduction to those individuals with multiple coronary risk factors or with marked elevation of serum cholesterol greatly improves cost-effectiveness (33–36). Hay et al. studied the cost-effectiveness of lovastatin for primary coronary prevention using a model of CHD incidence based on the Framingham risk equations and CHD mortality estimates derived from Framingham observations (35). Estimates of the costs of specific CHD events and procedures were based on expert consensus (73). They found that using lovastatin 20 mg/day to treat a 35-year-old man with serum cholesterol of 240–299 mg/dl and no other risk factors had a cost-effectiveness ratio of $34,000 (1990 dollars) per life-year gained. The cost-effectiveness ratio improved to $13,000 per life-year gained, however, when the same treatment was applied to a 35-year-old male smoker with hypertension. Using a similar model, Oster and Epstein found that for a 50-year-old man with a cholesterol level of 290 mg/dl but no additional risk factors, the cost-effectiveness ratio for lifetime treatment with cholestyramine was $135,000 per year of life gained (1985 dollars), while the addition of three coronary risk factors (smoking, hypertension, and diabetes mellitus) improved the cost-effectiveness ratio to $58,000 per year of life gained (33).

The difference in cost-effectiveness ratios between these two studies is partly based on the fact that Hay and colleagues studied lovastatin, one of the most effective agents for reducing the serum cholesterol level, while Oster and Epstein studied the less effective but equally expensive drug cholestyramine. But this difference alone cannot explain the difference fully. Hay and colleagues used a consensus panel's estimates of the costs of treating CHD, while Oster and other investigators based their cost estimates on data from cohort studies. The much higher treatment cost estimates in the former analysis may have led to an overestimate of any financial savings from the prevention of CHD. Finally, some of this difference may be due to other assumptions, whose impact cannot be measured from the published methods.

The higher absolute risk of CHD for men compared with women implies that if other risk factors are equal, interventions to reduce serum cholesterol levels will be more effective on an absolute basis and hence more cost-effective in men than in women (34–36). Goldman et al. examined the relation of age, sex, other risk factors, and prior CHD on the cost-effectiveness of cholesterol reduction with lovastatin (36). For women aged 35–64, cost-effectiveness ratios were generally 3–5 times higher than for men with equivalent risk profiles (see Table 4). For women younger than 54 years or older than 75 years, lovastatin was estimated to have a cost-effectiveness ratio above $60,000 per year of life saved regardless of the levels of other risk factors. Even for women ages 55–74, the cost-effectiveness of lifetime treatment with lovastatin 20 mg/day was below $60,000 per year of life saved only in subgroups with multiple other risk factors. Since the influence of gender on CHD incidence declines with age, the cost-effectiveness ratios for men and women were comparable above age 65 (36).

Impact of Drug Cost and Effect on Serum Lipid Levels on the Cost-Effectiveness of Lipid-Lowering Therapy

Several studies have demonstrated that the cost-effectiveness of drug treatment to lower serum cholesterol depends on both the cost and effectiveness of the drug used. Kinosian and Eisenberg examined the cost-effectiveness of cholestyramine, colestipol, or oat bran for primary coronary prevention in moderately hyper-cholesterolemic, middle-aged men (i.e., the Lipid Research Clinics Primary Prevention Trial population) (64). Effect estimates (CHD cases prevented, CHD deaths prevented) were based on the primary results of the Lipid Research Clinics Primary Prevention Trial (26). They found that a 50% reduction in the cost of cholestyramine (by substituting bulk drug for packets) improved the cost-effectiveness ratio for cholestyramine from $108,000 to $58,000 per year of life saved (1987 dollars) (64). Similarly, in their analysis of lovastatin treatment, Goldman et al. found that the cost-effectiveness ratios within each age, sex, and risk factor subgroup would decline by approximately 30% if the price of lovastatin fell by 40% in 1997 due to the availability of generic formulations (36). Whether

Table 4 Cost-Effectiveness Ratio (Dollars per Year of Life Saved) of 20 mg/day of Lovastatin for Primary Prevention of Coronary Heart Disease

	Age (yr)				
	35–44	45–54	55–64	65–74	75–84
Men, pretreatment cholesterol > 300 mg/dl					
High risk	24,000	13,000	15,000	23,000	66,000
Moderate risk	130,000	49,000	29,000	32,000	92,000
Low risk	330,000	110,000	58,000	58,000	150,000
Women, pretreatment cholesterol > 300 mg/dl					
High risk	195,000	62,000	34,000	39,000	67,000
Moderate risk	480,000	140,000	62,000	46,000	87,000
Low risk	1,500,000	320,000	130,000	68,000	110,000

High risk = diastolic blood pressure >105 mmHg, smoker, weight >130% of ideal;
Moderate risk = diastolic blood pressure 95–104 mmHg, nonsmoker, weight 110–129% of ideal;
Low risk = diastolic blood pressure <95 mmHg, nonsmoker, weight <110% of ideal.
Source: Adapted from Ref. 36.

such price reductions are likely to occur is difficult to predict, however, as such price cuts have not occurred for cholestyramine.

Schulman and colleagues used a different approach to explore the role of the specific lipid-lowering medication in determining the cost-effectiveness of cholesterol reduction (10). They assumed that cholesterol-lowering medications are effective only to the extent that they alter serum lipid levels. Thus, the cost-effectiveness ratio for each specific agent was calculated as the incremental cost of treatment divided by the incremental change in serum cholesterol (or LDL, HDL, or a weighted average of LDL and HDL cholesterol levels). Estimates of the effect of each drug regimen on serum lipid levels were derived from randomized trial data and adjusted for declining compliance over a 5-year time horizon. Costs included the direct costs of medications and physician services as well as the costs of monitoring and treating drug side effects, with adjustments for medication changes and noncompliance. By examining the cost per unit change in serum lipid levels rather than cost per year of life saved, this analysis is not directly comparable to any other cost-effectiveness analysis. Nonetheless, this approach provided important insight into those treatment regimens that are "efficient" ways to treat hypercholesterolemia. In their analysis of cost per unit decrement in overall coronary risk, Schulman and colleagues found that only nicotinic acid and lovastatin (20 mg, 40 mg, or 80 mg/day) were cost-effective treatment strategies (10). Nicotinic acid appeared to be relatively cost-effective because it produced modest decreases in LDL cholesterol level and modest increases in HDL cholesterol level at a comparatively low cost. Although lovastatin was the most expensive agent, it was also potentially cost-effective because it produced the greatest expected reduction in overall coronary risk. The other drugs considered (cholestyramine, gemfibrozil, colestipol, and probucol) did not appear to be as cost-effective as either niacin or lovastatin in their model because the latter two drugs were estimated to produce more favorable lipid alterations at a lower cost. If long-term compliance with niacin therapy was slightly lower than their initial estimates, however, gemfibrozil replaced niacin as a potentially cost-effective alternative.

Since the absolute risk of developing CHD for an individual with moderate hypercholesterolemia is low—except in the presence of multiple other risk factors—many people must be treated with lipid-lowering agents in order to prevent a single case of CHD. Consequently, most of the costs in the cost-effectiveness equation represent the cost of cholesterol screening and treatment (35,64). The costs of treatment for CHD and its complications have comparatively less impact on the cost-effectiveness ratios for cholesterol lowering.

Cost-Effectiveness of Population-Based Interventions to Lower Cholesterol Levels

Population-based interventions to lower serum cholesterol levels are appealing for several reasons. Most individual treatments (such as dietary counseling or drug therapy) are targeted at high-risk individuals; but because CHD is so prevalent, most cases develop in individuals who are, in fact, "low risk." Simulations based on the Coronary Heart Disease Policy Model suggest that 50–70% of new cases of CHD will occur in individuals with serum cholesterol levels below 250 mg/dl (74). These simulations also suggest that life expectancy gains from an ambitious policy of targeted intervention (i.e., reducing serum cholesterol to 250 mg/dl in all individuals with current cholesterol levels > 250 mg/dl) could also be obtained by a more modest populationwide intervention that reduced populationwide serum cholesterol levels by 10 mg/dl in men and 23 mg/dl in women (74). Moreover, population-based interventions are relatively inexpensive on a per capita basis. For example, populationwide education programs in the Stanford Five-Community Study and in the North Karelia (Finland) Study reduced serum cholesterol levels by 3–4% (along with other favorable changes in risk factor profile) at an estimated annual cost of only $4–10 per person (75,76).

Based on these data, analyses suggest that population-based interventions to lower serum cholesterol can be highly cost-effective. Kristiansen and colleagues estimated the cost-effectiveness of the Norwegian cholesterol-lowering program in the population of Norwegian males ages 40–49 and compared these values to the incremental cost-effectiveness of individual dietary and drug therapy for the same population (65). Based on the costs of administering the Norwegian Nutrition Counsel, they estimated a cost-effectiveness ratio of approximately $20 per year of life saved for a populationwide educational program. In contrast, individual dietary therapy and targeted drug treatment were estimated to have incremental cost-effectiveness ratios of about $20,000 and $150,000 (1991 U.S. dollars) per year of life saved, respectively.

Preliminary studies by Tosteson and colleagues also suggest that population-wide interventions are highly cost-effective compared with targeted drug and dietary therapy. Using the Coronary Heart Disease Policy Model, they estimated that a nationwide program with similar costs and effects as the North Karelia Study would cost perhaps $10,000 per year of life saved (1989 dollars) (77). Moreover, a program with the same costs and effects as the Stanford Five-Community Study was estimated to save both money and lives. The true cost-effectiveness of such programs may be more dependent on their ability to alter smoking rates, salt intake, and the rates of detection, treatment, and control of blood pressure than on any improvements in serum cholesterol levels, however. Nonetheless, these highly favorable cost-effectiveness ratios suggest that populationwide interventions should be included in any comprehensive national strategy to lower cholesterol levels.

LIMITATIONS OF EXISTING STUDIES

The major limitation of the existing analyses of the cost-effectiveness of choles-terol reduction is their failure to incorporate quality-of-life considerations. Despite increasing emphasis on evaluating quality of life in clinical trials and in incorporat-ing quality-of-life considerations into both clinical decision-making and cost-effectiveness analyses, most studies of cholesterol reduction have focused solely on life expectancy as the outcome measure. However, recent studies demonstrate that patients with coronary disease have a poorer quality of life compared with healthy individuals (78) and would be willing to "give up" some of their life expectancy to live in excellent health (79,80). Moreover, outcomes research suggests that patients with chronic conditions (e.g., hypertension, prior myocar-dial infarction) who were traditionally considered "asymptomatic" have a lower quality of life compared with their healthy counterparts (78,81). In a comparison of the first National Cholesterol Education Program's treatment guidelines with the more conservative Canadian Task Force recommendations, Krahn and colleagues demonstrated that even trivial decrements in quality of life associated with dietary and medical therapy for hypercholesterolemia negate the life expectancy gains of the more aggressive NCEP strategy (82). Thus, future cost-effectiveness analyses should explicitly consider the impact of symptomatic coronary disease as well as treatment for asymptomatic hypercholesterolemia on quality of life.

A further limitation of the existing studies is the failure to consider the effects of drug therapy on the levels of both LDL and HDL cholesterol. Most of the published cost-effectiveness analyses have considered only the effect of treatment on total serum cholesterol and its relation to CHD risk. As such, these studies do not consider the impact of HDL cholesterol as an independent coronary risk factor on the cost-effectiveness of cholesterol lowering. Since many drugs both lower LDL and raise HDL cholesterol levels, ignoring lipid subfractions may substantially underestimate the benefits and cost-effectiveness of lipid modification.

Another potential source of error in the existing analyses is their failure to consider regression-dilution bias. Because many of the epidemiological analyses have been based on a single cholesterol level, it has been postulated that the strength of the relationship between serum cholesterol level and coronary risk may be underestimated; a 1% change in cholesterol may actually correspond to a 3% change in coronary risk (31,32). Since the cost-effectiveness analyses are largely based on the relationship between coronary risk and cholesterol level derived from these epidemiological investigations, regression-dilution bias may have led to an underestimate of the benefits and cost-effectiveness of cholesterol reduction for a given pretreatment cholesterol level.

Finally, none of the current studies has considered the possibility that cholesterol reduction may increase non-CHD mortality. Although this possibility is highly controversial, there is increasing evidence that the relationship between serum cholesterol level and mortality may be "U-shaped" (42). If cholesterol reduction does increase non-CHD mortality even slightly, the existing studies would have overestimated the life expectancy gains from cholesterol reduction, and the cost-effectiveness ratios would increase—primarily for those individuals whose absolute risk of CHD is relatively small compared with their risk of noncoronary mortality. Future clinical trials will serve to clarify the effect of lowering cholesterol levels on both coronary and noncoronary mortality. Until then, cost-effectiveness analyses should consider the possible adverse effects of cholesterol lowering, at least in sensitivity analyses.

SUMMARY

Although many of the existing analyses are not directly comparable, certain consistent findings among the various studies suggest several general conclusions about the cost-effectiveness of interventions to lower cholesterol levels. The preponderance of data suggest that both dietary and drug therapy to reduce serum cholesterol levels are highly cost-effective when used for secondary coronary prevention in patients with established coronary heart disease. Populationwide interventions to improve nutrition through educational programs and mass media also appear to be highly cost-effective and should be included in any comprehensive national program to reduce cholesterol and coronary heart disease.

Although preventive care is intuitively appealing and is often advocated as a means to reduce health care costs, formal economic analyses demonstrate that, similar to most preventive care (59,83), cholesterol lowering for primary coronary prevention does not "pay for itself." Nonetheless, most analyses suggest that drug treatment for young and middle-aged men with moderate-to-severe elevations of serum cholesterol (>240 mg/dl) and multiple other risk factors for CHD has a cost-effectiveness ratio below $40,000 per year of life saved—similar to federally funded programs such as outpatient hemodialysis and many other widely practiced medical interventions.

The appropriateness of cholesterol reduction for other populations, including young men with isolated mild hypercholesterolemia, women, and the elderly is less certain, however. The costs of screening and treatment for hypercholesterolemia in these populations are considerable and are currently of unproven benefit. Even if cholesterol reduction in these populations reduces the risk of developing CHD to the level that would be predicted from epidemiological studies, the cost-effectiveness ratios generally exceed $100,000 per year of life saved. Whether these cost-effectiveness ratios would improve significantly if the impacts of CHD and drug therapy on quality of life were incorporated in the analysis is currently subject to debate. Clearly, given the potential economic impact of treating hypercholesterolemia in female and elderly patients, further studies are needed to document the efficacy of cholesterol reduction in these populations. If cholesterol reduction has even minor adverse effects on noncoronary mortality, the impact on effectiveness and cost-effectiveness would be most noteworthy in persons with lower coronary risks or higher noncoronary risks.

Regardless of the population under consideration, economic analyses suggest that the specific cholesterol-lowering agent can have an important impact on cost-effectiveness. Based on current prices, nicotinic acid, the HMG-CoA reductase inhibitors, and possibly gemfibrozil appear to be cost-effective cholesterol-lowering agents. On the other hand, cholestyramine, colestipol, and probucol are estimated to be more expensive and less effective drugs for lowering serum cholesterol than available alternatives. If the price of cholesterol-lowering drugs, particularly the HMG-CoA reductase inhibitors, falls sharply in the future, then important improvements in the cost-effectiveness ratios could be realized.

REFERENCES

1. The Expert Panel: Report of the National Cholesterol Education Program Expert Panel on Detection, Evaluation, and Treatment of High Blood Cholesterol in Adults. Arch Intern Med 1988; 148:36–69.
2. American Heart Association. 1990 Heart and Stroke Facts. Dallas, TX: American Heart Association, 1990.
3. Task Force on Cholesterol Issues, American Heart Association. The cholesterol facts: a summary of the evidence relating dietary fats, serum cholesterol, and coronary heart disease. Circulation 1990; 81:1721–1733.
4. Sempos CT, Cleeman JI, Carroll MD, Johnson CL, Bachorik PS, Gordon DJ, Burt

VL, Briefel RR, Brown CD, Lippel K, Rifkind BM. Prevalence of high blood cholesterol among U.S. adults: an update based on guidelines from the Second Report of the National Cholesterol Education Program Adult Treatment Panel. JAMA 1993; 269:3009–3014.

5. Wilson PWF, Christiansen JC, Anderson KM, Kannel WB. Impact of national guidelines for cholesterol risk factor screening: the Framingham Offspring Study. JAMA 1989; 262:41–44.

6. Eddy DM. What do we do about costs? JAMA 1990; 264:1161–1170.

7. Leaf A. Cost effectiveness as a criterion for Medicare coverage. N Engl J Med 1989; 321:898–900.

8. Weinstein MC, Stason WB. Foundations of cost-effectiveness analysis for health and medical practices. N Engl J Med 1977; 296:716–721.

9. Eisenberg JM. Clinical economics: a guide to the economic analysis of clinical practices. JAMA 1989; 262:2879–2886.

10. Schulman KA, Kinosian B, Jacobson TA, Glick H, Willian MK, Koffer H, Eisenberg JM: Reducing high blood cholesterol levels with drugs: cost-effectiveness of pharmacologic management. JAMA 1990; 264:3025–3033.

11. Garber AM, Littenberg B, Sox HC, Wagner JL, Gluck M. Costs and health consequences of cholesterol screening for asymptomatic older Americans. Arch Intern Med 1991; 151:1089–1095.

12. Canadian Consensus Conference on Cholesterol. Final report: The Canadian Consensus Conference on the Prevention of Heart and Vascular Disease by Altering Serum Cholesterol and Lipoprotein Risk Factors. Can Med Assoc J 1988; 139(11, suppl):1–8.

13. Grover SA, Coupal L, Fahkry R, Suissa S. Screening for hypercholesterolemia among Canadians: how much will it cost? Can Med Assoc J 1991; 144:161–168.

14. Toronto Working Group on Cholesterol Policy. Asymptomatic hypercholesterolemia: a clinical policy review. J Clin Epidemiol 1990; 43:1021–1112.

15. Keys A. Coronary heart disease in seven countries. Circulation 1970; 41(suppl I): I-1–I-211.

16. Keys A, Menotti A, Karvonene MJ, Arvanis C, Blackburn H, Buzina R, Kjorkjevic BS, Dontas AS, Fidanza F, Keys MH. The diet and 15-year death rate in the Seven Countries Study. Am J Epidemiol 1986; 124:903–915.

17. Kagan A, Harris BR, Winkelstein W Jr, Johnson KG, Kato H, Syme SL, Rhoads GG, Gay ML, Nichaman MZ, Hamilton HB, Tillotson J. Epidemiologic studies of coronary heart disease and stroke in Japanese men living in Japan, Hawaii, and California: demographic, physical, dietary, and biochemical characteristics. J Chron Dis 1974; 27:345–364.

18. Kannel WB, Castelli WP, Gordon T, McNamara PM: Serum cholesterol, lipoproteins, and the risk of coronary heart disease: the Framingham Study. Ann Intern Med 1971; 74:1–12.

19. Multiple Risk Factor Intervention Trial Research Group. Multiple Risk Factor Intervention Trial: risk factor changes and mortality results. JAMA 1982; 248:1465–1477.

20. Bush TL, Barrett-Connor E, Cowan LD, Criqui MH, Wallace RB, Suchindran CM, Tyroler HA, Rifkind BM. Cardiovascular mortality and noncontraceptive use of

estrogen in women: results from the Lipid Research Clinics Program Follow-Up Study. Circulation 1987; 75:1102–1109.

21. Gordon DJ, Probstfield JL, Garrison RJ, Neaton JD, Castelli WP, Knoke JD, Jacobs KR, Bangdiwala S, Tyroler HA. High-density lipoprotein cholesterol and cardiovascular disease: four prospective American studies. Circulation 1989; 79:8–15.

22. Gordon DJ, Rifkind BM. High-density lipoprotein—the clinical implications of recent studies. N Engl J Med 1989; 321:1311–1316.

23. Gordon T, Castelli WP, Hjortland MC, Kannel WB, Dawber TR. High density lipoprotein as a protective factor against coronary heart disease: the Framingham study. Am J Med 1977; 62:707–714.

24. Gordon DJ, Knoke J, Probstfield JL, Superko R, Tyroler HA. High-density lipoprotein cholesterol and coronary heart disease in hypercholesterolemic men: the Lipid Research Clinics Coronary Primary Prevention Trial. Circulation 1986; 74:1217–1225.

25. Castelli WP, Garrison RJ, Wilson PWF, Abbott RD, Kalousdian S, Kannel WB. Incidence of coronary heart disease and lipoprotein levels: The Framingham Study. 1986; 256:2835–3838.

26. Lipid Research Clinics Program. The Lipid Research Clinics Coronary Primary Prevention Trial results: I. Reduction in incidence of coronary heart disease. JAMA 1984; 251:351–364.

27. Frick MH, Elo O, Haapa K, Heinonen OP, Heinsalmi P, Helo P, Huttunen JK, Kaitaniemi P, Koskinen P, Manninen V, Maenpaa H, Malkonen M, Manttari M, Norola S, Pasternack A, Pikkarainen J, Romo M, Sjoblom T, Nikkila EA. Helsinki Heart Study: primary prevention trial with gemfibrozil in middle-aged men with dyslipidemia. N Engl J Med 1987; 317:1237–1245.

28. Lipid Research Clinics Program. The Lipid Research Clinics Coronary Primary Prevention Trial results: II. The relationship of reduction in incidence of coronary heart disease to cholesterol lowering. JAMA 1984; 251:365–374.

29. McGee D, Gordon T. The results of the Framingham study applied to four other U.S.-based epidemiological studies of cardiovascular disease. In: Kannel WB, Gordon T, eds. The Framingham Study: an epidemiologic investigation of cardiovascular disease. NIH Publication No. 76-1083. Bethesda, MD: U.S. Government Printing Office, 1976.

30. Tyroler HA: Review of lipid-lowering clinical trials in relation to observational epidemiologic studies. Circulation 1987; 76:515–522.

31. Davis CE, Rifkind BM, Brenner H, Gordon DJ. A single cholesterol measurement underestimates the risk of coronary heart disease. JAMA 1990; 264:3044–3046.

32. MacMahon S, Peto R, Cutler J, Collins R, Sorlie P, Neaton J, Abbott R, Godwin J, Dyer A, Stamler J. Blood pressure, stroke and coronary heart disease: Part 1. Prolonged differences in blood pressure: prospective observational studies corrected for the regression dilution bias. Lancet 1990; 335:765–774.

33. Oster G, Epstein AM. Cost-effectiveness of antihyperlipemic therapy in the prevention of coronary heart disease: the case of cholestyramine. JAMA 1987; 258:2381–2387.

34. Martens LL, Rutten FFH, Erkelens DW, Ascoop CA. Clinical benefits and cost-

effectiveness of lowering serum cholesterol levels: the case of simvastatin and cholestryamine in the Netherlands. Am J Cardiol 1990; 65(suppl F):27F–32F.

35. Hay JW, Wittels EH, Gotto AM. An economic evaluation of lovastatin for cholesterol lowering and coronary artery disease reduction. Am J Cardiol 1991; 67:789–796.

36. Goldman L, Weinstein MC, Goldman PA, Williams LW. Cost-effectiveness of HMG-CoA reductase inhibition for primary and secondary prevention of coronary heart disease. JAMA 1991; 265:1145–1151.

37. Dorr AE, Gunderson K, Schneider JC, Spencer TW, Martin WB. Colestipol hydrochloride in hypercholesterolaemic patients—effect on serum cholesterol and mortality. J Chron Dis 1978; 31:5–14.

38. Committee of Principal Investigators. A co-operative trial in the primary prevention of ischaemic heart disease using clofibrate. Br Heart J 1978; 40:1069–1118.

39. Dayton S, Pearce ML, Hashmoto S, Dixon WJ, Tomiyasu U. A controlled clinical trial of a diet high in unsaturated fat in preventing complications of atherosclerosis. Circulation 1969; 39:1–63.

40. Muldoon MF, Manuck SB, Matthews KA. Lowering cholesterol concentrations and mortality: a quantitative review of primary prevention trials. BMJ 1990; 301:309–314.

41. Yusuf S, Wittes J, Friedman L. Overview of results of randomized clinical trials in heart disease: II. Unstable angina, heart failure, primary prevention with aspirin and risk factor modification. JAMA 1988; 260:2259–2263.

42. Jacobs D, Blackburn H, Higgins M, et al. Report of the Conference on Low Blood Cholesterol: mortality associations. Circulation 1992; 86:1046–1060.

43. Maniolio TA, Pearson TA, Wenger NK, Barrett-Connor E, Payne GH, Harlan WR. Cholesterol and heart disease in older persons and women. Review of an NHLBI workshop. Ann Epidemiol 1992; 2:161–176.

44. Pekkanen J, Lin S, Heiss G, Suchindran C, Leon A, Rifkind BM, Tyroler HA. Ten-year mortality from cardiovascular disease in relation to cholesterol level among men with and without preexisting cardiovascular disease. N Engl J Med 1990; 322:1700–1707.

45. Coronary Drug Project Research Group: Natural history of myocardial infarction in the Coronary Drug Project: long-term prognostic importance of serum lipid levels. Am J Cardiol 1978; 42:489–498.

46. Stamler J, Wentworth D, Neaton JD. Is relationship between serum cholesterol and risk of premature death from coronary heart disease continuous and graded? Findings in 356,222 primary screenees of the Multiple Risk Factor Intervention Trial (MRFIT). JAMA 1986; 256:2823–2828.

47. Carlson LA, Roenhamer G. Reduction of mortality in the Stockholm Ischaemic Heart Disease Secondary Prevention Study by combined treatment with clofibrate and nicotinic acid. Acta Med Scand 1988; 223:405–418.

48. Kallio V, Hamalainen H, Hakkila J, Luurila OJ. Reduction in sudden deaths by a multifactorial intervention programme after acute myocardial infarction. Lancet 1979; ii:1091–1094.

49. Roussouw JE, Lewis B, Rifkind BM. The value of lowering cholesterol after myocardial infarction. N Engl J Med 1990; 323:1112–1119.

50. Breniske JF, Levy RI, Kelsey SF, Passamani ER, Richardson J, Loh IK, Stone NJ, Aldrich RF, Battaglini JW, Moriarty DJ. Effects of therapy with cholestyramine on progression of coronary arteriosclerosis: results of the NHLBI Type II Coronary Intervention Study. Circulation 1984; 69:313–324.

51. Arntzenius AC, Kromhout D, Barth JD, Reiber JH, Bruschke AV, Buis B, van Gent CM, Kempen-Voogd N, Strikwerda S, van der Veide EA. Diet, lipoproteins, and the progression of coronary atherosclerosis: the Leiden Intervention Trial. N Engl J Med 1985; 312:805–812.

52. Blankenhorn DH, Nessim SA, Johnson RL, Sanmarco ME, Azen SP, Cashin-Hemphill L. Beneficial effects of combined colestipol-niacin therapy on coronary atherosclerosis and coronary venous bypass grafts. JAMA 1987; 257:3233–3240.

53. Brown G, Albers JJ, Fisher LD, Schaefer SM, Lin JT, Kaplan C, Zhao XQ, Bisson BD, Fitzpatrick VF, Dodge HT. Regression of coronary artery disease as a result of intensive lipid-lowering therapy in men with high levels of apolipoprotein B. N Engl J Med 1990; 323:1289–1298.

54. Canner PL, Berge KG, Wenger NK, Stamler J, Friedman L, Prineas RJ, Friedewald W. Fifteen year mortality in Coronary Drug Project patients: long-term benefit with niacin. J Am Coll Cardiol 1986; 8:1245–1255.

55. Detsky AS, Naglie IG. A clinician's guide to cost-effectiveness analysis. Ann Intern Med 1990; 113:147–154.

56. Doubilet P, Weinstein MC, McNeil BJ. Use and misuse of the term "cost effective" in medicine. N Engl J Med 1986; 314:253–256.

57. Goldman L, Sia STB, Cook EF, Rutherford JD, Weinstein MC. Costs and effectiveness of routine therapy with long-term beta-adrenergic antagonists after acute myocardial infarction. N Engl J Med 1988; 319:152–157.

58. Weinstein MC, Stason WB. Cost-effectiveness of coronary artery bypass surgery. Circulation 1982; 66:2.

59. Weinstein MC, Stason WB. Cost-effectiveness of interventions to prevent or treat coronary heart disease. Annu Rev Public Health 1985; 6:41–63.

60. Drummond MF. Survey of cost-effectiveness and cost-benefit analyses in industrialized countries. World Health Stat Q 1985; 38:383–401.

61. Stason WB, Weinstein MC. Allocation of resources to manage hypertension. N Engl J Med 1977; 296:732–739.

62. Kupperman M, Luce BR, McGovern B, Podrid PJ, Bigger JT, Ruskin JN: An analysis of the cost-effectiveness of the implantable defibrillator. Circulation 1990; 81: 91–100.

63. Larsen GC, Manolis AS, Sonnenberg FA, Beshansky JR, Estes NAM. Pauker SG. Cost-effectiveness of the implantable cardioverter-defibrillator: effect of improved battery life and comparison with amiodarone therapy. J Am Coll Cardiol 1992; 19:1323–1334.

64. Kinosian BP, Eisenberg JM. Cutting into cholesterol: cost-effective alternatives for treating hypercholesterolemia. JAMA 1988; 259:2249–2254.

65. Kristiansen IS, Eggen AE, Thelle DS. Cost-effectiveness of incremental programmes for lowering serum cholesterol concentration: is individual intervention worth while? BMJ 1991; 302:1119–1122.

66. Weinstein MC, Coxson PG, Williams LW, Pass TM, Stason WB, Goldman L. Forecasting coronary heart disease incidence, mortality, and cost: the Coronary Heart Disease Policy Model. Am J Public Health 1987; 285:2381–2387.

67. Kannel WB, Wolf PA, Garrison RF, eds. The Framingham Study: An Epidemiological Investigation of Cardiovascular Disease. Section 35. Survival Following Initial Cardiovascular Events: Thirty Year Follow-Up. NIH Publication No. 88-2969. Washington, DC: Government Printing Office, 1988.

68. Shaper AG, Pocock SJ, Walker M, Phillips AN, Whitehead TP, Macfarlane PW. Risk factors for ischaemic heart disease: the prospective phase of the British Regional Heart Study. J Epidemiol Commun Health 1985; 39:197–209.

69. Gordon T, Kannel WB. Multiple risk functions for predicting coronary heart disease: the concept, accuracy, and application. Am Heart J 1982; 103:1031–1039.

70. The Pooling Project Research Group. Relationship of blood pressure, serum cholesterol, smoking habit, relative weight, and ECG abnormalities to incidence of major coronary events: final report of the Pooling Project. J Chron Dis 1978; 31:201–306.

71. Brand JRI, Rosenman RH, Sholtz RI, Friedman M. Multivariate prediction of coronary heart disease in the Western Collaborative Group Study compared to the findings of the Framingham Study. Circulation 1976; 53:348–355.

72. Taylor WC, Pass TM, Shepard DS, Komaroff AL. Cholesterol reduction and life expectancy: a model incorporating multiple risk factors. Ann Intern Med 1987; 106: 605–614.

73. Wittels EH, Hay JW, Gotto AM. Medical costs of coronary artery disease in the United States. Am J Cardiol 1990; 65:432–440.

74. Goldman L, Weinstein MC, Williams LW. Relative impact of targeted versus populationwide cholesterol interventions on the incidence of coronary heart disease: projections of the Coronary Heart Disease Policy Model. Circulation 1989; 80: 254–260.

75. Farquhar JW, Fortmann SP, Flora JA, Taylor CB, Haskell WL, Williams PPT, Maccoby N, Wood P. Effects of communitywide education on cardiovascular disease risk factors. The Stanford Five-City Project. JAMA 1990; 264:359–365.

76. Puska P, Salonen JT, Nissinen A, Tuomilehto J, Vartiainen EK, Tanskanen A, Ronnqvist P, Koskela K, Huttunen J. Change in risk factors for coronary heart disease during 10 years of a community intervention programme (North Karelia project). BMJ 1983; 287:1840–1844.

77. Goldman L, Gordon DJ, Rifkind BM, Hulley SB, Detsky AS, Goodman DS, Kinosian B,Weinstein MC. Cost and health implications of cholesterol lowering. Circulation 1992; 85:1960–1968.

78. Stewart A, Greenfield S, Hays RD, Wells K, Rogers WH, Berry SD, McGlynn EA, Ware JE. Functional status and well-being of patients with chronic conditions: results from the Medical Outcomes Study. JAMA 1989; 262:907–913.

79. Tsevat J, Goldman L, Lamas GA, Pfeffer MA, Chapin CC, Connors KF, Lee TH. Functional status versus utilities in survivors of myocardial infarction. Med Care 1991; 29:1153–1159.

80. Miyamoto JM, Eraker SA. Parameter estimates for a QAYL utility model. Med Decis Making 1985; 5:191–213.

81. Fryback DG, Dasbach EJ, Klein R, Klein BE, Dorn N, Peterson K, Martin PA. The Beaver Dam Health Outcomes Study: initial catalog of health-state quality factors. Med Decis Making 1993; 13:89–102.
82. Krahn M, Naylor CD, Basinski AS, Detsky AS. Comparison of an aggressive (U.S.) and a less aggressive (Canadian) policy for cholesterol screening and treatment. Ann Intern Med 1991; 115:248–255.
83. Russell LB. The role of prevention in health reform. N Engl J Med 1993; 329: 352–354.

15

Emerging Opportunities for Atherosclerosis Prevention: Beyond Cholesterol-Lowering Therapy

Daniel Steinberg and Joseph L. Witztum

University of California, San Diego, La Jolla, California

INTRODUCTION

The proposition that appropriate treatment of hypercholesterolemia reduces risk of coronary heart disease (CHD) is no longer controversial. As this book amply demonstrates, there is every reason to believe that the current intensive effort to improve our management of hypercholesterolemia has begun to have its impact on morbidity and mortality from CHD. Undoubtedly we will see further improvements in exactly how we manage diet and drug treatment of hypercholesterolemia. However, there are reasons to doubt that the approaches currently in use, even with further refinements, will be enough to eliminate CHD.

Our current goal for the general population—a total cholesterol level of 200 mg/dl or less—while realistic, is probably not low enough if we want total war against CHD. What is the best we possibly could do? What should be our goals, if we set aside the issue of feasibility? The Japanese death rate from CHD in the 1950s was only about 10% of that in the United States—a huge difference—and serum cholesterol levels averaged about 160 mg/dl (1). It could be argued that some or most of that difference might be on a genetic basis, but migration studies speak against that (2). When Japanese migrate to the United States, their cholesterol levels rise and their CHD death rates rise to approach those of the American population. Although there may be some genetic component responsible for the striking differences in CHD death rates, it seems that the major factor must be the plasma cholesterol level as environmentally determined (or other environmental factors). What emerges from this speculative analysis is that we may have to set a goal for total cholesterol level of 160–170 mg/dl if we want to see a 90% reduction in CHD mortality.

Is such a reduction feasible? With intensive drug therapy, yes. With the drugs currently available, we can reduce total cholesterol levels to 160–170 mg/dl, although that generally requires intervention with more than one drug. Under current guidelines, intensive drug treatment is only indicated for patients at very high risk or those with established CHD (3). Ironically, the patients who are at high risk are in a way better off in the sense that we can bring their levels down (and do) to much lower levels than those reached by persons at lower risk, in whom treatment is limited to diet! However, this is not a plea for drug treatment for low-risk patients—at least not yet!

What can be accomplished with diet? Grundy has probably done as much as anyone to analyze in detail the impact of diet on plasma cholesterol levels. He presented a superb scholarly discussion of the issue in his 1990 Duff Lecture (4). The currently recommended AHA Phase I diet generally reduces plasma cholesterol levels by about 50 mg/dl. This reduction would be good enough to reach the 200 mg/dl mark in a large part of our population (those with levels below 250 mg/ dl), but it will not reach the 160–170 mg/dl mark in most people. The simple fact is that the Japanese diet of the 1950s was almost devoid of saturated fat, extremely low in total fat, and extremely low in cholesterol—more so than even the Phase II AHA diet. We are slowly moving people toward an AHA Phase I diet, but there still is a lot to be done to achieve even that relatively minor change in dietary habits. It will probably take generations to get the American people to switch completely to AHA-type diets. Can we really ever expect to get the entire country on AHA-type diet? At the moment it seems that the Japanese cannot even maintain their own population on a Japanese-style diet! McDonald's has invaded the Ginza Strip, and the more Kobe beef you eat, the higher your status in Japan (and the higher your cholesterol level!). The point, of course, is that if we have not already hit a ceiling (floor?) on what we can accomplish with diet, we soon will. This is one reason it behooves us to explore "beyond cholesterol," i.e., to look at interventions not necessarily directed at correcting hypercholesterolemia but at blocking the downstream consequences of hypercholesterolemia within the artery wall. As we learn more about the interactions between lipoproteins and the cells of the artery wall and about the sequence of events that leads to atherosclerotic lesions, we are beginning to see novel ways to intervene and slow the atherogenic process at any given cholesterol level. In this chapter, we will review the current status of our understanding of how hypercholesterolemia leads to the vascular lesion and speculate about the new modes of intervention suggested by new insights into pathogenesis.

A second reason to look "beyond cholesterol" is that the correlation between hypercholesterolemia and CHD risk, while highly significant, is by no means absolute. While most patients with homozygous familial hypercholesterolemia have clinically evident disease before they reach puberty, there are instances in which they have survived into their thirties and forties without manifest disease (5). Conversely, we have all seen patients with three-vessel disease at an early age but with total cholesterol levels below 200. These"exceptions that prove the rule" are not all explained by classical risk factors (smoking, hypertension, diabetes, etc.). They represent tr..e exceptions and tell us that there must be other factors— unknown risk factors—that modulate the impact of hypercholesterolemia on the artery wall.

The purpose of this chapter, then, is to look into the future and to try to anticipate what qualitatively new modes of intervention are likely to emerge from current intensive research efforts directed at understanding the cellular and molecular mechanisms of atherogenesis.

PATHOGENESIS OF ATHEROSCLEROSIS

Despite the fact that the cause-and-effect relationship between hypercholesterolemia and atherosclerosis has been recognized for many years, the detailed mechanisms by which hypercholesterolemia brings about the observed changes in the artery wall remained quite obscure until recently. Over the past couple of decades, thanks to the remarkable advances in modern cell biology and molecular biology, a consensus has emerged with respect to some of the key elements involved in the initiation of atherosclerosis. We can now identify at least a few well-defined processes in the artery wall induced by hypercholesterolemia, processes that are probably central to lesion initiation. From the discussion of pathogenesis outlined below, one can immediately see a number of theoretical options for intervention. The most thoroughly studied of these options is interference with the oxidative modification of LDL. The hypothesis that oxidative modification is centrally important in the atherogenic process has received very strong support from work with animal models of atherosclerosis, and several clinical intervention trials using natural antioxidant vitamins have been funded and are in progress. Results should be available in 3–5 years and are anxiously awaited. Meanwhile, it is important to recognize that much of this chapter relates to provocative but clinically untested hypotheses.

A CURRENT "CONSENSUS SEQUENCE" OF CELLULAR AND MOLECULAR EVENTS LEADING TO THE INITIAL LESION IN ATHEROSCLEROSIS—THE FATTY STREAK

The schema in Figure 1, based on the work of a large number of investigators, summarizes a generally accepted view of the sequence of events by which hypercholesterolemia induces the earliest lesion in atherosclerosis—the fatty

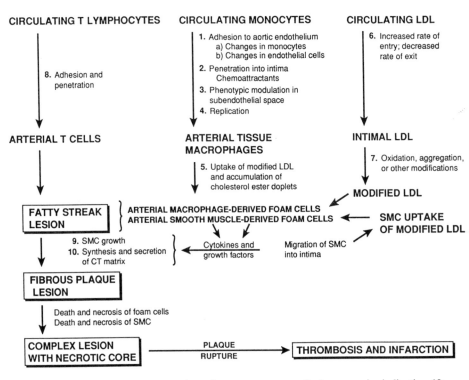

Figure 1 Schematic representation of current concepts of atherogenesis, indicating 10 potential sites at which one might intervene to slow the progression of the disease to the fibrous plaque stage. As discussed in the text, some of these interventions, while theoretically capable of slowing the development of fatty streaks and fibrous plaques, might have unacceptably deleterious side effects. For example, interference with the adhesion and penetration of monocytes, if it occurred globally, would embarrass defense mechanisms against infection. The only one of these theoretically feasible modes of intervention that has been clearly shown to prevent hypercholesterolemia-induced atherosclerosis—and then only in experimental animals—is inhibition of the oxidative modification of LDL.

streak. This is the earliest definable gross lesion in both human and in experimental atherosclerosis. It consists primarily of an accumulation of so-called "foam cells," i.e., cells that are heavily loaded with lipid droplets, mainly cholesterol esters. These cells lie between an intact, unbroken endothelium and the internal elastic membrane. The notion that some endothelial cells need to be stripped away with exposure of underlying connective tissue matrix before a lesion can develop (6) has been largely abandoned. Careful studies fail to reveal any breaks in the endothelial lining overlying the fatty streak (7,8). The foam cells are so numerous and so enlarged that they protrude slightly into the lumen of the vessel. However, the fatty streak never significantly interferes with blood flow. Indeed, the fatty streak is a clinically silent lesion. It develops very early in life, covering 20–40% of the surface of the abdominal aorta by age 35 (9). The intense interest in the origin of this clinically silent lesion, of course, is that it is the precursor of the later clinically significant lesions. The fatty streak is the father of the fibrous plaque and the complex lesion that eventually leads to myocardial infarction and/or stroke. The presumption is that if we understood the pathogenesis of the fatty streak and could inhibit its formation, we might not ever have to concern ourselves with the pathobiology of the fibrous plaque and the complex lesion. They would simply never develop.

Most of the foam cells in a fatty streak represent circulating monocytes that have penetrated between endothelial cells and have taken up residence in the subendothelial space. There they take up lipoproteins at a sufficient rate to become heavily loaded with lipid, primarily cholesterol esters. A small but significant fraction of foam cells is derived also from smooth muscle cells that have migrated from the media up through breaks in the internal elastic membrane into the subendothelial space and also acquired lipids. The fatty streak also contains a few T lymphocytes but these do not contain large quantities of lipid. Their role in pathogenesis is unclear.

Examining the sequence of events in Figure 1, we can immediately see points at which intervention might, in principle, slow down the atherogenic process. As will become apparent, however, intervention in some of these processes might also have serious adverse side effects.

ADHESION OF MONOCYTES TO ENDOTHELIAL CELLS

One of the earliest consequences of feeding an atherogenic diet to experimental animals is an increase in the adherence of monocytes to arterial endothelium (10,11). This reflects the effects of hypercholestrolemia both on circulating monocytes and on endothelial cells. Some of the adhesion molecules involved have already been identified, characterized, and cloned (12), and there are undoubtedly others yet to be defined. Knowing the structure of these adhesion molecules, it should be possible in principle to design drugs that would compete and prevent the adhesion of monocytes to arterial endothelium. This might reduce the number of foam cells and slow the development of the fatty streak. However, one would want to be very sure that the intervention did not at the same time interfere with the adherence of leukocytes to capillary endothelium, a first step in combating infection or dealing with trauma. In other words, a generalized interference with leukocyte adhesion might lead to the very serious consequences seen in some of the disorders in which white cells fail to exercise their sentinel duties (e.g., chronic granulomatous disease). If one could selectively interfere with the adherence of monocytes to arterial endothelium without inhibiting adherence to capillary and venule endothelium, then this mode of intervention might be considered. It is still not known exactly how hypercholesterolemia induces the changes in monocytes and/or endothelial cells. There is some evidence that oxidation of the LDL may be necessary before it acts on endothelial cells (13). If so, this represents still another reason why antioxidants may be effective in blocking the early stages in atherogenesis (see below).

PENETRATION OF ADHERENT MONOCYTES INTO THE SUBENDOTHELIAL SPACE

Once the cells have adhered tightly enough to the endothelial surface to resist the shear forces of blood flow, they can "roll" toward endothelial cell junctions and penetrate, presumably in response to one or more chemoattractants being generated in the intima. Here we have another potential point for intervention— blocking the generation of the chemoattractant molecules or interfering with the response of the leukocytes to them. Oxidized LDL itself has been identified as a potential chemoattractant, both for monocyte/macrophages (14) and for T lymphocytes (15). Navab and coworkers (16), using an elegant cell culture model of the arterial intima, have shown that addition of LDL in such a system increases the adherence and penetration of monocytes but only if the cells are capable of oxidatively modifying the added LDL. The same group showed that minimally oxidized LDL stimulates the release from endothelial cells of monocyte chemoattractant protein-1 (17). Moreover, in situ hybridization studies showed that this protein is strongly expressed in fatty streak lesions (18). Further research will probably uncover still other chemoattractants, thus providing additional potential sites of intervention. However, the same caveat discussed above with regard to monocyte adherence would obviously apply also to interfering with penetration of monocytes (unless that interference again were selective).

PHENOTYPIC MODULATION, GROWTH, AND REPLICATION OF MONOCYTES

Circulating monocytes have a relatively short lifetime in the vascular space, no more than a day or two. As soon as they enter the extravascular, extracellular spaces, they begin to undergo a remarkable series of changes in biological properties. The array of genes they express changes radically and they undergo a Jekyll-and-Hyde transformation to become "tissue macrophages." We are just beginning to understand some of the factors that regulate this shift in the pattern of gene expression. To cite one example relevant to our current discussion, the level of expression of the acetyl LDL receptor or "scavenger" receptor is very low in the circulating monocyte but increases dramatically when the monocyte differentiates into a macrophage—in culture (19) or in the arterial wall. This receptor, as we shall discuss in a moment, appears to be responsible for much of the uptake of lipoprotein that converts the macrophage to a foam cell. Recent studies show that macrophages in atherosclerotic lesions grow and divide. One of the critical factors regulating monocyte differentiation and growth is macrophage colony-stimulating factor (M-CSF). Minimally oxidized LDL stimulates the release of this growth factor (20) and the gene for M-CSF is strongly expressed in arterial lesions (21). Already, then, we recognize several "players" in the phenotypic modulation of macrophages, their growth, and their replication that could become targets for intervention. We will discuss the suitability of these as targets when we discuss the oxidative modification hypothesis below.

As mentioned above, some of the lipid-laden cells in fatty streaks are derived from smooth muscle cells. Presumably these represent cells from the media that have been induced to change their properties and to migrate into the subendothelial space, either as a primary response to hypercholesterolemia or secondary to changes in the intima induced by hypercholesterolemia. Intervention to block the migration and growth of these smooth muscle cells is an obvious target for inhibiting progressive enlargement of the fibrous plaque. It is not clear how important the formation of smooth muscle cell–derived foam cells is in the development of the fatty streak. While the number of such cells is small compared to the number of macrophage-derived foam cells, they could nevertheless play a critical role in processes responsible for progression. Certainly, smooth muscle cell replication and synthesis of connective tissue matrix play a major role in the progression of the fatty streak to the fibrous plaque, which we will discuss below.

UPTAKE AND STORAGE OF EXCESS LIPID IN FOAM CELLS

For many years it has been recognized by pathologists that the lipid-laden foam cell is the hallmark of the fatty streak lesion. However, until quite recently relatively little was known about the mechanisms by which these cells take up lipoproteins. The discovery of the LDL receptor by Goldstein and Brown (22), for which they received the Nobel Prize in 1985, ushered in a new era. They showed that the LDL receptor functioned to tightly regulate the cholesterol content of cells by increasing or decreasing the uptake of LDL according to the cell's needs for cholesterol. Thus, cells growing in culture take up LDL at a rate sufficient to supply their needs, but, once the cell cholesterol content has reached an appropriate level, the cell downregulates the expression of the LDL receptor and ceases to take up LDL from the medium. Consequently, it is not possible to build up the cholesterol content of cells beyond a certain level by increasing the concentration of LDL in the medium. Importantly, Brown and Goldstein showed that this applies also to monocyte/macrophages, i.e., it is not possible to build up their cholesterol content above a ceiling imposed by this regulation of the LDL receptor (23). The same is true with regard to smooth muscle cells; their cholesterol content cannot be increased beyond a certain point even in the presence of very high concentrations of LDL in the medium (24). These findings posed a paradox: How can monocytes and smooth muscle cells be the precursors of cholesterol-loaded foam cells when they appear to protect themselves against engorgement with cholesterol beyond a rather low ceiling? Another paradox was posed by the fact that fatty streak lesions in patients or in animals that totally lack the LDL receptor look pretty much the same as lesions in those that have normal LDL receptors. This similarity included the presence of cholesterol-laden foam cells just like those in individuals with normal LDL receptors. In other words, the absence of the LDL receptor was responsible for the high plasma LDL concentrations, but the uptake of LDL into the foam cells probably was occurring by some pathway other than the LDL receptor pathway.

These two paradoxes could be resolved by postulating that it is not the native, circulating LDL that is being taken up by monocyte/macrophages but rather some modified form of that LDL. Goldstein et al. (25) showed that chemical acetylation of LDL modified it to a form that was taken up rapidly enough by macrophages to generate foam cells. They showed further that that uptake occurred via a specific, saturable receptor—the acetyl LDL receptor. Unlike the LDL receptor, this receptor did not down-regulate when the cholesterol content of the macrophages increased. However, there was no evidence that a modification equivalent to chemical acetylation could occur in vivo. It was at this point that studies by Henriksen et al. (26) showed that endothelial cells in culture are capable of drastically modifying the biological and chemical properties of LDL, converting it to a form that is taken up rapidly enough by monocyte/macrophages to account for foam cell formation.

THE OXIDATIVE MODIFICATION HYPOTHESIS FOR ATHEROGENESIS

The discovery that cells were capable of modifying LDL to a form recognized by the acetyl LDL receptor seemed to resolve the paradoxes surrounding foam cell formation from monocytes. It was shown that any of the major cell types characterizing an atherosclerotic lesions—the endothelial cells, the smooth muscle cells, and the monocyte/macrophages—could carry out this modification. Later studies showed that the modification was strongly inhibited by the addition of antioxidants or even by the addition of relatively low concentrations of serum (27). The hypothesis was therefore greeted with understandable skepticism. How could oxidative modification be important when certainly serum (and probably extracellular fluids to an only slightly lesser extent) contain large numbers of antioxidants at relatively high concentrations? It seemed necessary to propose that the oxidative modification must take place in a sequestered environment in which the LDL was at least temporarily exposed to pro-oxidant conditions without the benefit of the protection afforded by antioxidants. In fact, cells create such microenvironments. For example, killer cells act by grappling with their target cells, creating a closed space between them into which active forms of oxygen or lytic enzymes are then secreted in order to "finish the job." Wright and Silverstein have directly demonstrated microenvironments between macrophages and the surface on which they are plated (28). We, to this day, cannot describe in great detail the conditions in which LDL undergoes oxidative modification in vivo, but we are now sure that it happens. The lines of evidence for this include the following:

1. A gentle extraction of atherosclerotic lesions yields LDL with all the chemical and biological properties of oxidized LDL (29).
2. Immunohistochemical methods show that antibodies against oxidized LDL react with antigens present in atherosclerotic lesions but not in normal artery (30).
3. Autoantibodies that react with oxidized LDL (but not native LDL) have been demonstrated in the plasma of rabbits and of humans (30), and there is preliminary evidence that the titer of such autoantibodies correlates positively with rates of progression of atherosclerotic lesions (31).
4. Antioxidants of several different kinds have been demonstrated to significantly slow the rate of progression of early atherosclerotic lesions in seven animal studies, two in the LDL receptor–deficient rabbit (32,33), three in the cholesterol-fed rabbit (34–36), and two in the cholesterol-fed monkey (37,38). Negative results have been reported in one cholesterol-fed rabbit study (39); in one LDL receptor–deficient rabbit study in older animals, probucol arrested progression but did not induce regression (40).

The animal data are impressive, but a few words of caution are in order. There has been a tendency to assume that antioxidants are basically interchangeable, differing only in their relative potency as antioxidants. Also, it has been assumed by many that the antioxidants work at the level of the LDL particle itself. In other words, the inhibition of atherogenesis by probucol could be attributed to its residence within the LDL particle and the protection it afforded by virtue of that. Indeed, probucol is a highly effective antioxidant and it does reside in the lipoproteins, including LDL. However it also enters cells, and there is a good possibility that some of its effectiveness is due to actions within cells in addition to its action as a resident antioxidant in the LDL particle (41). Probucol has several biological properties that could make it antiatherogenic over and above its antioxidant properties. Considerations like these must be kept firmly in mind as we think about extrapolating from the available animal evidence to the human disease.

Another limitation of the available evidence in experimental animals is that it relates almost exclusively to fatty streak lesions. In almost all of the studies reported thus far, atherosclerosis has been assessed by Sudan IV staining and planimetry, i.e., simply measuring the total surface area covered by lipid-rich fatty streak lesions. There is no evidence as to whether antioxidants will slow the progression of fibrous plaques or whether they will have any effect on complex lesions.

Finally, it should be noted that with one exception, the animal model studies have not used natural antioxidants. The one study using natural antioxidants is a small study in non-human primates (37) and the results were marginally significant.

ADDITIONAL PROPERTIES OF OXIDIZED LDL THAT MAY ENHANCE ITS ATHEROGENIC POTENTIAL

The role of oxidized LDL in foam cell formation was the first property to be discovered but was by no means the last. The fact that oxidized LDL is itself chemotactic (14) and that it also acts indirectly by stimulating the release of MCP-1 to recruit additional monocytes (17) sets the stage for a geometric rate of growth of the lesion once it has been initiated. Monocytes can oxidize LDL, and oxidized LDL can recruit monocytes. Not only that, but oxidized LDL also inhibits any tendency for monocytes to migrate back out of a developing lesion (14). Another "vicious cycle" is set up because oxidized LDL stimulates the release of M-CSF, which facilitates the differentiation and growth of the monocyte/ macrophage (16). That differentiation includes the expression of scavenger receptors, increasing further the rate of uptake of oxidized LDL and the generation of foam cells. We will not take time to go into all of these properties of oxidized LDL, but one deserves particular attention. Endothelial cells incubated for several days in the presence of LDL show severe cytotoxicity, and Morel et al. (42) discovered that this can be blocked by the addition of antioxidants. In other words, the oxidized LDL generated during the incubation, not the native LDL, was cytotoxic. Exactly how the properties of the endothelium are altered by oxidized LDL under in vivo conditions we do not know. What is of interest here is that macrophages are present not only in the developed fatty streak lesion but throughout the history of the atheroma. Complex lesions show high concentrations of macrophages, particularly at the shoulder, which is the site at which rupture commonly occurs. By virtue of its cytotoxicity, oxidized LDL could conceivably play a role in the thinning of the cap that antecedes fatal thrombosis. The macrophage itself also is capable of secreting a large number of other cytotoxic materials and lytic enzymes. One could readily visualize that the high concentration of active macrophages at the shoulder region might contribute importantly to the erosion of a thinned cap and the plaque rupture, which is now accepted as responsible for most fatal thromboses.

THE CURRENT STATUS OF ANTIOXIDANTS IN THE PREVENTION OF HUMAN ATHEROSCLEROSIS

To the extent that the rabbit and nonhuman primate models parallel the human disease, one might expect that antioxidants would also be effective in humans. It is not unreasonable to assume that the steps involved in the pathogenesis of atherosclerosis induced by elevated LDL levels would be pretty much the same in animals and in humans. There is a very high level of interest in testing this explicitly, but so far the proof is not in hand.

Epidemiological studies show a negative correlation between antioxidant vitamin intake or antioxidant vitamin levels in plasma, on the one hand, and CHD events, on the other (43–45). However, epidemiological correlations do not, of course, establish cause-and-effect relationships. They are only suggestive, but they increase our interest in doing the definitive intervention trials necessary to prove the benefit of antioxidants. A preliminary report has appeared from the Harvard Physician's Health Survey, which was begun more than 8 years ago to test whether aspirin would reduce CHD risk and whether β-carotene would reduce risk of bowel cancer (46). The cohort consisted of over 20,000 physicians who took aspirin, β-carotene, both, or neither every other day for an initially planned 7 years. Aspirin quickly proved to be effective, and that part of the study was discontinued early on. At 7 years there were still evidently not enough events to allow a firm decision about the efficacy of β-carotene against cancer. However, because of the increasing interest in the oxidative modification hypothesis, the investigators decided to analyze the data in a subset of the physicians who had had evidence of CHD at the onset of the study. Patients with CHD were supposedly excluded, but somehow 333 men were randomized who had CHD at the time the study began. That is the subset that was analyzed by Gaziano et al. (46) with respect to coronary events. Their presentation at the 1990 meeting of the American Heart Association showed a significant ($p < 0.04$) 45% reduction in pooled coronary events (myocardial infarction, angioplasty, coronary bypass surgery, or sudden death). A full report of these fascinating findings has yet to appear because the decision was made to extend the β-carotene study and the investigators do not want to publish the results in detail until the major part of the study is stopped.

A preliminary verbal report was made in September 1993 of the PQRST Study, a Swedish study to determine whether treatment with probucol would slow the progression of femoral atherosclerosis (47). Men with peripheral vascular disease were first tested to be certain they would respond to treatment with cholestyramine. Those showing a satisfactory drop in cholesterol on cholestyramine (8%) were then tested for their response to the addition of probucol. Those who showed a further 8% decrease in cholesterol level on probucol were considered eligible for randomization. The whole group was randomized to treatment either with cholestyramine alone or cholestyramine plus probucol. Femoral angiograms were performed at repeated intervals over 3 years and quantified by an objective computer technique. The primary endpoint was a change in the volume of a defined segment of the femoral artery. Those treated with cholestyramine showed a small, marginally significant increase in femoral volume over the 3-year period of the study. Those being treated with both drugs showed a result that did not differ significantly from that of those treated with cholestyramine alone. In other words, there was no evidence in this study that probucol, either through its cholesterol-lowering effect or its antioxidant effect (or any other effect), improved prognosis over and above that obtained by treatment with cholestyramine alone.

What it comes down to is that we do not have a single fully reported, large-scale, double-blind study to tell us whether or not antioxidants will be valuable in prevention of *human* atherosclerosis. Several additional studies are either under way or in the planning stages. The two studies referred to above will, hopefully, present their full results for evaluation in the near future. In the meantime, we simply cannot answer the question "Does an increase in intake of antioxidant vitamins or treatment with antioxidant drugs improve prognosis of human coronary artery disease?"

ALTERNATIVE POINTS OF INTERVENTION—THEIR MERITS AND DEMERITS

Examining Figure 1, one sees a number of processes that could conceivably be blocked that would slow the progression of lesion development. We can lump together for discussion all of the steps involved in the initial adhesion, penetration, and phenotypic modulation of the circulating monocyte to become the arterial tissue macrophage. If we could interfere at any point along this chain of events, we would presumably reduce the rate of accumulation of foam cells: no monocyte penetration, no foam cells. Again, if we could interfere with the expression or functioning of the scavenger receptors, we might get penetration and phenotypic modulation, but the macrophages would accumulate lipids at a slower rate, again slowing the development of the fatty streak. This might be beneficial, but, as discussed below, there are reasons to question that.

First, if the interference with monocyte adhesion and penetration were global, that could interfere with the normal protective mechanisms against bacterial invasion. The usual response to introduction of foreign materials or bacteria is the entrance into the affected area of white blood cells, initially neutrophils and then monocytes. Interference with this would in a sense create a patient with the same problems facing patients with chronic granulomatous disease—an inability to control inflammation. If the intervention could be made tissue-specific—interfering only with the penetration of monocytes into arterial wall—then we might have a better chance of obtaining the desired result without unexpected side effects. But even this kind of highly selective intervention may not be desirable. Thus, it has been suggested that one of the reasons for the persistence of the scavenger receptors in evolution is that the removal of oxidized LDL may be protective (48). Oxidized LDL is cytotoxic, and if it were not "cleared" by the macrophages it might induce endothelial damage sufficient to cause thrombosis and death. Oxidation of LDL is a process that may go on in all animal species, including those that do not develop atherosclerosis (because their lipid levels are too low and their lifespan too short). Obviously, there would be natural selection for a process that protects young animals from fatal thromboses. Uptake of oxidized LDL by scavenger receptors could be such a process.

The scavenger receptors may have an even more general purpose. They may be responsible for recognizing damaged cells that need to be removed before they damage adjacent, healthy cells. Cells that have been damaged and are dying increase their production of free radicals, and their plasma membrane lipids undergo oxidative damage. Sambrano et al. (49) have shown that oxidized red blood cell membranes are in fact recognized by macrophage scavenger receptors, mainly the oxidized LDL receptor. Whether the phenomenon is general or not remains to be determined, but there is the real possibility that scavenger receptors are required for many kinds of "mopping-up operations." Therefore any interference with the expression of functioning of those receptors might have unacceptable side effects.

Before leaving this issue, it should be noted that simply because a process is of central importance in homeostasis does not necessarily mean that it is not a legitimate target for pharmacological intervention. A striking example is the enormous success of treatment of hypercholesterolemia by inhibition of HMG-CoA reductase. When that was first proposed, it was met with highly vocal skepticism because it seemed to be tampering with one of the most essential homeostatic mechanisms—the provision of cholesterol and other essential polyisoprenoids. As it turns out, the judicious use of these inhibitors allows us to control hypercholesterolemia without any important side effects.

Studies in experimental animals show that calcium channel blockers can significantly slow the progression of experimental atherosclerosis (50). A number of mechanisms have been proposed, but none is clearly established as yet. Based on the empirical findings, clinical trials have already been carried out. It seems that some calcium channel blockers may inhibit the development of new coronary lesions but have little or no effect on the progression of established lesions (51). These findings certainly suggest that the site of action lies early in the pathogenetic schema, possibly on endothelial cell function or monocyte/macrophage function, but we simply do not know. This continues to be a promising area for further study.

FIBROUS PLAQUES, COMPLEX LESIONS, PLAQUE RUPTURE, AND THROMBOSIS

The transition from a fatty streak to a fibrous plaque is due mainly to cell replication and an increase in synthesis and secretion of connective tissue matrix.

Remarkable progress has been made in recent years in delineating the growth factors and cytokines produced by cells in the atherosclerotic lesion that may contribute to the development of the fibrous plaque (52–54). The first of these to be identified was platelet-derived growth factor (PDGF). We now know that PDGF is by no means limited to platelets but is synthesized and secreted also by endothelial cells, by macrophages, and probably by smooth muscle cells themselves. Studies over the past decade have identified a large and still growing list of other growth factors and cytokines that could, in principle, contribute to the evolution of the stenotic lesion. We know that many of these are expressed in the lesions themselves, but we still cannot say with certainty which are the important players in the process. A good deal of research has gone into evaluating factors that may play a role in restenosis after angioplasty or after deendothelialization in animal models. However, it is not clear that these are the same factors responsible for the development of the fibrous plaque in the ordinary course of evolution of an atherosclerotic lesion. In any case, once we know which factors are key players, it is conceivable that we could intervene with inhibitors—chemical or biological—that would slow the growth of the lesion. Recent advances in structural biology may make it possible to define the "contours" of the receptors involved and synthesize low molecular weight, absorbable molecules that could act competitively, i.e., the "designer drug" approach.

Most myocardial infarctions are the result of thrombosis, which usually occurs as the result of plaque rupture. As first proposed by Constantinides (55) and extended by Davies (56), the fatal thrombosis almost always overlies a plaque that shows a thin fibrous cap that has actually ruptured, allowing blood to make contact with the interior of the atheroma. That is presumably what initiates the thrombosis by exposing blood to tissue factor and to other prothrombogenic materials within the atheroma. It is now appreciated that the fatal thrombosis occurs seldom on very highly stenotic lesions but with a much higher frequency on lesions at 40–60% stenosis. Moreover, these lesions tend to have a large lipid pool in the necrotic center. Finally, and perhaps most important, these lesions still show a significant population of macrophages, mainly at the shoulder, which is the point at which the rupture most commonly occurs. It is quite possible, then, that the final erosion of the fibrous cap could be the result of release from macrophages of proteases, lipases, or other lytic enzymes or the release of reactive oxygen species. These macrophages could also contribute by producing oxidized LDL, which is cytotoxic. Once we understand more about the mechanisms leading to plaque rupture, we may be in a position to "stabilize" such plaques and prevent the fatal thrombosis. If so, inhibitors of those enzymes might be therapeutic and could be given to patients with unstable angina, which often precedes myocardial infarction. Finally, if we could prevent the fatal thrombosis even after plaque rupture, that could forestall the fatal event and allow time for repair. Obviously, research on anticoagulants continues to be a major thrust.

SUMMARY

In this chapter we have tried to give an overview of current research on atherosclerosis to indicate the many new approaches made possible by our increasing understanding of the pathogenesis at the level of the artery wall. Which of these will prove feasible and find their way into the clinic nobody can predict at the moment. The approach closest to being used in practice is the use of antioxidants to protect LDL against oxidative modification. Clinical trials are in progress at several centers, and data should be available within a few years. Inhibitors of smooth muscle cell growth are being intensively studied in restenosis, but the results in clinic trials have been disappointing. Some of the other approaches have not yet even been tested in experimental animals, but they have been discussed for completeness' sake to give an idea of what lies ahead. In the meanwhile, we should intensify our efforts to control the established risk factors, including hypercholesterolemia.

ADDENDUM

After completion of this chapter, results were reported from a large prospective trial in 29,133 Finnish male smokers, 50–69 years of age at time of entry, treated with vitamin E alone (50 mg/day), β-carotene alone (20 mg/day), both, or placebo only (57). Follow-up lasted 5–8 years. The primary endpoint of the trial was lung cancer, and secondary endpoints were other cancers. The data showed no reduction in lung cancer or other cancers. In fact, in the β-carotene group there may have been a small increase in lung cancer. Although not a primary or secondary endpoint, mortality from cardiovascular disease was recorded. Neither vitamin significantly affected mortality due to cardiovascular disease. As the accompanying editorial pointed out (58), "There was an apparently higher risk of hemorrhagic stroke among those treated with vitamin E, although this association may also have been due to chance."

While this study has some limitations, particularly with respect to the cardiovascular disease results, the results are sobering and should cause us to look more carefully at the nature of clinical trials designed to test the antioxidant hypothesis with respect to coronary heart disease. In this study, only 50 mg of vitamin E were given. That was only enough to raise plasma vitamin E levels by about one third and probably insufficient to confer much protection against LDL oxidation. The intakes found to be protective according to the prospective epidemiological studies of Stampfer et al. (44) and Rimm et al. (45) were hundreds of mg per day above the usual dietary intake, i.e., protection was seen only in those taking large supplements. The dose of β-carotene was adequate, however, comparable to that used by Gaziano and coworkers (46) in their prospective intervention trial. The reason for the difference in results remains to be established. Points that will need to be considered include the fact that all of these participants had been very heavy smokers for about 30 years and were 50–69 years old when they entered the study. Presumably they must have had fairly advanced coronary artery disease at the beginning of the intervention.

These recent results underscore our concern that intervention trials be designed to focus on early lesions. The only experimental animal evidence we have is limited to fatty streak lesions. We suggest that intervention trials should always include as one parameter the rate of progression of carotid intimal/medial thickness determined by ultrasound, presumably a marker for early stages of atherosclerosis. If event trials are to be conducted, every effort should be made to start patients at a younger age and to extend the time of the study to the extent possible. Only in that way may it be possible to see the impact of prevention of early lesions on the clinical events triggered by late lesions.

REFERENCES

1. Kimura N. Analysis of 100,000 postmortem examinations in Japan. In: Keys A, White PD, eds. World Trends in Cardiology I. Cardiovascular Epidemiology. New York: Harper and Row, 1965:22–33.
2. Marmot MG, Syme SL, Kagan A, Kato H, Cohen JB, Belsky J. Epidemiologic studies of coronary heart disease and stroke in Japanese men living in Japan, Hawaii, and California: prevalence of coronary and hypertensive heart disease and associated risk factors. Am J Epidemiol 1975; 102(6):514–525.
3. Expert Panel on Detection, Evaluation, and Treatment of High Blood Cholesterol in Adults. Summary of the Second Report of the National Cholesterol Education Program (NCEP) (Adult Treatment Panel II). JAMA 1993; 269(23):3015–3023.
4. Grundy SM. Multifactorial etiology of hypercholesterolemia: implications for prevention of coronary heart disease. Arterio Thromb 1991; 11(6):1619–1635.
5. Seftel HC, Baker SG, Sandler MP, Forman MB, Joffe, LI, Mendelsohn D, Jenkins T, Mieny CJ. A host of hypercholesterolaemic homozygotes in South Africa. Br Med J 1980; 281:633–636.
6. Ross R, Glomset J. The pathogenesis of atherosclerosis. N Engl J Med 1976; 295: 369–377, 420–425.
7. Goode TB, Davies PF, Reidy MA, Bowyer DE. Aortic endothelial cell morphology observed in situ by scanning electron microscopy during atherogenesis in the rabbit. Arteriosclerosis 1977; 27:235–251.
8. Ross R. The pathogenesis of atherosclerosis—an update. N Engl J Med 1986; 314: 488–500.
9. Pathological Determinants of Atherosclerosis in Youth (PDAY) Research Group. Natural history of aortic and coronary arteriosclerotic lesions in youth. Arterio Thromb 13:1291–1298.
10. Poole JCF, Florey HW. Changes in the endothelium of the aorta and the behavior of macrophages in experimental atheroma of rabbits. J Pathol Bacteriol 1958; 75: 245–252.

11. Gerrity RG. The role of the monocyte in atherogenesis, I. Transition of blood-borne monocytes into foam cells in fatty lesions. Am J Pathol 1981; 103:181–190.

12. Cybulsky MI, Gimbrone MA. Endothelial expression of a mononuclear leukocyte adhesion molecule during atherosclerosis. Science 1991; 261:788.

13. Berliner JA, Territo MC, Sevanian A, Ramin S, Kim JA, Bamshad B, Esterson M, Fogelman AM. Minimally modified low density lipoprotein stimulates monocytes endothelial interactions. J Clin Invest 1990; 85:1260–1266.

14. Quinn MT, Parthasarathy S, Fong LG, Steinberg D. Oxidatively modified low density lipoproteins: a potential role in recruitment and retention of monocyte/macrophages during atherogenesis. Proc Natl Acad Sci USA 1987; 84:2995–2998.

15. McMurray HF, Parthasarathy S, Steinberg D. Oxidatively modified low density lipoprotein is a chemoattractant for human T-lymphocytes. J Clin Invest 1993; 92:1004–1008.

16. Navab M, Imes SS, Hama SY, Hough GP, Ross LA, Bork RW, Valente AJ, Berliner JA, Drinkwater DC, Laks H, et al. Monocyte transmigration induced by modification of low density lipoprotein in cocultures of human aortic wall cells is due to induction of monocyte chemotactic protein 1 synthesis and is abolished by high density lipoprotein. J Clin Invest 1991; 88(6):2039–2046.

17. Cushing SD, Berliner JA, Valente AJ, Territo MC, Navab M, Parhami F, Gerrity R, Schwartz CJ, Fogelman AM. Minimally modified low density lipoprotein induces monocyte chemotactic protein 1 in human endothelial cells and smooth muscle cells. Proc Natl Acad Sci USA 1990; 87:5134–5138.

18. Ylä-Herttuala S, Lipton BA, Rosenfeld ME, Säkioja T, Yoshimura T, Leonard EJ, Witztum JL, Steinberg D. Expression of monocyte chemoattractant protein 1 in macrophage-rich areas of human and rabbit atherosclerotic lesions. Proc Natl Acad Sci USA 1991; 88:5252–5256.

19. Fogelman AM, Haberland ME, Seager J, Holcom M, Edwards PA. Factors regulating the activities of the low density lipoprotein receptor and the scavenger receptor in human monocyte/macrophages. J Lipid Res 1981; 22:1131–1141.

20. Rajavashisth TB, Andalibi A, Territo MC, Berliner JA, Navab M, Fogelman AM, Lusis AJ. Induction of endothelial cell expression of granulocyte and macrophage colony-stimulating factors by modified low density lipoproteins. Nature (Lond) 1990; 344:254–257.

21. Rosenfeld ME, Ylä-Herttuala S, Lipton BA, Ord VA, Witztum JL, Steinberg D. Macrophage colony-stimulating factor mRNA and protein in atherosclerotic lesions of rabbits and man. Am J Pathol 1992; 140:291–300.

22. Goldstein JL, Brown MS. The low density lipoprotein pathway and its relation to atherosclerosis. Annu Rev Biochem 1977; 46:897–930.

23. Brown MS, Goldstein JL. Lipoprotein metabolism in the macrophage: implications for cholesterol deposition in atherosclerosis. Annu Rev Biochem 1983; 52:223–261.

24. Weinstein DB, Carew TE, Steinberg D. Uptake and degradation of low density lipoprotein by swine arterial smooth muscle cells with inhibition of cholesterol biosynthesis. Biochim Biophys Acta 1976; 424:404–421.

25. Goldstein JL, Ho YK, Basu SK, Brown MS. Binding site on macrophages that mediates uptake and degradation of acetylated low density lipoprotein, producing massive cholesterol deposition. Proc Natl Acad Sci USA 1979; 76:333–337.

26. Henriksen T, Mahoney EM, Steinberg D. Enhanced macrophage degradation of low density lipoprotein previously incubated with cultured endothelial cells: recognition by receptor for acetylated low density lipoproteins. Proc Natl Acad Sci USA 1981; 78:6499–6503.

27. Steinbrecher UP, Parthasarathy S, Leak DS, Witztum JL, Steinberg D. Modification of low density lipoprotein by endothelial cells involves lipid peroxidation and degradation of low density lipoprotein phospholipids. Proc Natl Acad Sci USA 1984; 83: 3883–3887.

28. Wright SD, Silverstein SC. Phagocytosing macrophages exclude proteins from the zones of contact with opsonized targets. Nature 1984; 309:359–361.

29. Ylä-Herttuala S, Palinski W, Rosenfeld ME, Parthasarathy S, Carew TE, Butler S, Witztum JL, Steinberg D. Evidence for the presence of oxidatively modified low density lipoprotein in atherosclerotic lesions of rabbit and man. J Clin Invest 1989; 84:1086–1095.

30. Palinski W, Rosenfeld ME, Ylä-Herttuala S, Gurtner GC, Socher SA, Butler SW, Parthasarathy S, Carew TE, Steinberg D, Witztum JL. Low density lipoprotein undergoes oxidative modification in vivo. Proc Natl Acad Sci USA 1989; 86:1372–1376.

31. Salonen JT, Ylä-Herttuala S, Yamamoto R, Butler S, Korpela H, Salonen R, Nyyssonen K, Palinski W, Witztum JL. Autoantibody against oxidized LDL and progression of carotid atherosclerosis. Lancet 1992; 339(8798):883–887.

32. Carew TE, Schwenke DC, Steinberg D. Antiatherogenic effect of probucol unrelated to its hypocholesterolemic effect: evidence that antioxidants in vivo can selectively inhibit low density lipoprotein degradation in macrophage-rich fatty streaks and slow the progression of atherosclerosis in the Watanabe heritable hyperlipidemic rabbit. Proc Natl Acad Sci USA 1987; 84:7725–7729.

33. Kita T, Nagano Y, Yokode M, et al. Probucol prevents the progression of atherosclerosis in Watanabe heritable hyperlipidemic rabbit, an animal model for familial hypercholesterolemia. Proc Natl Acad Sci USA 1987; 84:5928–5931.

34. Daugherty A, Zweifel BS, Schonfeld G. Probucol attenuates the development of aortic atherosclerosis in cholesterol-fed rabbits. Br J Pharmacol 1989; 98:612–618.

35. Bjorkhem I, Henriksson-Freyschuss AH, Breuer O, Diczfalusy U, Berglund L, Henriksson P. The antioxidant butylated hydroxytoluene protects against atherosclerosis. Arterior Thromb 1991; 11:15–22.

36. Sparrow CP, Doebber TW, Olaszewski J, Wu MS, Ventre J, Stevens KA, Chao YS. Low density lipoprotein is protected from oxidation and the progression of atherosclerosis is slowed in cholesterol-fed rabbits by the antioxidant N,N′-diphenylphenylene-diamine. J Clin Invest 1992; 86:1885–1891.

37. Verlangieri AJ, Bush MJ. Effects of d-α-tocopherol supplementation on experimentally induced primate atherosclerosis. J Am Coll Nutr 1992; 11:131–138.

38. Sasahara M, Raines EW, Carew TE, Steinberg D, Wahl PW, Chait A, Ross R. Inhibition of hypercholesterolemia-induced atherosclerosis in *Macaca nemestrina* by probucol: I. Intimal lesion are correlates inversely with resistance of lipoproteins to oxidation. J Clin Invest 1994; 94:155–164.

39. Stein Y, Stein O, Delplanque B, Fesmire JD, Lee DM, Alaupovic P. Lack of effect of

probucol on atheroma formation in cholesterol-fed rabbits kept at comparable plasma cholesterol levels. Arteriosclerosis 1989; 75:145–155.

40. Daugherty A, Zweifel BS, Schonfeld G. The effects of probucol on the progression of atherosclerosis in mature Watanabe heritable hyperlipidemic rabbits. Br J Pharmacol 1991; 103(1):1013–1018.

41. Parthasarathy S. Evidence for an additional intracellular site of action of probucol in the prevention of oxidative modification of low density lipoprotein. Use of a new water-soluble probucol derivative. J Clin Invest 1992; 89:1618–1621.

42. Morel DW, DiCorleto PE, Chisolm GM. Endothelial and smooth muscle cells alter low density lipoprotein in vitro by free radical oxidation. Arteriosclerosis 1994; 4:357–364.

43. Gey KF. Plasma vitamins E and A inversely correlated to mortality from ischemic heart disease in cross-cultural epidemiology. Ann NY Acad Sci 1989; 570:268–282.

44. Stampfer MJ, Hennekens CH, Manson JE, Colditz GA, Rosner B, Willett WC. Vitamin E consumption and the risk of coronary heart disease in women. N Engl J Med 1993; 328:1444–1449.

45. Rimm EB, Stampfer MJ, Ascherio A, Giovannucci E, Colditz GA, Willett WC. Vitamin E consumption and the risk of coronary heart disease in men. N Engl J Med 1993; 328:1450–1456.

46. Gaziano MJ, Manson JE, Ridker PM, Buring JE, Hennekens CH. Beta carotene therapy for chronic stable angina. Circulation 1990; 82:III–201.

47. Walldius G, Carlson LA, Erikson U, Olsson AG, Johansson J, Molgaard J, Nilsson S, Stenport G, Kaijser L, Lassvik C. Development of femoral atherosclerosis in hypercholesterolemic patients during treatment with cholestyramine an probucol/ placebo: Probucol Quantitative Regression Swedish Trial (PQRST): a status report. Am J Cardiol 1988; 62(3):37B–43B.

48. Steinberg D. Arterial metabolism of lipoproteins in relation to atherogenesis. Ann NY Acad Sci 1990; 598:125–135.

49. Sambrano GR, Parthasarathy S, Steinberg D. Recognition of oxidatively damaged erythrocytes by a macrophage receptor with specificity for oxidized low density lipoprotein. Proc Natl Acad Sci USA 1994; 91:3265–3269.

50. Henry PD, Bentley KI. Suppression of atherogenesis in cholesterol-fed rabbits treated with nifedipine. J Clin Invest 1981; 68:1366–1369.

51. Lichtlen PR, Hugenholtz PG, Rafflenbeul W, Hecker H, Jost S, Deckers JW. Retardation of angiographic progression of coronary artery disease by nifedipine. Results of the International Nifedipine Trial on Antiatherosclerotic Therapy (INTACT). INTACT Group Investigators. Lancet 1990; 335(8698):1109–1113.

52. Li H, Cybulsky MI, Gibrone MA, Libby P. An atherogenic diet rapidly induces VCAM-1, a cytokine-regulatable mononuclear leukocyte adhesion molecule, in rabbit aortic endothelium. Arterio Thromb 1993; 13(2):197–204.

53. Clinton SK, Libby P. Cytokines and growth factors in atherogenesis. Arch Path Lab Med 1992; 116(12):1292–1300.

54. Ross R. The pathogenesis of atherosclerosis: a perspective for the 1990's. Nature 1993; 362(6423):801–809.

55. Constantinides P. Plaque fissures in human coronary thrombosis. J Athero Res 1966; 6:1.

56. Davies MJ. A macro and micro view of coronary vascular insult in ischemic heart disease. Circulation 1990; 82(3 suppl):I138–146.
57. The Alpha-Tocopherol, Beta Carotene Cancer Prevention Study Group. The effect of vitamin E and beta carotene on the incidence of lung cancer and other cancers in male smokers. N Engl J Med 1994; 330:1029–1035.
58. Hennekens CH, Buring JE, Peto R. Antioxidant vitamins—benefits not yet proved. N Engl J Med 1994; 330:1080–1081.

16

The Scandinavian Simvastatin Survival Study (4S)
Editor's Summary

Basil M. Rifkind

National Heart, Lung, and Blood Institute, National Institutes of Health, Bethesda, Maryland

As this book was going to press, the results of the Scandinavian Simvastatin Survival Study (4S) were reported (1). Given their importance in being derived from the first, large-scale clinical trial of a drug from the HMG-CoA reductase inhibitor class (sometimes called statins) and their contribution to resolving the controversial issue of total mortality (see Chapter 2), a brief summary and commentary are provided here by the editor.

The 4S was a multicentered, randomized, double-blind, placebo-controlled clinical trial of the effect of simvastatin therapy in patients with a history of coronary heart disease (CHD), either angina pectoris or myocardial infarction, followed for a median duration of 5.4 years.

The prespecified primary endpoint for the 4S was total mortality because of the uncertainty as to whether cholesterol lowering adversely affects the noncoronary mortality rate, thereby negating some or all of the reduction in coronary mortality that it has been shown to produce. Total mortality has been difficult to evaluate in individual trials because of the large sample sizes that are required to develop a trial design with sufficient statistical power. The 4S was designed with requisite power, because of the greater degree of cholesterol lowering that was anticipated through use of a statin drug, leading to a greater expected decrease in CHD mortality rates. Also, a high mortality rate could be expected in patients with a history of CHD. The study recruited 4444 patients (81% male) aged 35–70 at entry, in excess of the calculated sample size, with a total of 440 deaths being specified as the target, a goal that was virtually achieved.

A cholesterol-lowering diet was prescribed for all patients prior to randomization. The lipid entry criteria for the study were serum cholesterol levels 213–310 mg/dl (5.5–8.0 mmol/liter) on the diet. Serum triglyceride levels were 222 mg/dl or less (2.5 mmol/liter).

The baseline lipid values for the placebo- and simvastatin-treated groups are show in Table 1 and are similar, as are the levels of other important CHD risk factors such as blood pressure, smoking, and body weight. In particular, the serum cholesterol was 262 mg/dl (5.70 mmol/liter) in the placebo group, and 262 mg/dl (5.74 mmol/liter) in the active drug group. Thus the randomization process appears to have been successful in producing two comparable groups.

Simvastatin was prescribed at an initial dose of 20 mg/day, which was increased to 40 mg/day if the LDL-C level did not drop below 202 mg/dl (5.2 mmol/ liter). Thirty-seven percent of the patients had their dose raised. Substantial changes in lipid levels occurred as a result of treatment and were maintained throughout the duration of the trial, with mean total and LDL-C cholesterol levels falling by 25% and 35%, respectively. Triglyceride levels fell by 10% and HDL-C levels rose by 8%, typical responses to statin therapy. These findings reflect the high degree of compliance recorded by the study.

Some of the mortality findings are given in Table 2. A total of 438 patients died—256 in the placebo group and 182 in the simvastatin group, a highly significant 30% reduction in total mortality ($p < 0.0003$). Most of this reduction was due to a 42% reduction in coronary deaths (placebo 189, simvastatin 111). In contrast, there were no significant changes in the number of deaths from noncardiovascular causes including the categories of violent deaths (7 vs 6) and cancers (35 vs. 33).

The large reduction in events was not confined to deaths, there being corresponding reductions in several nonfatal coronary endpoints. In all, 620 placebo-treated patients compared with 431 simvastatin-treated patients had one or more major coronary events, a 34% decrease ($p < 0.00001$). Of special note is the reduction in the number of revascularization procedures in the simvastatin group; the patients' risk of undergoing coronary bypass surgery or angioplasty was decreased by 37% (placebo 383 vs. simvastatin 252; $p < 0.00001$). Interestingly,

Table 1 Lipid Levels at Entry

	Placebo	Simvastatin
Cholesterol (mg/dl)		
Total	262	262
HDL	46	46
LDL	189	189
Triglyceride (mg/dl)	134	132

Table 2 Selected Endpoints

Deaths	Placebo (n = 2223)	Simvastatin (n = 2221)	Relative risk (95% C.I.)
All causes	256	182	0.70 (0.58–0.85)
Coronary	189	111	0.58 (0.46–0.73)
All cardiovascular	207	138	0.65 (0.52–0.80)
All noncardiovascular	49	46	
Cancer	35	33	
Suicide	4	5	
Trauma	3	1	
Other	7	7	

there were also fewer fatal or nonfatal cerebrovascular events in the active treatment group, including fewer nonembolic strokes and transient ischemic attacks.

Additional analyses were performed on prespecified subgroups of particular interest, namely older patients and women. There were 420 women in the placebo group and 407 in the simvastatin-treated group. Only 25 of placebo-treated women and 27 of simvastatin-treated women died, and of these 17 and 13, respectively, were due to CHD. However, 91 versus 59 major coronary events were found in the placebo and simvastatin groups, respectively, a statistically significant 35% reduction ($p < 0.010$) with a relative risk similar to that observed in men.

The influence of age was explored by comparing the event rates in patients under age 60 with those above this age. The 4S report indicates that, although the observed relative risk reductions produced by simvastatin were somewhat less in the older patients, they were statistically significant ($p < 0.01$ in both age groups for mortality and $p < 0.0001$ for major coronary events), and the absolute differences between treatment groups were similar.

As indicated, specific categories of deaths, such as all fatal (and nonfatal) or all gastrointestinal cancers, or suicides and accidents and violence, were unaltered. Myopathy, an occasional known adverse effect of statin drugs, occurred in only one woman and led to rhabdomyolysis; she recovered after cessation of treatment. An addditional six simvastatin patients and one on placebo had raised levels of creatine kinase but remained asymptomatic. The frequency of lens changes was not reported.

COMMENT

The 4S findings confirm that drugs of the HMG-CoA reductase inhibitor class are highly effective cholesterol-lowering agents, with the added possible advantages of reducing triglyceride levels while moderately elevating HDL-C levels. These attributes, and the potential for a high level of compliance, as was observed over the duration of the study, make simvastatin and related statin drugs extremely attractive for cholesterol lowering.

In contrast to the use of resins or nicotinic acid, alternative cholesterol-lowering agents, the patients treated with simvastatin were free of troublesome side effects that can interfere with therapy. Previous studies of statin drugs have also found them to have low toxicity. Taken in the aggregate, the safety data, including the impressive reduction in total mortality and no increase in noncoronary mortality, suggest that there should be little hesitation in prescribing simvastatin or related drugs when they are indicated, provided patients are alerted to the possibility of myopathy and are monitored, especially initially, for hepatic toxicity. Previous suggestions of possible lens toxicity have not been confirmed in other studies. The 4S findings should go a long way to allay concerns about the possible toxicity of cholesterol lowering, either as a general consequence of such therapy irrespective of the agent, or as a specific consequence of therapy.

The results of the 4S are a major contribution to the field of cholesterol lowering, especially by showing it to lower total mortality. In addition, the 4S results dramatically confirm and extend the findings from meta-analyses of secondary prevention trials that cholesterol lowering reduces the rates of fatal and nonfatal coronary events (2–4). They are complemented by the results of several recently reported smaller angiographic studies of statin drugs, which have shown a beneficial effect on coronary lesions and in which large reductions in event rates have often been seen. The 4S describes a much greater reduction in coronary disease rates than has generally been obtained in previous large studies, presumably attributable to the much greater cholesterol reductions obtained in the prior trials, and possibly to the other lipid changes. The findings are consistent with the rule of thumb that each 1% reduction in cholesterol levels produces a 2% fall in the rate of fatal and nonfatal coronary events (5). Reductions of this magnitude are of considerable clinical and public health import. They should not be difficult to obtain in the practice setting, given the ease of administration of the statin drugs and their freedom from troublesome side effects.

The case for secondary prevention of CHD has recently been presented (4). The Adult Treatment Panel of the National Cholesterol Program has emphasized its importance and has issued guidelines for its implementation (6). The 4S results strongly reinforce the potential value of this approach and are a powerful stimulus to cardiologists, internists, general practitioners, and others involved in the care of patients with manifestations of CHD to routinely evaluate their lipid status, and to treat the majority with diet and, often, drugs to achieve levels of LDL-C below 100 mg/dl. As pointed out by the 4S investigators, of 100 patients similar to those in the study and treated for 6 years with simvastatin, 4 of 9 would be saved from a coronary death, 7 of 21 would avoid a nonfatal heart attack, and 6 of 19 would not need coronary bypass surgery or angioplasty.

The 4S results still leave a number of questions in the cholesterol-CHD field partially unresolved. For example, it remains to be seen, given the still considerable number of coronary events that were not prevented, whether still greater degrees of cholesterol lowering will further reduce the CHD rate, as is suggested by observational data on cohorts that show a gradient of risk for CHD, even at quite low cholesterol levels (7). Subsequent analysis of 4S data relating lipid changes to event rates will presumably speak to this issue. Greater cholesterol reductions could be achieved by higher doses of statin drugs and by combination drug therapy (e.g., statin + resin).

Cholesterol levels as low as 213 mg/dl were eligible for the 4S and so would be considered for treatment. The gradient of risk that is still evident at even lower levels also suggests that therapy would be useful in CHD (and healthy) subjects with low LDL-C levels if their aggregate risk status (presence of CHD or multiple risk factors) was high enough to justify such treatment.

While the 4S does not speak directly to the use of statin drugs for primary prevention of CHD, its findings are relevant. Cholesterol lowering reduces the rate of CHD in both primary and secondary intervention patients, as seen in meta-analyses or in individual trials. The 4S findings are in agreement and show the more marked reduction in cholesterol to result in a greater fall in CHD rates; if anything, similar reductions can be expected in primary prevention populations. Noncoronary mortality was unaltered and no trends occurred in cancer or violent death rates. There is little or no biological reason to expect a different adverse event experience in primary prevention patients—rather the opposite. It can be argued that the greater numbers of noncoronary deaths that would be expected in the setting of a primary prevention study would allow a more powerful test of safety than was possible in 4S. The major issue for primary prevention is whether a commensurate reduction in relative risk of CHD and its consequences would translate into sufficient benefit to justify the treatment of this lower risk group and the resulting substantial costs. However, even if ongoing primary prevention trials produce similar reductions in CHD without toxicity, contrary to the situation in secondary prevention, they will not necessarily lead to such widespread application. Cost-benefit considerations and the possibility that very long-term use of statin drugs might still be toxic will still confine their use to those at highest risk along the lines defined by the National Cholesterol Education Program (6).

REFERENCES

1. Scandinavian Simvastatin Survival Study Group. Randomized trial of cholesterol lowering in 4444 patients with coronary heart disease; the Scandinavian Simvastatin Survival Study (4S). Lancet 1994; 344:1383–1389.
2. Holme I. Relation of coronary heart disease incidence and total mortality to plasma cholesterol reduction in randomized trials: use of meta-analysis. Br Heart J 1993; 69(S):S42–S47.

3. Law MR, Wald NJ, Thompson SG. By how much and how quickly does reduction in serum cholesterol concentration lower risk of ischemic heart disease? Br Med J 1994; 308:367–373.
4. Rossouw JE, Lewis B, Rifkind BM. The value of lowering cholesterol after myocardial infarction. N Engl J Med 1990; 323:1112–1119.
5. Lipid Research Clinics Program. The Lipid Research Clinics Coronary Primary Prevention Trial Results. I. Reduction in incidence of coronary heart disease. JAMA 1984; 251:351–364.
6. Report of the National Cholesterol Education Program Expert Panel on Blood Cholesterol Levels in Children and Adolescents. NIH Publication 91-2732, September 1991.
7. Neaton JD, Blackburn H, Jacobs D, Kuller L, Lee D-J, Sherwin R, Shih J, Stamler J, Wentworth D. Serum cholesterol level and mortality findings for men screened in the Multiple Risk Factor Intervention Trial. Arch Intern Med 1992; 152:1490–1500.

Index

Adaptive intimal hyperplasia, 73
Adolescence
    cholesterol levels and, 120, 133–144
    lipoprotein metabolism and, 101
Adult Treatment Panel (NCEP)
    cardiovascular risk factors and, 5–6
    controversies of, 9–23
    follow-up and, 3–4
    guidelines of, 2–23, 104, 272, 311
        original vs. new, 9
    rationale for recommendations of, 9–23
    screening of high-risk adults and, 2
    therapeutic decisions and, 3–4
Advertising
    cholesterol and, 158
Age issues, 10–11. *See also* Elderly
    dyslipoproteinemia and, 99, 100,
        101–104
    ischemic heart disease and, 100
    lipoprotein metabolism and, 101–104
Alcohol, 193
Alzheimer's disease, 113
American Heart Association, 183, 188, 232
Androgen, 119

Android obesity, 15, 122
Angina, 85
Angiographic trials, 59–65
    inclusion criteria of, 51
    meta-analysis of, 50, 63–64
    narrative analysis of, 50, 53, 63–64
Antioxidants, 344–348
Apo B risk factor, 18, 121, 305–306
Apolipoproteins, 305–306
    apo B risk factor, 18, 121, 305–306
    Lp(a) risk factor, 18–19, 81, 121–
        122, 305–306
Arterial lesions in the young
    serum cholesterol and, 134–135
Arterial wall
    lipid accumulation and, 72, 339
    monocyte/macrophage recruitment
        into, 72–73, 341–344
Aspirin and coronary heart disease, 347
Atherosclerosis, 19–20
    adaptive intimal hyperplasia and, 73
    children and, 134, 135
    evolution of atherosclerotic plaque
        and, 69–87

[Atherosclerosis]
fatty streak and, 339–341
incrustation hypothesis, 70
initiation of, 70–73
lipid accumulation and, 13, 72
lipid hypothesis, 70
lipoproteins and, 121–122
localization of, 70–73
oxidative modification hypothesis for, 344–348
pathogenesis of, 339
pathology of in youth, 134
prevention of, 337–352
progression of, 73–85
regression of, 86–87
in various countries, 149
vascular response-to-injury hypothesis, 70–72
women and, 121–122

β-carotene, 347
Beta-quantification technique, 301
Bile-acid resins, 274–276
Bile-acid sequestrants, 143
combined with gemfibrozil, 284–285
combined with HMG-CoA reductase inhibitors, 282–284
combined with niacin, 282
Body fat distribution
gender and, 122–123
Body mass index, 229, 230

Cafeterias
cholesterol and, 158–159, 177–178
Calcium channel blockers, 349
Calorie balance, 192
Calorie-modified food consumption, 219
Calorie needs, 223–225
Canadian Consensus on Cholesterol Lowering, 313–314
Canadian Task Force on Prevention, 100
Carbohydrates, 192
Cardiovascular risk factors, 5–6, 18–19, 137–138, 142, 271, 324–325. *See also* Risk assessment

[Cardiovascular risk factors]
exogenous gonadal hormones and, 123–125
lipoproteins and, 121–122
negative risk factors, 5
positive risk factors, 6
CASS study, 74
CDC Cholesterol Network program, 295
CDP trial, 38, 39, 52
CHD. *See* Coronary heart disease
Chef training, 175
Chemical precipitation, 299
Childhood Panel (NCEP), 2, 138
Children
cholesterol levels and, 133–144
diet therapy and, 140, 142, 256–258
drug therapy and, 142–143
Cholesterol
content of various foods, 233
dietary recommendations, 155, 191
food labeling and, 244–245, 246, 247
reduction of in diet, 154–159
Cholesterol definitive method, 295
Cholesterol diagnostic products, 296
Cholesterol levels. *See also* HDL cholesterol levels; LDL cholesterol levels; Serum cholesterol levels; VLDL cholesterol levels
adolescence and, 133–144
atherosclerosis regression and, 87
children and, 133–144
cutpoints for, 11, 126, 142, 292–293
dangers of low blood cholesterol, 22–23
high-risk approach to lowering, 1–23
international studies and, 135–136
measuring, 291–306
men vs. women, 101–103
mortality and, 33–46
pregnancy and, 120
stratification of individuals by, 3
tracking of, 136–137
in women, 119–128
Cholesterol-Lowering Atherosclerosis Study. *See* CLAS trial

Cholesterol-lowering trials, 34–46, 38–
  39. *See also* Primary prevention
  (population strategy); Secondary
  prevention (high-risk strategy);
  specific trials
  angiographic trials, 49–65
  control groups and, 34–35
  criteria for relevancy of, 34–36
  duration of trials, 36
  hormone-based interventions, 36, 43
  "intent-to-treat" and, 34–35
  interventions and, 43
  meta-analysis of, 36–46
  primary prevention, 40
  randomization and, 34–35, 38–39
  secondary prevention, 40
  successes of, 42
  as unconfounded test, 35–36
Cholesterol Reduction in Seniors
  Program (CRISP) trial, 111
Cholesterol reference method, 294,
  295, 296
Cholesterol screening, 2, 139, 157–158
  costs of, 312–314
Cholesteryl ester hydrolase enzyme, 294
Cholestyramine, 8, 275, 347
  costs of, 313, 320, 321, 324, 325
CLAS trial, 53, 60, 272
Clofibrate, 8, 53–54, 55, 280–281
Colestipol, 8, 275
  costs of, 313, 320, 325
Community approach, 167–180
  intervention and, 174–180
  rationale of, 168–170
  versus primary prevention (population
  strategy), 154
Compact analyzers, 304–305
Complex lesions, 349–350
Congestive heart failure, 113
Corn oil, 52, 56
Coronary Drug Project, 54, 316
Coronary heart disease. *See also*
  Ischemic heart disease
  ability to change rate of, 152–153
  cholesterol levels and, 33–46, 106,
  109, 314–317

[Coronary heart disease]
  clinical outcomes of prevention trials,
  59
  costs of, 311–330
  death of men and women of, 124
  diet and, 65. *See also* Diet therapy
  drugs and, 65. *See also* Drug therapy
  dyslipoproteinemia and, 99–113
  HDL levels and, 13–14, 106
  high-risk approach to, 1–23
  lipids and, 13, 104–106. *See also*
  Lipids
  lipoproteins and, 106–109. *See also*
  Lipoproteins
  multiple pregnancy and, 120–121
  primary prevention of. *See* Primary
  prevention (population approach)
  secondary prevention of, 49–65. *See
  also* Secondary prevention (high-
  risk strategy)
  in various countries, 149
  weight loss and, 109
Coronary Heart Disease Policy Model,
  322, 328
Coronary thrombi, 85. *See also*
  Ischemic heart disease
Cost-effectiveness, 311–330
  diet therapy and, 198, 251
  drug costs and, 325–327
  drug therapy and, 313
  incremental cost-effectiveness
  analysis, 318
  limitations of studies on, 328–329
  methodological considerations of,
  317–319
  primary prevention (population
  strategy) and, 324–325, 327–328
  secondary prevention (high-risk
  strategy) and, 1–23, 322–324
  serum cholesterol levels and, 319–
  322
  treatment and screening of
  hypercholesterolemia and, 312–314

D-thyroxine, 39
DART trial, 52, 57

Diabetes, 5, 17–18, 253–254
   CHD mortality and, 124
   women and, 123
Diet and Reinfarction Trial. *See* DART
   trial
Diet Heart Feasibility Study, 196
Diet therapy, 4, 6–7, 20–21, 39, 43,
   56, 57, 172–173, 183–199, 338.
   See also Eating patterns
   children and, 140, 142
   coexisting disorders and, 251–254
   cost-effectiveness and, 198, 251
   eating patterns and, 155–156
   effectiveness of, 194–197
   elderly and, 110
   guidelines for food selection, 238–
      240
   HDL cholesterol levels and, 197–198
   high blood cholesterol and, 212
   individual responsiveness to, 195–197
   LDL cholesterol levels and, 197–198
   number of adults who require, 210
   population responsiveness to, 194–
      195
   primary prevention (population
      strategy) and, 154–159, 184–187
   professional groups, 265
   resources for patients, 261–265
   resources for professionals, 258–261
   scientific basis for, 183–188
   secondary prevention (high-risk
      strategy) and, 187–188
   side effects of, 197–198
   triglyceride levels and, 197–198
   VLDL cholesterol levels and, 198
   women and, 122
Dietary assessment, 228–230
Dietitians, diet therapy and, 249–251
Discounting principle, 319
Drug costs, 325–327
Drug therapy, 4, 7–9, 21–22, 54, 62–
   63, 271–286
   categorizing the patient, 273–274
   children and, 142–143
   combination therapy, 281–286
   cost-effectiveness of, 313, 328–329

[Drug therapy]
   elderly and, 111
   lipid-lowering drugs, 8
   monotherapy for
      hypercholesterolemia, 274–279
   monotherapy for
      hypertriglyceridemia, 279–281
   women and, 127
Dyslipidemia, 5
Dyslipoproteinemia
   elderly and, 99–113
   treatment issues and, 109–113

Eating patterns, 155–156, 170, 171. *See
   also* Diet therapy
   changes in, 161–163
   guidelines for food selection, 238–
      240
   implementing change, 209–265
   strategies to change, 157–160
   theoretical model of, 172–174
Edinburgh trial, 52, 53–54, 125
EDTA plasma, 298
Education. *See* Nutrition education
Elderly, 11–13
   cost issues, 112
   diet and, 110
   drug therapy and, 111
   dyslipoproteinemia and, 99–113
   lipid levels and, 104–106
   lipoprotein metabolism and, 104
Electrophoretic methods, 303
Endothelial cells, 76–78
Endothelial injury, 70–71
Energy needs assessment, 223–228
Environmental issues, 133, 183, 337
Enzymic reagents, 294, 303
Estrogen, 39, 119, 120
Estrogen-replacement therapy (ERT),
   127. *See also* Hormone-
   replacement therapy
Ethnic issues
   diet therapy and, 21
   lipids and, 104–105
Exchange lists, 232
Exercise, 21, 123, 227

Exogenous gonadal hormones
    as cardiovascular risk factor, 123–125
Expert Panel on Detection, Evaluation
    and Treatment of High Blood
    Cholesterol in Adults, 110
Expert Panel on Detection, Evaluation
    and Treatment of High Blood
    Cholesterol in Children, 256

Familial Atherosclerosis Treatment
    Study. *See* FATS trial
Familial hyperlipidemia, 134
Family history. *See* Genetic issues
Fat-reduction techniques, 232–237
Fats
    counting grams of, 231–232
    dietary recommendations, 155, 188–
        189, 235
    food labeling and, 242, 246, 247
    food sources of, 217–223
    intake in United States, 162
    reduction of in diet, 154–159
FATS trial, 53, 60–61
Fatty streak, 339–341, 344
Femoral artery, 348
Fiber, 192
    contents of various foods, 216
Fibrates, 38, 43, 280–281
Fibrin(ogen), 85
Fibrin(ogen) degradation products, 85
Fibrous plaques, 349–350
Finland, cholesterol levels and, 135,
    136, 137
Finnish Mental Hospital studies, 125,
    185
Fluvastatin, 8
Foam cells, 340–341
    lipids in, 343–344
Food. *See* Eating patterns
Food costs, 159, 160, 161
Food groups, serving per day, 234
Food industry, cholesterol and, 158
Food institutions, cholesterol and, 158–
    159, 175–179
Food labeling, 158, 159–160, 240–247
Frame size, 226

Framingham Heart Study, 271, 314,
    315, 319
Framingham Offspring Study, 313
Friedewald estimation, 302

Gemfibrozil, 8, 38, 55, 280–281
    combined with bile-acid sequestrants,
        284–285
    combined with HMG-CoA reductase
        inhibitors, 285–286
    costs of, 313, 320
Gender issues, 5, 11–13. *See also*
    Women
    cholesterol levels and, 119–128
    ischemic heart disease and, 100
    mean lipoproteins in males and
        females, 119
Genetic issues, 14
    children and, 134, 135, 139
    hyperlipidemia and, 14
    lipoprotein metabolism and, 14
    risk assessment and, 140
Genzyme Corporation, 302
Ghana, cholesterol levels and, 135, 136
Glucose intolerance, 5
Gynoid obesity, 15, 122

HDL cholesterol levels, 3–4, 105
    age trends and, 120
    children and, 133–144
    cutpoints for, 292–293
    diabetes and, 17
    diet therapy and, 197–198
    drug therapy and, 7, 22
    gender issues and, 119–128
    in initial evaluations, 13
    measuring, 299–301
    men vs. women, 101–102
    as negative risk factor, 13–14
    in therapeutic decisions, 14
HDL reference method, 300–301
Health professionals, diet therapy and,
    249–251, 257
Healthy People 2000, 154, 217
Helsinki trial, 38, 52, 55–56, 279, 281,
    315

Hexane extraction, 294
High-risk approach. *See* Secondary
    prevention (high-risk strategy)
HMG-COA reductase inhibitors, 278–
    279, 349
  combined bile-acid sequestrants,
    282–284
HMG-CoA reductase inhibitors
  combined with gemfibrozil, 285–286
  combined with niacin, 284
Hormone-based interventions, 36, 38,
    39, 43
Hormone-replacement therapy, 123, 125,
    127
Hungary
  cholesterol levels and, 137
  food costs in, 161
Hypercholesterolemia, 125. *See also*
    Cholesterol levels
  diets for, 209
  drug therapy for, 274–279
  in various countries, 149
Hyperlipidemia. *See also* Cholesterol
    levels
  high-risk approach to, 1–23
Hyperplasia, intimal adaptive, 73
Hypertension, 5, 17, 252–253, 317
Hypertriglyceridemia, 5, 16–17, 252
  drug therapy for, 279–281

Incrustation hypothesis, 70
Individual approach. *See* Secondary
    prevention (high-risk strategy)
Infancy, lipoprotein metabolism and,
    101
Injury
  atherosclerosis and, 70–72
  types of, 70
Insulin resistance
  obesity and, 5, 15–16
Intervention levels, 171–172
Intimal hyperplasia, 73
Ischemic heart disease. *See also*
    Coronary heart disease
  age and, 100
  clofibrate and, 53–54

[Ischemic heart disease]
  corn oil and, 56
  gender and, 100
  plaque progression and, 85–86
Israel, cholesterol levels and, 135, 137
Italy, cholesterol levels and, 135, 136,
    137

Japan, coronary heart disease and, 149,
    183, 337, 338

LDL cholesterol levels, 10, 105
  age trends and, 120
  apo B and, 18
  children and, 133–144
  cutpoints for, 11, 142, 292
  diabetes and, 17
  diet therapy and, 197–198
  drug therapy and, 271–286
  gender issues and, 119–128
  LDL particle size and density, 15
  Lp(a) and, 18–19
  measuring, 301–303
  men vs. women, 101–102
  postmenopausal rise of, 119
  stratification of individuals by, 3
LDL oxidation, 87
Lieden trial, 187
Lifestyle Heart Trial, 187–188
Lipid hypothesis, 70
Lipid-lowering drugs, 8
Lipid Research Clinics Coronary
    Primary Prevention Trial. *See*
    LRC-CPPT trial
Lipids, 13, 23, 104–106. *See also*
    Hyperlipidemia
  body mass and, 105
  elderly and, 104–106
  health status and, 105
  measuring, 291–306
  race and, 104
Lipoprotein disorders
  treatment of, 125–127
Lipoprotein metabolism, 101–104
Lipoproteins, 3–4, 23, 106–109. *See*
    also LDL cholesterol level

[Lipoproteins]
  atherosclerosis and, 121–122
  and female life cycle, 120–121
  gender issues and, 119–123
  measuring, 291–306
Los Angeles Veterans Domiciliary trial,
  39
Lovastatin, 8, 62–63, 111, 278–279
  costs of, 313, 320, 321, 322, 323,
  324, 325, 326, 327
Lp(a) risk factor, 18–19, 81, 121–122,
  305–306
LRC-CPPT trial, 38, 271, 314, 315,
  325

Macrophages, 73, 78–79, 83
MARS trial, 53, 62–63
Mass media, cholesterol awareness and,
  157, 160
MCA low fat trial, 39
Measurement issues, 291–306
  beta-quantification technique, 301
  chemical precipitation, 299
  cutpoints, 11, 126, 142, 292–293
  enzymic reagents, 294, 303
  Friedewald estimation, 302
  precision, 297
  quality assurance, 297
  sources of preanalytic variation, 298
Meats, fat in, 218–219
MEDFICTS, 229
Medical Research Council. *See* MRC
  trial
Menopause
  hormonal replacement therapy and,
  125
  lipoprotein metabolism and, 101
  postmenopausal rise of LDL, 119
Menstrual cycle, cholesterol levels and,
  120
Meta-analysis
  angiographic trials and, 50, 63–64
  of cholesterol-lowering trials, 36–46,
  315
  of mortality, 36–46
  secondary prevention and, 50, 58–59

MI. *See* Myocardial infarction
Migrating population studies, 133, 183,
  337
Minnesota Coronary Survey, 185–186
Minnesota Heart Health Program, 157,
  174–180
Monitored Atherosclerosis Regression
  Study. *See* MARS trial
Monocyte/macrophage recruitment into
  arterial wall, 72–73, 341–344
Monosaturated fat, 189–190
Mortality
  cholesterol levels and, 33–46, 107,
  315, 323
  clinical outcomes of prevention trials,
  59, 315, 316
  from coronary heart disease, 1, 109,
  323
  meta-analysis of, 36–46
  non-CHD, 329
MRC trial, 39, 52, 56
MRFIT. *See* Multiple Risk Factor
  Intervention Trial
Multiple Risk Factor Intervention Trial,
  236
Myocardial infarction, cholesterol levels
  and, 33–46

Narrative analysis
  angiographic trials and, 50, 53, 63–
  64
  secondary prevention and, 50, 52
National Academy of Sciences, Nutrition
  Committee of, 171
National Cholesterol Education
  Program, 125–126, 161, 171, 183,
  272, 296, 298. *See also* Adult
  Treatment Panel; Childhood Panel
  analytic performance goals, 297, 300
National health policy vs. national
  agricultural policy, 160
National Heart Lung and Blood
  Institute, 1
National Institutes of Health, 1
NCEP. *See* National Cholesterol
  Education Program

Netherlands
  cholesterol levels and, 135, 136, 137
Newcastle trial, 38, 52, 53–54, 125
NHLBI Type II trial, 38, 53, 59–60, 272
Niacin, 38, 43, 55, 276–278, 280
  combined with bile-acid sequestrants,
    282
  combined with HMG-CoA reductase
    inhibitors, 284
  costs of, 313
Nicotinic acid, 111, 326
non-Q-wave infarction, 85–86
Northern Ireland, coronary heart disease
  and, 149
Northwest Lipid Research Clinic's Fat
  Intake Scale, 229
Norwegian cholesterol-lowering
  program, 327
Nuclear magnetic resonance
  spectroscopy (NMR), 302
Nutrition education, 158, 179–180
Nutrition Labeling and Education Act,
  240
Nutrition labels. *See* Food labeling

Obesity, 5, 23, 122–123, 192
  insulin resistance and, 15–16
Oleic acid, 189
Oliver, Boyd trial, 39
Oral contraceptives, 123, 125
Oslo trial, 39, 52, 56, 187, 317
Osteoarthritis, 113
Osteoporosis, 113
Oxidative modification hypothesis for
  atherosclerosis, 344–348
Oxidized LDL, 344–349

Pathobiological Determinants of
  Atherosclerosis in Youth study, 134
Pawtucket Heart Health Program, 174–
  180, 229
PDAY study. *See* Pathobiological
  Determinants of Atherosclerosis in
  Youth study
Philippines
  cholesterol levels and, 135, 136

Physicians, diet therapy and, 249–251
Plaque evolution, 69–85. *See also*
  Atherosclerosis
Plaque regression, 85–86
Plaque rupture, 85, 349–350. *See also*
  Plaque evolution
Platelets, 78, 83–84
Poland, cholesterol levels and, 137
Polyunsaturated fat, 190–191
Population approach. *See* Primary
  prevention (population strategy)
Population-based approach. *See* Primary
  prevention (population strategy)
Population Panel (NCEP), 1
Portugal, cholesterol levels and, 135, 137
POSCH trial, 39, 52, 53
PQRST study, 347
Pravastatin, 8, 278–279
  costs of, 313
Pregnancy, cholesterol levels and, 120
Price supports, 159
Primary prevention (population
  strategy), 1, 3, 149–164
  children and, 138
  cholesterol-lowering trials, 40, 314–
    316
  clinical outcomes of, 59
  versus community intervention
    approach, 154
  confusions about, 159–160
  cost-effectiveness of, 324–325, 327–
    328
  definition of, 49, 150, 152
  diet therapy and, 154–159, 184–187,
    254–256
  rationale for, 152–154
  versus secondary prevention (high-
    risk strategy), 153
  successes of, 161–163
Primate studies, 121
Probucol, 87, 347
  costs of, 313, 320
Processed food, 171
Progesterone, 120
Protein, 191
Puberty. *See* Adolescence

Q-wave infarction, 86
Quality-adjusted life years, 317, 318
Quality of life, cost-effectiveness and, 328

Rapid thrombus-mediated atherosclerosis progression, 80–85
Rate-Your-Plate, 229
Resins, 38, 43
Restaurants, 171, 175–176, 239–240
Risk assessment, 140, 291–306
Risk factors. *See* Cardiovascular risk factors
Rose Corn Oil trial, 52

Saturated fats
  coronary heart disease and, 169
  counting grams of, 231–232
  dietary recommendations, 155, 189, 235
  food labeling and, 243, 246, 247
  food sources of, 217–223
  reduction of in diet, 154–159
Scandinavian Simvastatin Survival Study, 357–362
Scavenger receptors, 349
SCOR trial, 53
Scotland, coronary heart disease and, 149
Scottish physician trial, 38
Secondary prevention (high-risk strategy), 1–23, 3, 9
  cardiovascular risk factors and, 5–6
  children and, 138–143
  cholesterol-lowering trials, 40, 51–59, 316–317
  clinical outcomes of, 59, 316–317
  of coronary heart disease, 49–65
  cost-effectiveness of, 1–23, 3, 9, 322–324
  definition of, 49
  diet therapy and, 187–188
  follow-up and, 3–4
  inclusion criteria of trials, 51
  meta-analysis of, 50, 58–59
  narrative analysis of, 50, 52, 58–59
  versus primary prevention (population strategy), 153

[Secondary prevention (high-risk strategy)]
  screening and, 2, 3–4, 139
Serum cholesterol levels, 170
  age-adjusted mean, 163
  arterial lesions in the young and, 134–135
  cost-effectiveness and, 319–322
  diet therapy and, 20
  levels in Japan, 151
  measuring, 294
  stratification of individuals by, 3
  in United States, 150, 151
Sex. *See* Gender issues
Shelf-labeling, 178, 238
Simvastatin, 8, 111, 278–279
  costs of, 313, 320
Smooth muscle cells, 79–80, 344
Sodium, 193–194
  food labeling and, 247
Soybean oil, 39, 52, 56
Specialized Center of Research Study, 61
St. Thomas Atherosclerosis Regression Study. *See* STARS trial
Stanford Five-Community Study, 328
STARS trial, 38, 39, 53, 61–62, 188
Stary morphological studies, 74–76, 80, 82
Step I and Step II diet
  components of, 188–194
  description of, 211–216
  drug therapy and, 273
  implementing, 230–238, 247–249, 254–258
  recommended daily intake, 184
Stockholm trial, 38, 52, 55, 279, 316–317
Subendothelial space, 342
Supermarkets, 178–179, 238–239
Surgery, 39, 43, 57–58, 62
Sweden, coronary heart disease and, 152, 153
Sydney trial, 39

Tamoxifen, 36
Taxation, 159

Thrombin, 84–85
Thrombosis, 346, 349–350
Triglyceride levels
  cutpoints for, 292–293
  diet therapy and, 197–198
  gender issues and, 119–128
  measuring, 303–304
  niacin and, 276
  pregancy and, 120
  reference method, 304

United States, cholesterol levels and,
  135, 136, 137
USVA Drug-Lipid Trial, 39

Vascular response-to-injury hypothesis,
  70–72
Veterans Administration Cardiology
  Drug-Lipid study, 51–53

Vitamins, 193
VLDL cholesterol levels, 3–4, 105, 301
  diet therapy and, 198

Weight, women and, 122
Weight control. *See* Diet therapy;
  Exercise
Weight status, 226
WHO trial, 38, 54, 281
Women. *See also* Gender issues
  cholesterol levels in, 119–128
  lipoproteins and female life cycle,
    120–121
  treatment of lipoprotein disorders
    and, 125–127
  weight and, 122
Women's Health Trial, 186–187, 237
World Health Organization. *See* WHO
  trial

About the Editor

BASIL M. RIFKIND is Senior Scientific Advisor to the Vascular Research Program, Division of Heart and Vascular Diseases of the National Heart, Lung, and Blood Institute (NHLBI), National Institutes of Health (NIH), Bethesda, Maryland. The editor of the book *Drug Treatment of Hyperlipidemia* (Marcel Dekker), Dr. Rifkind is a Fellow of the Royal College of Physicians and Surgeons of Glasgow and a member of the Royal College of Physicians of Edinburgh. He received the M.D. degree (1972) from the University of Glasgow, Scotland.